W9-BYT-065

Enzyme kinetics

Cambridge Chemistry Texts

Enzyme kinetics

D. V. ROBERTS

Department of Physical Biochemistry
The John Curtin School of Medical Research
The Australian National University

CAMBRIDGE UNIVERSITY PRESS

CAMBRIDGE

LONDON · NEW YORK · MELBOURNE

Published by the Syndics of the Cambridge University Press
The Pitt Building, Trumpington Street, Cambridge CB2 1RP
Bentley House, 200 Euston Road, London NW1 2DB
32 East 57th Street, New York, NY 10022, USA
296 Beaconsfield Parade, Middle Park, Melbourne 3206, Australia

First published 1977

Printed in Great Britain at The Spottiswoode Ballantyne Press
by William Clowes & Sons Limited, London, Colchester and Beccles

Library of Congress Cataloguing in Publication Data

Roberts, D. V.

Enzyme kinetics.

(Cambridge chemistry texts)

Bibliography: p.

Includes index.

1. Enzymes. 2. Chemical reaction, Rate of.
I. Title.
QP601.R67 574.1'925 76-11091
ISBN 0 521 21274 X hard covers
ISBN 0 521 29080 5 paperback

Contents

Preface

The study of the mechanism of action of enzymes can be undertaken using a multitude of different experimental approaches, only one of which is a study of the kinetics of the enzyme-catalysed reaction. Other important techniques include amino acid sequencing, X-ray crystallography, chemical modification and nuclear magnetic resonance studies. No one method can supply sufficient information to describe adequately the mechanism of action of an enzyme; indeed even the sum of all the information available from a number of different experimental studies is not enough to explain the detailed action of even the simpler enzymes, never mind the more complicated multi-subunit enzymes or enzyme–enzyme complexes. The initial studies of an enzyme-catalysed reaction using kinetic techniques does, however, provide a firm basis upon which to plan other more specific experiments.

The object of this book is to cover in a fairly comprehensive manner the detailed kinetic analysis of a number of different enzyme models. The contents of the book cover not only the enzyme kinetics that would be presented to an honours student in biochemistry as part of a molecular enzymology or physical biochemistry course, but also include topics that would be of more interest to postgraduate students and research workers. In this respect, the book contains sections on coupled enzyme systems, oscillatory kinetics, computer simulation of biochemical systems and statistical analysis of enzyme kinetic data which are not normally found in textbooks of comparable size. The initial plan of the book was based on the various enzyme kinetics courses presented to students at Queen's University of Belfast during their three-year honours course. It was noted that many students with limited mathematical backgrounds had problems with differential equations, determinants, statistics and other mathematical methods used in the course. For this reason, it was decided that in this book there should be sufficient steps in the development of kinetic equations that most readers would be able to follow the derivations. The mathematical equations are, however, complicated enough to prevent the reader from simply attempting to memorise the algebraic manipulations.

It is a great pleasure to acknowledge my gratitude to a number of friends. In particular, I am indebted to Professor D. T. Elmore of Queen's University of Belfast for his assistance in the initial planning, for many suggestions and helpful criticisms during the preparation of the manuscript; to Professor L. W. Nichol of the Australian National University for his encouragement and helpful suggestions during the period of my Research Fellowship in his department; to Dr P. W. Kuchel for many stimulating discussions and for reading the entire manuscript and suggesting a number of alternative mathematical derivations. Finally I would like to thank my wife, Jeanette, for her patience and to apologise to my son, David, for the lack of attention he received during the writing of the manuscript.

1 Introduction to kinetics

1.1. Introduction

Chemical reactions can be studied in two general ways. The direction that a reaction is likely to take and the concentrations at equilibrium can be predicted from the free energies of the reactants and products of the reaction. This thermodynamic approach to the study of a chemical reaction gives no information about the rate at which the reaction proceeds to equilibrium. The kinetic approach, however, is concerned with the rates at which reactions occur and with the factors that affect the rate of a reaction, such as pH, temperature, presence of catalysts, and is therefore essentially an experimental investigation. The interpretation of the kinetic results obtained for a chemical reaction and their dependence on other factors can lead to a fuller understanding of the mechanism of the reaction. The difference between the thermodynamic and kinetic approaches to the study of a chemical reaction can be seen more fully by considering the reaction between hydrogen and oxygen. Thermodynamic calculations indicate that this reaction should proceed spontaneously since it is accompanied by a large decrease in free energy. No reaction is detectable at room temperature and atmospheric pressure, however, even over a period of years. The introduction of a catalyst or spark causes the reaction to proceed at a considerable rate. An explanation of this will be given later.

1.2. Reaction rates

Before discussing the effects of enzymes on the rates of reactions, it is necessary to introduce a number of terms and relationships that frequently occur in chemical kinetics. Chemical reactions may be classified in two ways. They may be described on the basis of the number of molecules that react to form products (molecularity). Thus reactions such as

$$A \xrightarrow{k} B \tag{1.1}$$

are *unimolecular* whereas reactions such as

$$A + B \xrightarrow{k} C$$

are *bimolecular*. These reactions may also be classified on a kinetic basis, i.e. by *reaction order*. Thus if the rate, v, of a reaction is proportional to the concentration of only one component,

$$v = kc, \tag{1.2}$$

then the reaction is said to be *first order*. The proportionality constant, k, is the *rate constant* for the reaction. In equation (1.1) the rate of the reaction, v, is proportional to the concentration of reactant A. The rate of the reaction can be described either by the rate of loss of A or by the rate of formation of B. Thus the velocity

$$v = -\frac{d[A]}{dt} = \frac{d[B]}{dt}.$$

Therefore

$$-\frac{d[A]}{dt} = \frac{d[B]}{dt} = k[A]. \tag{1.3}$$

An equation like (1.3) is called a *rate equation*. Rate equations are *differential equations* that may be integrated to give equations into which the experimental results may be substituted directly. An example of a first order reaction is the decomposition of ethane,

$$C_2H_6 \xrightarrow{k} C_2H_4 + H_2.$$

The rate of production of ethylene is proportional to the concentration of ethane present,

$$\frac{d[C_2H_4]}{dt} = \frac{d[H_2]}{dt} = k[C_2H_6].$$

A *second order* reaction can be of two types; the rate of reaction can depend on the square of a single concentration

$$v = kc^2$$

or it can depend on the product of the concentrations of two reacting species,

$$v = kc_a c_b. \tag{1.4}$$

In general the overall order, n, of a reaction is the sum of the powers of the concentrations of the reactants. Thus if

$$v = kc_a^{n_a}c_b^{n_b}$$

then the order, n, of the reaction is given by

$$n = n_a + n_b.$$

An example of the first type of second order reaction is the decomposition of hydrogen iodide into hydrogen and iodine.

$$2HI \xrightarrow{k_f} H_2 + I_2,$$

thus

$$v = \frac{d[H_2]}{dt} = \frac{d[I_2]}{dt} = k_f[HI]^2.$$

The reverse reaction is also second order and is an example of the second type:

$$H_2 + I_2 \xrightarrow{k_r} 2HI.$$

The rate of the reaction v is given by

$$v = k_r[H_2][I_2].$$

It must be stressed that the order of a reaction is an experimental quantity and cannot necessarily be deduced from the chemical equation. The reaction between hydrogen and bromine can be described by an equation similar to that for the reaction between hydrogen and iodine:

$$H_2 + Br_2 \xrightarrow{k} 2HBr. \tag{1.5}$$

At first sight the reaction might be thought to be second order, but an investigation of the experimental conditions that affect the rate of production of hydrogen bromide shows that it obeys a complex rate equation

$$\frac{d[HBr]}{dt} = \frac{k[H_2][Br_2]^{1/2}}{1 + k'([HBr]/[Br_2])}.$$

The fact that the reaction between hydrogen and bromine is not a simple second order reaction indicates that the reaction does not take place according to the simple reaction scheme described in equation (1.5).

Third order reactions, which depend on the product of three concentration terms, are relatively rare. Some chemical reactions have rates which are independent of the concentration of any reactant and the kinetics of these reactions are *zero order*. As will be seen later, many catalysed reactions are zero order with respect to the reactants; the rate of the reaction depends only on the concentration of the catalyst. Consider the reaction,

$$A + B \xrightarrow{k} C;$$

the rate equation that describes this reaction is

$$v = \frac{d[C]}{dt} = k[A][B].$$

The reaction is normally second order. If the concentration of one of the reactants is much higher than the other, e.g. $[B] \gg [A]$, then the concentration of B will not change appreciably during the course of the reaction. The rate will therefore appear to depend solely on the concentration of A. The reaction is said to obey *pseudo first order* kinetics. The apparent rate constant, k', is given by the equation

$$\frac{d[C]}{dt} = k'[A],$$

where $k' = k[B]$.
The reaction is therefore *zero order* with respect to reactant B.

The rate constant, k, is numerically equal to the rate of the reaction when the concentration of all the reactants is unity. Its *dimensions* can be deduced from the rate equation and will vary with the order of the reaction. For a first order reaction, equation (1.2), the units of k are those of velocity, v, (mol l^{-1} s^{-1}) divided by those of concentration, c, (mol l^{-1}) and are therefore s^{-1}. For a second order reaction, equation (1.4), the units of k will be those of velocity, v, (mol l^{-1} s^{-1}) divided by the product of c_a and c_b (mol l^{-1})^2 and are therefore l mol^{-1} s^{-1}. In general for a reaction of order n, the dimensions of k are $mol^{(1-n)}$ $l^{(n-1)}$ s^{-1}.

1.3. Analysis of kinetic results

In a kinetic study of a chemical reaction, the change in concentration with time of one or more reactants or products is followed. The change in concentration of a reactant or product may be determined directly or indirectly by following the change in some other parameter related to concentration, e.g. change in optical density, fluorescence, pH, or by measuring the acid or alkali uptake required to maintain a constant pH. In order that the rate constant and order of reaction may be determined from these results, it is necessary to convert the rate equation into more suitable forms. Since the rate equations are differential equations there are two methods of determining the rate constant and order of reaction.

1.3.1. Differential method. The differential method uses the differential equation directly. The term dx/dt (where x represents the concentration of product formed or reactant consumed in time t) refers to the rate of the reaction. From a plot of x against time the slope of the graph (dx/dt) at various values of x can be determined and these values are substituted directly into the rate equation (see fig. 1.1). Although apparently simple, it lacks accuracy owing to the difficulty in determining the slope of a curve, especially if the curve is not smooth. Alternatively, instead of determining the velocity at various points along a curve, the initial velocity of the reaction is measured at the very beginning of the curve, and this process is repeated at different initial concentrations of substrate(s) (see fig. 1.2). Determination of the initial rate of the

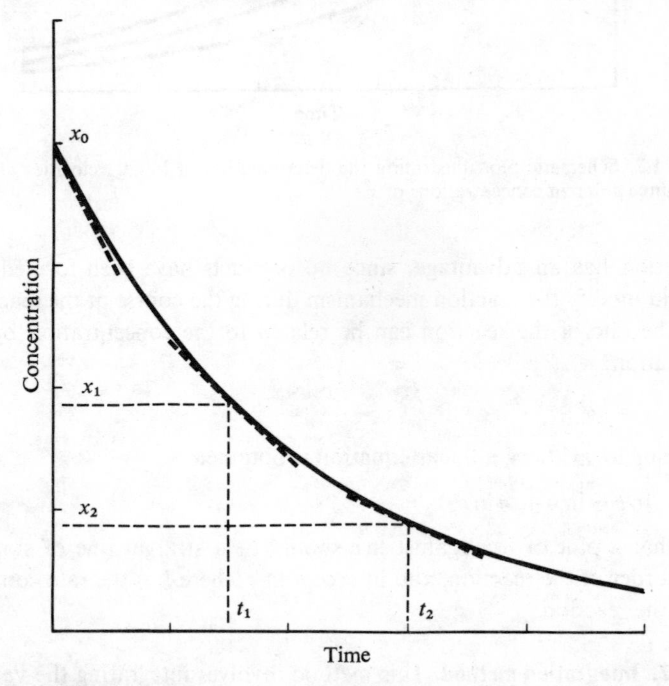

Fig. 1.1. Schematic plot of concentration against time. Tangents (dx/dt) are drawn at the initial concentration x_0 at $t=0$ and at concentrations x_1 and x_2 at times t_1 and t_2 respectively.

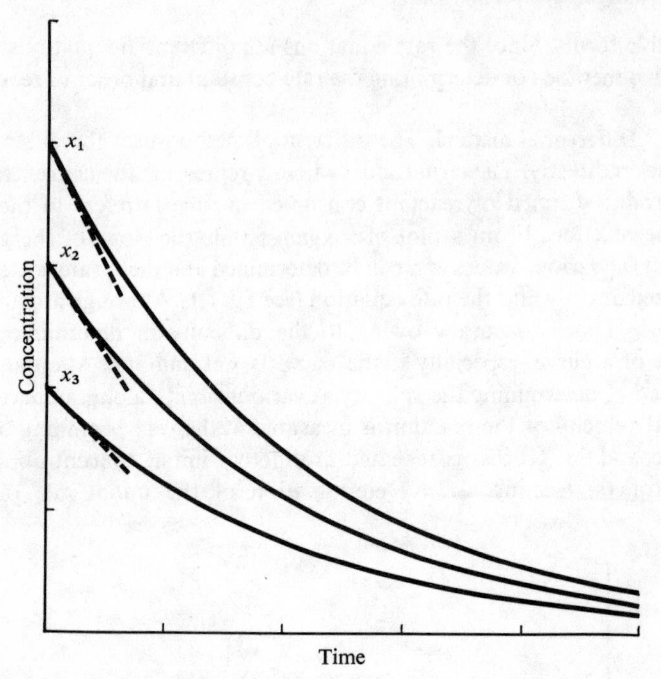

Fig. 1.2. Schematic plot illustrating the determination of initial velocities at $t = 0$ for three different concentrations of x.

reaction has an advantage, since no products have been formed that could modify the reaction mechanism during the course of the reaction.

The rate of the reaction can be related to the concentration by the equation,

$$v = kc^n.$$

Taking logarithms, a linear equation is obtained,

$$\ln v = \ln k + n \ln c,$$

so that a plot of $\ln v$ against $\ln c$ should be a straight line of slope n, the order of the reaction, and intercept $\ln k$ where k is the rate constant for the reaction.

1.3.2. Integration method. This method involves integrating the various rate equations and obtaining a solution for the variation of the concentration of a reactant x with time.

Zero order kinetics. Reactions of zero order are not common except in heterogeneous systems and in catalysed reactions in solution. The rate of the reaction is independent of the concentration of the reactants,

$$v = \frac{\mathrm{d}x}{\mathrm{d}t} = k,$$

where x is the concentration of the product formed. Integration of this equation with respect to time gives

$$x = kt + c.$$

Since $x = 0$ when $t = 0$, then $c = 0$. Therefore

$$x = kt,$$

which is the equation of a straight line of slope k passing through the origin. The dimensions of k are mol $\mathrm{l^{-1}\ s^{-1}}$. The *half life* of a reaction is the time taken for half of the reactant to be consumed or half of the product to be produced. For zero order kinetics, $t = t_{1/2}$ when $x = x_0/2$ (where x_0 is the initial concentration of the reactant or final concentration of the product). Therefore

$$t_{1/2} = \frac{x_0}{2k}.$$

The half life is proportional to the initial concentration of reactant or final concentration of product.

First order reactions. A first order reaction can be described by the equation

$$\mathrm{A} \xrightarrow{k} \mathrm{B}.$$

Let the initial concentration of A at $t = 0$ be a, while the concentration of B at this time is zero. After a time t, a quantity x has been transformed from A into B; therefore the new concentrations of A and B are respectively $(a - x)$ and x. The rate of formation of B is $\mathrm{d}x/\mathrm{d}t$ so that

$$\frac{\mathrm{d}x}{\mathrm{d}t} = k(a - x),$$

which can be integrated to give

$$-\ln(a - x) = kt + c.$$

Since $x = 0$ when $t = 0$, $c = -\ln a$ and hence the complete solution for the equation is

$$\ln[a/(a - x)] = kt. \tag{1.6}$$

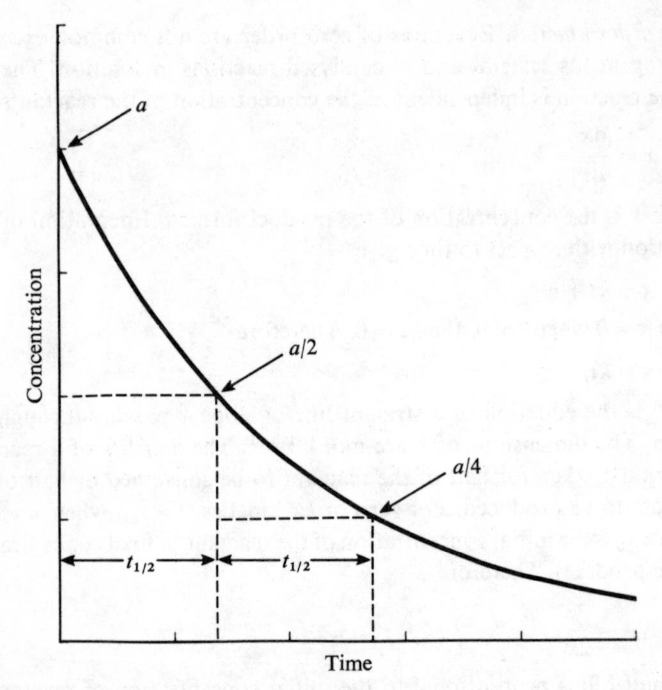

Fig. 1.3. Schematic plot of a first order exponential decay process. The half life ($t_{1/2}$) can be used to determine the rate constant k.

The equation can be written in the alternative exponential form as

$$(a-x) = a\mathrm{e}^{-kt}.$$

This shows that the concentration, $(a-x)$, of the reactant decays exponentially with time from an initial value a at $t=0$ to zero at $t=\infty$ (see fig. 1.3). From equation (1.6), it can be seen that the rate constant k depends on the ratio of two concentrations and has the dimensions of 1/time or s^{-1}. There are two important points regarding first order reactions. First, the rate constant, k, can be determined from the *ratio* of two concentrations determined at two times; the actual values of the concentrations are not needed. Secondly, the half life of the reaction is a constant and does not depend on the initial concentration. Since $t=t_{1/2}$ when $x=a/2$, substitution in equation (1.6) gives

$$kt_{1/2} = \ln 2$$

or $$t_{1/2} = \frac{0.693}{k}. \tag{1.7}$$

Provided a reaction follows first order kinetics, the time taken for 50% of the reaction to be completed cannot be altered by changing the concentration of the reactant A. From equation (1.7), it can be seen that the determination of the half life of the reaction gives the rate constant k directly. This is a useful semiquantitive way to determine the rate constant. For accurate determinations of the rate constant, a linear regression analysis (see appendix II) of $\ln(a/(a-x))$ or $\ln(a-x)$ against time t is recommended.

Second order reactions. This is the most common order of reaction that occurs. A second order reaction may be expected to occur if the reaction takes place in a single step and the concentrations of the reactants are of similar magnitudes. As mentioned previously, second order reactions can be of two types:

$$2A \longrightarrow B \tag{1.8}$$

or

$$A + B \longrightarrow C. \tag{1.9}$$

For equation (1.8), the rate equation is

$$\frac{dx}{dt} = k(a-x)^2,$$

where a is the initial concentration of A at $t = 0$ and x is the concentration of product at time t. This equation can be rearranged to give

$$\frac{dx}{(a-x)^2} = k dt$$

which can be integrated directly to give

$$\frac{1}{(a-x)} = kt + c.$$

Since at $t = 0$, $x = 0$, then $c = 1/a$ so that the complete solution becomes

$$\frac{1}{(a-x)} = kt + \frac{1}{a} \tag{1.10}$$

or

$$kt = \frac{x}{a(a-x)}.$$

From equation (1.10) a plot of $1/(a-x)$ against time t is a straight line of slope k and intercept $1/a$. The half life, $t_{1/2}$, can be evaluated by putting

$x = a/2$, hence

$$t_{1/2} = \frac{1}{ka}.$$

The half life of the reaction is now inversely proportional to the initial concentration of reactant A as well as to the rate constant k. This means that an increase in the concentration of A will reduce the time taken to reach 50% reaction.

For the reaction between two different substances, equation (1.9), the rate equation is

$$\frac{dx}{dt} = k(a-x)(b-x), \tag{1.11}$$

where a and b are the initial concentrations of reactants A and B and x is the concentration of product at time t. Rearranging equation (1.11) gives

$$\frac{dx}{(a-x)(b-x)} = k\,dt,$$

which can be integrated by the method of partial fractions giving

$$\frac{-1}{(b-a)}\ln(a-x) - \frac{1}{(a-b)}\ln(b-x) = kt + c, \tag{1.12}$$

and since $x = 0$ at $t = 0$, then

$$c = \frac{1}{(a-b)}\ln\frac{a}{b}.$$

Hence the complete solution of equation (1.12) is

$$\ln\frac{(a-x)}{(b-x)} = (a-b)kt + \ln\frac{a}{b}. \tag{1.13}$$

From equation (1.13), a plot of $\ln(a-x)/(b-x)$ against time is a straight line of slope $(a-b)k$ and an intercept $\ln a/b$. In this reaction, in which the initial concentrations of the reactants may be different, the evaluation of the half life is not a meaningful concept. Nevertheless it is apparent that an increase in concentration of either reactant will lead to a decrease in reaction time.

Equations with non-integral orders of reaction. There are many reactions in which the rate of reaction does not depend on the concentrations of the reactants raised to some *integral* power. These reactions usually have mechanisms which are more complex than the simple chemical equation that satisfies the overall chemical change that takes place. An example of this is the reaction between hydrogen and bromine mentioned earlier. A first order reaction in which the products act as inhibitors would show second order kinetics as the reaction proceeded. For these complex reactions the integration method cannot be used. It is more useful to use the differential method and determine only the initial rates of the reaction for different concentrations of reactants.

1.4. Equilibrium and consecutive reactions

The previous sections have dealt with simple reactions involving the direct conversion of reactant(s) into product(s). Two additional complications frequently occur; reactions may reach an equilibrium state or they may consist of two or more consecutive steps. The solutions of the rate equations for these systems are particularly relevant to the reactions involving enzymes. The simplest example of an equilibrium reaction is

$$\mathrm{A} \underset{k_{-1}}{\overset{k_1}{\rightleftharpoons}} \mathrm{B}$$

where k_1 is the rate constant for the forward reaction $\mathrm{A} \rightarrow \mathrm{B}$ and k_{-1} is the rate constant for the reverse reaction $\mathrm{B} \rightarrow \mathrm{A}$. (The use of k_j and k_{-j} for the forward and reverse rate constants for the jth equation is to be recommended.) The reaction will proceed until an equilibrium position is reached, at which point the rate of the forward reaction is equal to that of the reverse reaction.

Suppose the initial concentration of A at $t = 0$ is $[\mathrm{A}_0]$ and the concentrations of A and B after time t are [A] and [B] respectively; then the rate of formation of B, $\mathrm{d}[\mathrm{B}]/\mathrm{d}t$, will be given by

$$\frac{\mathrm{d}[\mathrm{B}]}{\mathrm{d}t} = k_1[\mathrm{A}] - k_{-1}[\mathrm{B}], \tag{1.14}$$

since $k_1[\mathrm{A}]$ is the rate of the forward reaction and $k_{-1}[\mathrm{B}]$ is the rate of the reverse reaction. It is useful to introduce here the concept of a *conservation equation* which will occur frequently in enzyme kinetics. The sum of the concentrations of A and B after a time t is equal to $[\mathrm{A}_0]$, the initial concentration of A at zero time:

$$[\mathrm{A}_0] = [\mathrm{A}] + [\mathrm{B}].$$

This is called a conservation equation. Substitution for [A] in equation (1.14) gives

$$\frac{d[B]}{dt} = k_1([A_0] - [B]) - k_{-1}[B]. \tag{1.15}$$

Since at equilibrium the net rate of the reaction is zero, $d[B]/dt = 0$ and

$$k_1([A_0] - [B_e]) = k_{-1}[B_e], \tag{1.16}$$

where $[B_e]$ is the concentration of product B at equilibrium. Substituting for k_{-1} from equation (1.16) in equation (1.15) gives

$$\frac{d[B]}{dt} = \frac{k_1[A_0]([B_e] - [B])}{[B_e]}.$$

Since k_1, $[A_0]$ and $[B_e]$ are all constants, this can be integrated to give

$$-\ln([B_e] - [B]) = k_1 \frac{[A_0]}{[B_e]} t + c.$$

When $t = 0$, $[B] = 0$, so that $c = \ln[B_e]$ and the complete solution is

$$\ln\left(\frac{1}{[B_e] - [B]}\right) = k_1 \frac{[A_0]}{[B_e]} t + \ln[B_e].$$

If $[B_e]$ is determined from equilibrium measurements, then k_1 can be determined from the variation of [B] with time. The reverse rate constant, k_{-1}, can then be evaluated from equation (1.16).

The added complication of an equilibrium system would not become apparent using the differential method because initially the reverse reaction rate is zero and this would have no effect on the initial rate of reaction.

Sequential reactions. The simple first order conversion of a reactant A into product B may be complicated by the further conversion of B into a new product C. This can be described by the equation

$$A \xrightarrow{k_1} B \xrightarrow{k_2} C.$$

Each step of the reaction will have a unique rate constant, k_1 for the conversion of A into B and k_2 for the conversion of B into C. If the initial concentration of A at zero time is $[A_0]$, then at any other time the conservation equation,

$$[A_0] = [A] + [B] + [C], \tag{1.17}$$

must be observed.

The rate of disappearance of A is given by the equation

$$-\frac{d[A]}{dt} = k_1[A],$$

the solution of which is

$$[A] = [A_0]e^{-k_1 t}. \tag{1.18}$$

The rates of production of B, $d[B]/dt$, and of C, $d[C]/dt$, are

$$\frac{d[B]}{dt} = k_1[A] - k_2[B] \tag{1.19}$$

and $$\frac{d[C]}{dt} = k_2[B].$$

Substitution for [A] from equation (1.18) into equation (1.19) gives

$$\frac{d[B]}{dt} + k_2[B] = k_1[A_0]e^{-k_1 t}.$$

This is a simple first order differential equation which can be integrated using the integrating factor $e^{\int k_2 dt}$ to give

$$[B] = \frac{k_1[A_0]}{(k_2 - k_1)}(e^{-k_1 t} - e^{-k_2 t}).$$

From equation (1.17),

$$[C] = [A_0] - [A] - [B];$$

hence

$$[C] = [A_0]\left(1 - \frac{1}{(k_2 - k_1)}(k_2 e^{-k_1 t} - k_1 e^{-k_2 t})\right).$$

The variation of A, B and C with time is shown in fig. 1.4. Reactant A follows a simple exponential decay; the rate of formation of C is slow at the beginning when there is no B, passes through a maximum when [B] reaches a maximum and falls to zero when [B] reaches zero. This sequential reaction is the simplest that can be described. It has an analytical mathematical solution, since each of the rate equations can be solved. Rate equations for more complex reaction schemes frequently cannot be

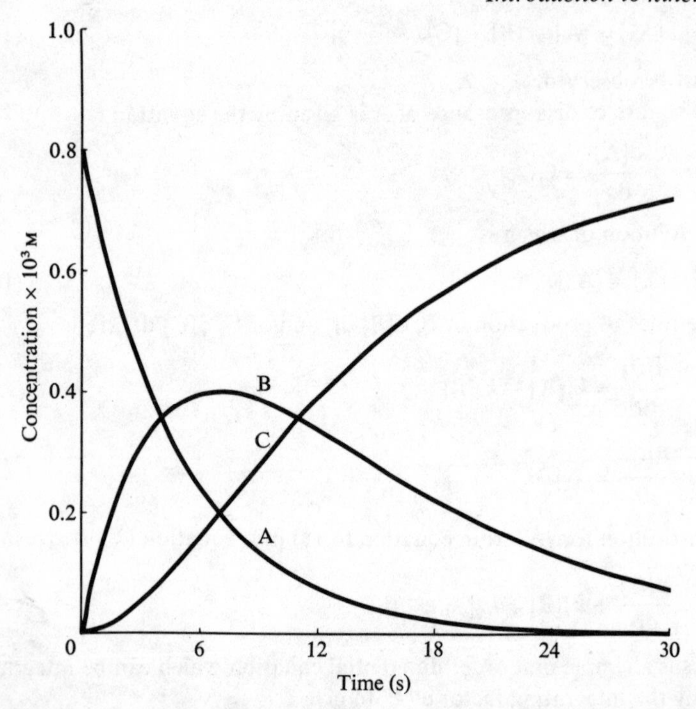

Fig. 1.4. Schematic plot of the variation in concentration with time of the reactants for the sequential reaction $A \xrightarrow{k_1} B \xrightarrow{k_2} C$ with $k_1 = 0.2\ s^{-1}$ and $k_2 = 0.1\ s^{-1}$.

integrated without making some assumptions regarding the variation of the concentrations of some reactants with time. Even the most complex systems, however, can usually be studied using computer techniques (see chapter 10).

1.5. The effect of temperature on reaction rates

Very early in the study of chemical reactions it was found empirically (Hood, 1878, 1885) that the variation of the rate constant k with the absolute temperature T could be described by an equation of the form

$$\ln k = A - \frac{B}{T}. \tag{1.20}$$

The variation of the equilibrium constant, K, with temperature can be deduced from the classical thermodynamic equations (van't Hoff, 1887);

$$\Delta G^\circ = \Delta H^\circ - T\Delta S^\circ,$$

whence

$$\frac{d(\Delta G^\circ)}{dT} = -\Delta S^\circ;$$

then

$$\Delta G^\circ = \Delta H^\circ + T\,\frac{d(\Delta G^\circ)}{dT}. \qquad (1.21)$$

This is the Gibbs–Helmholtz equation. Since

$$\Delta G^\circ = -RT \ln K,$$

then

$$\frac{d(\Delta G^\circ)}{dT} = -R \ln K - RT\,\frac{d(\ln K)}{dT};$$

hence substitution for $d(\Delta G^\circ)/dt$ in equation (1.21) gives

$$\frac{d(\ln K)}{dT} = \frac{\Delta H^\circ}{RT^2}, \qquad (1.22)$$

which is known as the van't Hoff isochore.

For a reversible reaction $A + B \rightleftharpoons C + D$, the equilibrium conditions can be formulated by equating the forward and reverse reactions,

$$k_1[A_e][B_e] = k_{-1}[C_e][D_e],$$

where k_1 and k_{-1} are the forward and reverse rate constants respectively. The equilibrium constant, K, is equal to k_1/k_{-1}. Van't Hoff proposed that equation (1.22) could be written as

$$\frac{d(\ln k_1)}{dT} - \frac{d(\ln k_{-1})}{dT} = \frac{\Delta H^\circ}{RT^2},$$

which may be split into two separate equations, one relating to the forward reaction and the other to the reverse reaction:

$$\frac{d(\ln k_1)}{dT} = \frac{E_1}{RT^2} + c,$$

$$\frac{d(\ln k_{-1})}{dT} = \frac{E_{-1}}{RT^2} + c,$$

where $\Delta H^\circ = E_1 - E_{-1}$ and c is a constant. Experimentally the constant, c, was found to be zero, so that the general relationship applicable to a rate constant is of the form

$$\frac{d(\ln k)}{dT} = \frac{E_a}{RT^2}, \tag{1.23}$$

which can be integrated to give

$$\ln k = \frac{-E_a}{RT} + \text{constant} \tag{1.24}$$

or $$k = Ae^{-E_a/RT}. \tag{1.25}$$

Equation (1.24) agrees with the earlier empirical relationship, equation (1.20), whereas the rearranged form, equation (1.25), is the well known Arrhenius equation. The constant A is called the frequency factor whilst the term $e^{-E_a/RT}$ is the Boltzmann expression for the fraction of molecules having an energy in excess of the value E_a. The energy E_a was referred to by Arrhenius as the *activation energy* for the reaction. It is the minimum energy that must be acquired by the reactant molecules before a reaction can take place.

1.5.1. The collision theory of reactions. One of the fundamental concepts of chemistry is that all molecules of a given compound are identical. In the simple first order reaction, A → B, the reaction proceeds at a finite rate and at the half life time, 50% of the molecules have reacted. Whilst these molecules have been reacting, what has prevented the remaining molecules, which are supposed to be identical, from reacting? There must obviously be something that differentiates between a molecule that is capable of undergoing a reaction and one that is not. In a second order reaction of the type, A + B → C, the molecules A and B must collide or approach very close. If every collision between A and B were to result in the formation of C, then all reactions would take place instantaneously since the frequency of collisions in gaseous and aqueous phases is extremely high. In order to obtain finite rates of reactions, only certain collisions must give rise to reaction. Arrhenius (1889) proposed that molecules should be divided into two categories; normal molecules which do not take part in reactions and molecules which have acquired a certain level of energy and have become activated. It is these activated molecules which are capable of undergoing reactions and the level of energy required to raise them to this excited state is called the activation energy. These activated molecules arise from random thermal collisions that impart energy to a level higher than average to a few molecules. An increase in temperature will lead to an increase in molecular motion and hence to an increase in the number of collisions that produce activated molecules. An increase in temperature will therefore lead to an increase in reaction rate. The collision between two activated molecules cannot by itself lead to the formation of a stable molecule no matter how great the affinity between the two colliding molecules. The new molecule would not only have the energy of the activated molecules but also that due to the kinetic energy gained on collision. This 'quasi-molecule' would therefore be unstable and vibrate apart unless the excess energy were removed. The excess energy must be removed by collision with a third body; this can be the walls of the reaction vessel, solvent molecules or other reactant molecules.

The collision theory proposes that the reaction rate is dependent on the number of collisions giving rise to activated molecules or complexes. The proportion of molecules whose energy is in excess of the minimum required for reaction is given by the Boltzmann term, $e^{-E/RT}$. If the total number of collisions in unit time is Z, then the number of collisions leading to reaction will be $Ze^{-E/RT}$. It is possible to calculate Z assuming that the molecules behave like hard spheres. For like molecules,

$$Z = 2n^2d^2 \sqrt{\left(\frac{\pi \boldsymbol{k}T}{M}\right)},$$

where n = number of molecules of molecular diameter d and molecular weight M, and $\boldsymbol{k}$ is the Boltzmann constant. For unlike molecules,

$$Z_{AB} = 2n_A n_B d_{AB}^2 \sqrt{\left(\frac{2\pi \boldsymbol{k}T}{M}\right)},$$

where n_A and n_B are the number of A and B molecules and $d_{AB} = (d_A + d_B)/2$, and $1/M = 1/M_A + 1/M_B$. The term Z is called the collision number and can be equated with the frequency factor in the Arrhenius equation. For simple reactions between small molecules, the theory gives results which agree reasonably well with the observed values. For reactions between more complex molecules that have more degrees of freedom involving vibrational, rotational and translational motion, the discrepancy between the calculated rate and the observed rate can be several orders of magnitude. Because of this discrepancy, the equation for the number of collisions leading to reaction was modified empirically to include a probability or steric factor, P, so that the rate constant $k = PZe^{-E/RT}$. The values of P usually range from unity to about 10^{-8},

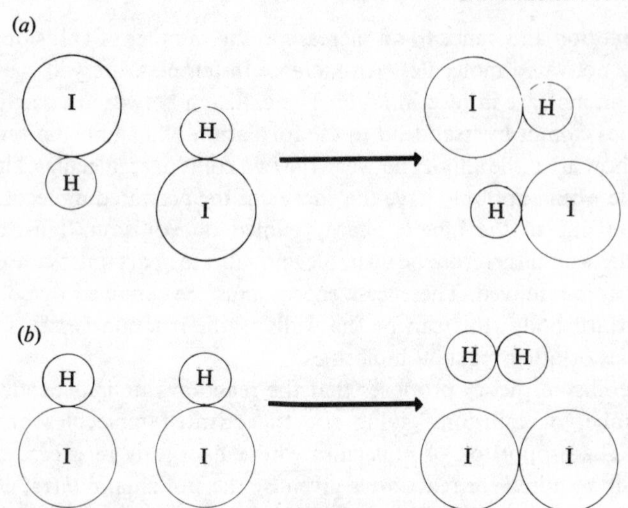

Fig. 1.5. A diagrammatic representation of the importance of correct orientation between interacting molecules. In (*a*) collision only leads to the formation of starting material, HI, whereas in (*b*) collision leads to the formation of products, H_2 and I_2.

although values as large as 10^6 have been observed for reactions between oppositely charged ions. In a large molecule, only a small part of the molecule may be susceptible to reaction, so that a collision between two molecules with the wrong orientation would not bring the two reactive areas of the molecules into juxtaposition. Such a collision would not bring about a reaction even though the molecules possessed sufficient energy as a specific orientation of the molecules is required. This is illustrated in fig. 1.5 for the reaction $2HI \rightarrow H_2 + I_2$. In (*a*) the molecules are incorrectly orientated so that any reaction that does take place will only lead to the formation of two new HI molecules. Only in (*b*) would the correct orientation lead to the formation of H_2 and I_2 molecules.

1.5.2. Transition state theory. In view of the anomalies in the collision theory, especially when $P \gg 1$, Eyring (1935*a*,*b*) formulated a new theory based on the concept of a transition state which exists as an intermediate stage in any chemical reaction. The transition state complex can be regarded as a molecule which has only a transient existence and breaks down at a definite rate to give the products of the reaction. The rate of the reaction would then be determined by the concentration of the transition state complex and the rate of its decomposition. The reaction, $A + B \rightleftharpoons C + D$, must now include the formation of the transition state complex (sometimes referred to as the activated complex). Hence

$$A + B \rightleftharpoons X^{\ddagger} \rightleftharpoons C + D$$

where $X^{\ddagger}$ is the transition state complex. The activation energy for the reaction can then be defined as the energy the reactant molecules must acquire in order to form the transition state complex. This is the difference in energy between the complex $X^{\ddagger}$ and the reactants and can be regarded as a barrier which prevents the reaction of A and B until sufficient energy is acquired. Fig. 1.6 is a diagrammatic example of the energy changes in the reaction $A + B \rightleftharpoons C + D$. It is possible from classical thermodynamic calculations to determine the free energies of the reactants A and B in the initial state and of the products C and D in the final state and hence to determine the overall free energy change for the reaction. In this reaction, $A + B \rightleftharpoons C + D$, classical or equilibrium thermodynamic measurements would indicate that there was an overall decrease in energy, ΔE, for the reaction and hence it should occur spontaneously. Classical thermodynamic measurements, however, do

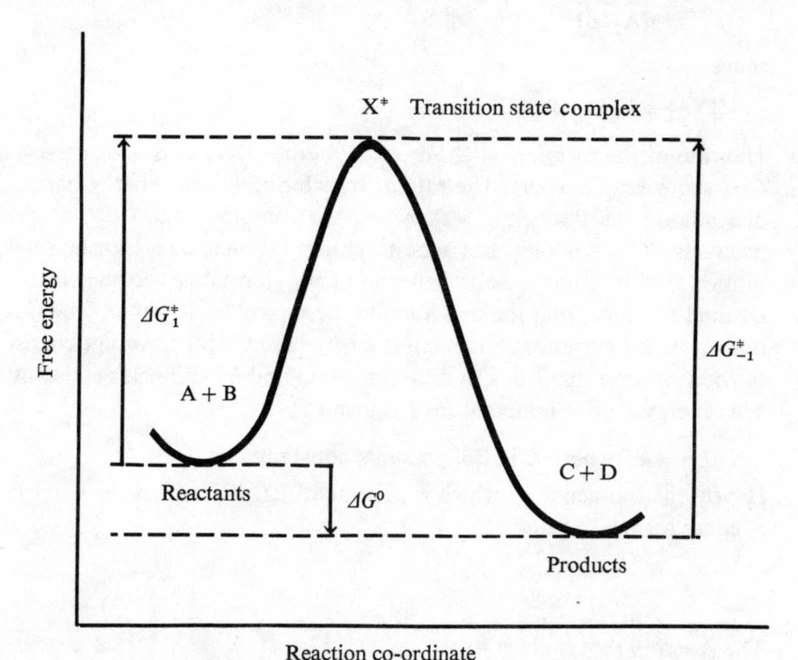

Fig. 1.6. Schematic representation of the free energy changes in the chemical reaction $A + B \rightleftharpoons C + D$.

not give any indication of the size of the activation energy barrier that must be overcome before any reaction between A and B can occur. The reaction between hydrogen and oxygen is a good example. There is a considerable decrease in free energy in the formation of water molecules, but under normal conditions (e.g. room temperature and atmospheric pressure), mixtures of hydrogen and oxygen are stable indefinitely. There is a high activation energy for the reaction and under these conditions molecules cannot achieve this energy by collisions. The passage of a spark through the mixture causes some of the molecules to react and the energy emitted is sufficient to cause a chain reaction amongst the rest of the molecules. The addition of a catalyst (e.g. palladium on charcoal) has the effect of lowering the activation energy for the reaction. This catalytic effect will be dealt with in chapter 5.

If the complex, $X^{\ddagger}$, is in equilibrium with the reactants then the equilibrium constant, $K^{\ddagger}$, for the formation of the complex is given by

$$K^{\ddagger} = \frac{[X^{\ddagger}]}{[A][B]};$$

hence

$$[X^{\ddagger}] = K^{\ddagger}[A][B].$$

The rate of the reaction $-d[B]/dt = -d[A]/dt = [X^{\ddagger}] \times$ (rate of passage over the energy barrier). The rate of transfer across the energy barrier depends on the frequency with which the complex breaks down into products. The complex has vibrational, rotational and translational motion and it decomposes when one of its vibrations becomes translational motion along the line holding the molecules together. The frequency of this vibration, ν, is equal to $E^{\ddagger}/h$ where $E^{\ddagger}$ is the average energy of the vibration that leads to decomposition and h is Planck's constant. This energy $E^{\ddagger}$ at a temperature T is given by

$E^{\ddagger} = \boldsymbol{k}T$ where $\boldsymbol{k}$ is Boltzmann's constant.

Hence, the frequency, ν, which is the rate of passage across the barrier, is equal to

$$\nu = \frac{\boldsymbol{k}T}{h}.$$

The reaction rate is given by

$$-\frac{d[A]}{dt} = -\frac{d[B]}{dt} = K^{\ddagger}\frac{\boldsymbol{k}T}{h}[A][B],$$

where

$$K^{\ddagger}\frac{\boldsymbol{k}T}{h} = k,$$

the rate constant for the reaction. If the assumption is made that the transition state complex is in true equilibrium with the reactants, then it is possible to equate the classical thermodynamic constants pertaining to an equilibrium to the transition state equilibrium. This assumption is true provided $E^{\ddagger}/RT$ has a value of 5 or more; below this value the reaction occurs so rapidly that no true equilibrium can exist. Therefore

$$\Delta G^{\ddagger} = \Delta H^{\ddagger} - T\Delta S^{\ddagger} \tag{1.26}$$

and $$\Delta G^{\ddagger} = -RT \ln K^{\ddagger} \tag{1.27}$$

where $\Delta G^{\ddagger}$ is the standard free energy change of formation of the transition state, $\Delta H^{\ddagger}$ is the standard enthalpy (heat content) of the transition state and $\Delta S^{\ddagger}$ is the standard entropy of activation change for the transition state. From equations (1.26) and (1.27)

$$\ln K^{\ddagger} = \frac{\Delta S^{\ddagger}}{R} - \frac{\Delta H^{\ddagger}}{RT}$$

or $$K^{\ddagger} = \mathrm{e}^{\Delta S^{\ddagger}/R}\mathrm{e}^{-\Delta H^{\ddagger}/RT}$$

The rate constant, k, for the reaction is then given by the equation

$$k = \frac{\boldsymbol{k}T}{h}\mathrm{e}^{\Delta S^{\ddagger}/R}\mathrm{e}^{-\Delta H^{\ddagger}/RT}. \tag{1.28}$$

Taking logarithms gives

$$\ln k = \ln\frac{\boldsymbol{k}}{h} + \ln T + \frac{\Delta S^{\ddagger}}{R} - \frac{\Delta H^{\ddagger}}{RT} \tag{1.29}$$

and differentiating equation (1.29) gives

$$\frac{\mathrm{d}(\ln k)}{\mathrm{d}T} = \frac{1}{T} + \frac{\Delta H^{\ddagger}}{RT^2}. \tag{1.30}$$

From equations (1.23) and (1.30), the energy of activation $E_a = \Delta H^{\ddagger} + RT$ and hence equation (1.28) becomes

$$k = \frac{\mathrm{e}\boldsymbol{k}T}{h}\mathrm{e}^{\Delta S^{\ddagger}/R}\mathrm{e}^{-E_a/RT}. \tag{1.31}$$

It is of interest to note that the quantity $\mathrm{e}\boldsymbol{k}T/h$ is similar in magnitude to the collision number Z and that the probability factor P can be equated with the term $e^{\Delta S^{\ddagger}/R}$. The overall rate is dependent on two exponential terms. Obviously, the higher the energy of activation, E_a, the slower will be the reaction. The effect of the entropy term is not so obvious. It is in fact this entropy term that causes the large discrepancies when the simple collision theory is applied to complex molecules. When the reacting molecules contain a large number of atoms, the loss of translational and rotational freedom on the formation of the transition state is accompanied by a decrease in entropy and hence the entropy of activation $\Delta S^{\ddagger}$ is negative and the term $e^{\Delta S^{\ddagger}/R}$ will be small. This entropy term is very significant in explaining the rate enhancements achieved in enzyme-catalysed reactions compared to the non-enzymic reaction. Both the energy of activation and the entropy of activation are therefore very important in determining reaction rate.

2 Simple enzyme-catalysed reactions

2.1. The concept of an enzyme–substrate complex

In the previous chapter a brief mention was made of the reaction between hydrogen and oxygen. If a suitable catalyst, such as palladium on charcoal, is added to the mixture, the reaction will proceed very rapidly and usually explosively. How has this dramatic change been brought about? An examination of the Arrhenius equation, equation (1.25), shows that, at a given temperature, the enhancement of the rate of a reaction can be explained by an increase in the term $e^{-E_a/RT}$ (by a reduction in E_a), by an increase in the value of A or by a combination of the two effects. In terms of the transition state theory, equation (1.28),

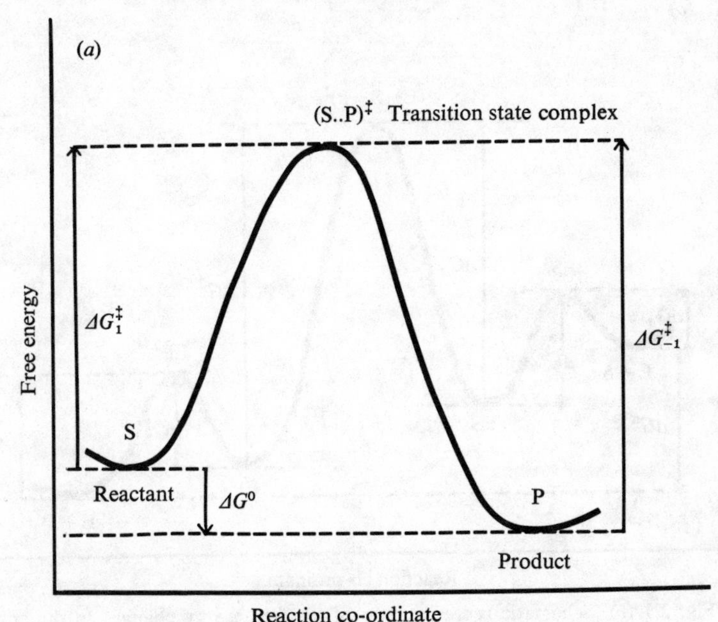

Fig. 2.1 (*a*). Schematic representation of the free energy changes in the chemical reaction $S \rightleftharpoons P$.

these effects can be described by either a reduction in the enthalpy of activation or by a lessening in the decrease in entropy required to form the transition state complex compared with the uncatalysed reaction. The lowering of the free energy of the transition state in an enzyme-catalysed reaction is achieved by the introduction into the reaction pathway of a number of new reaction intermediates (fig. 2.1*b*) and the number of these reaction intermediates will depend on the particular enzyme mechanism under consideration. The behaviour depicted in fig. 2.1(*b*) is typical of many enzyme-catalysed reactions. The original transition state complex for the uncatalysed reaction (fig. 2.1*a*) is frequently stabilised by the enzyme so that it no longer occupies a position of maximum energy in the reaction pathway (fig. 2.1*b*). The purpose of the enzyme is to provide groups in the binding site that constrain the substrate into the transition state in juxtaposition with other catalytic groups in the active site. For this reason, compounds that mimic the structure of the transition state (transition state analogues) and are therefore already

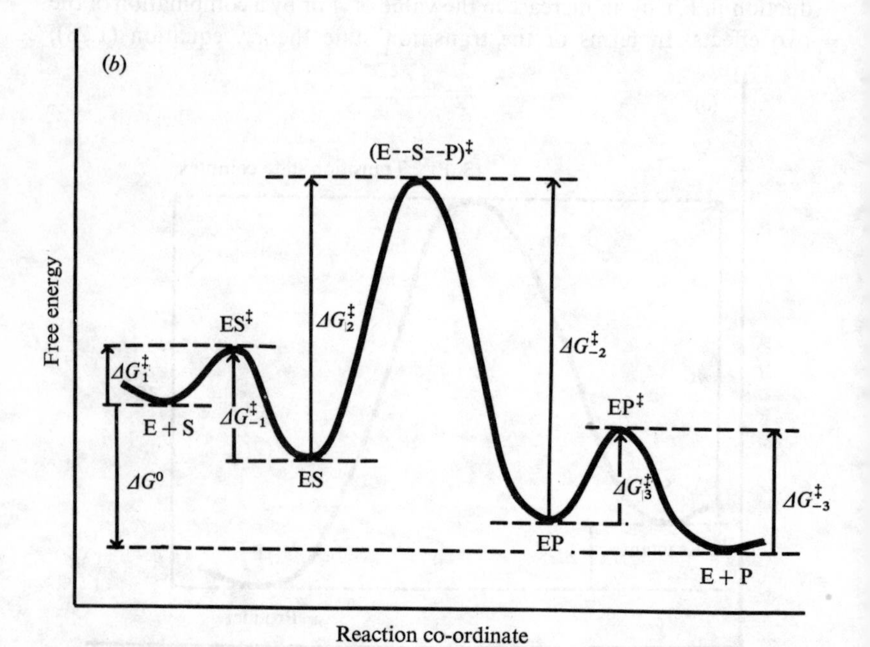

Fig. 2.1 (*b*). Schematic representation of the free energy changes in the enzyme-catalysed reaction $S \rightleftharpoons P$. Note in this case the formation of a number of enzyme intermediates as part of the reaction mechanism.

in a form that require a minimal free energy change upon binding to the enzyme, are often potent inhibitors of the enzyme-catalysed reaction (Wolfenden, 1972; Lienhard, 1972, 1973). In some group transfer reactions catalysed by mutase enzymes the transition state is believed to be the same in both the enzymic and non-enzymic catalysed reaction. An example of this type of enzyme is chorismic acid-mutase prephenate dehydrogenase, which catalyses the intramolecular conversion of chorismic acid to prephenic acid and also catalyses the conversion of prephenic acid to 4-hydroxyphenyl pyruvic acid, the latter step requiring NAD (Andrews, Smith & Young, 1973).

Whichever type of reaction pathway is used in an enzyme-catalysed reaction, there can be no difference in the standard free energy change ΔG° for the reaction and hence there can be no change in the equilibrium constant K for the reaction ($\Delta G^\circ = -RT \ln K$). This means that the effect of the enzyme or catalyst must be to increase the rates of the forward and reverse reactions proportionately. The catalyst is unchanged chemically after the reaction but may be changed physically so that it no longer behaves as a catalyst.

What are enzymes and how do they differ from the inorganic catalysts normally found in chemical laboratories? From the work carried out in the nineteenth century it had been shown that (*a*) the rate of an enzyme-catalysed reaction was usually faster than the same reaction catalysed by non-biological catalysts, (*b*) enzymes were very specific and one enzyme would usually catalyse only one particular type of reaction under very limited chemical and physical conditions, and (*c*) there existed a vast number of enzymes each catalysing a specific reaction which frequently formed links in a metabolic pathway. More recent work has shown that enzymes are proteins with a large number of amino acid residues linked together in a specific sequence. This linear sequence of amino acids is called the primary structure of the protein. The protein does not normally exist as an extended polypeptide but is coiled or folded into a specific arrangement or secondary structure which is stabilised by hydrogen bonds between the amido nitrogen and carbonyl oxygen of the polypeptide chain. The tertiary structure results from further folding of the molecule under the influence of ionic and hydrophobic interactions between the side chains of the amino acid residues. It is during this latter folding process that an area of the enzyme called the *active site*, which is responsible for the catalytic activity of the enzyme, is generated by the juxtaposition of certain functional groups such as carboxyl, amino, sulphydryl, hydroxyl and imidazolyl in the protein.

The various groups required for the catalytic activity are usually found in different areas of the linear amino acid sequence and are brought together by the folding of the molecule. The functioning of many enzymes requires the presence of small amounts of chemical agents other than the reactants of the catalysed reaction. These compounds are called co-factors and may be simply metal ions, more complex organic molecules or even other proteins (e.g. lactose synthetase system in which the protein α-lactalbumin modifies the reaction catalysed by the enzyme, *N*-acetyl lactosamine synthetase, to produce lactose (Hill & Brew, 1975)). The purpose of these co-factors is usually to supply a specific chemical function that is not possible with the enzyme alone. Thus a metal co-factor may undergo a cyclic oxidation–reduction process or act as a chelating agent for the substrate. Not only are enzymes highly specific for a particular reaction but they are also highly stereospecific and, in a reaction involving either an L or D isomer, only one form reacts. The other isomer frequently acts as an inhibitor for the reaction. The active site is therefore a highly organised, stereospecific, three-dimensional region of the molecule. Studies of the mechanism of action of enzymes involve the use of a multitude of different techniques such as kinetic measurements, chemical modification of the enzyme, X-ray structure analysis, electron spin resonance and nuclear magnetic resonance spectroscopy, which provide complementary and sometimes overlapping information about the nature of the active site.

2.2. Single-intermediate mechanism

Early studies on enzyme kinetics date back to 1902 when Brown published work on the enzyme-catalysed hydrolysis of sucrose by invertase. The non-enzymic hydrolysis of sucrose to glucose and fructose under acid conditions had earlier been reported (Wilhelmy, 1850). He found that the rate of decrease of sucrose concentration in acid solution was proportional to the sucrose concentration at any time, i.e. $-d[A]/dt = k[A]$. The non-enzymic process is therefore a first order process. In reality, since the hydrolysis depends on the concentration of water and hydronium ion, it is a third order reaction; since, however, the concentrations of the latter do not alter during the experiment, the hydrolysis appears to be first order. It is in fact a case of pseudo first order kinetics. Brown showed that the enzyme-catalysed hydrolysis of sucrose differed markedly from the acid-catalysed reaction. He found that the rate of reaction was independent of sucrose concentration, i.e. it obeyed zero order kinetics with respect to sucrose. The rate of the reaction depended solely on the quantity of enzyme present. These results cannot be explained by the simple equation,

$$\text{Enzyme [E]} + \text{sucrose [S]} \rightarrow \text{enzyme [E]} + \text{product [P]}.$$

Brown's interpretation of these results has formed the basis for all subsequent enzyme mechanisms. He suggested that an essential step in the enzyme-catalysed reaction was the formation of a complex between the enzyme and sucrose and it was the breakdown of this complex into products that was observed. Thus, an additional step has to be included in the reaction mechanism:

$$\text{Enzyme [E]} + \text{sucrose [S]} \underset{k_{-1}}{\overset{k_1}{\rightleftharpoons}} \text{complex [ES]} \xrightarrow{k_2} \text{enzyme [E]} + \text{products [P]}.$$

If the equilibrium between the enzyme, sucrose and complex is attained rapidly, the rate-limiting step will be the decomposition of the enzyme–sucrose complex and hence if $[S] \gg [E]$ the rate of the reaction will appear to be independent of [S]. Later work by Michaelis & Menten (1913) on the same system led them to formulate in precise terms the concept of an enzyme–substrate complex. They proposed the scheme

$$E + S \underset{k_{-1}}{\overset{k_1}{\rightleftharpoons}} ES \xrightarrow{k_2} E + P \tag{2.1}$$

where [ES] is the intermediate complex (the Michaelis–Menten complex). The Michaelis constant, K_S, is the apparent *dissociation* constant of the enzyme–substrate complex:

$$K_S = \frac{[E][S]}{[ES]}. \tag{2.2}$$

Since

$$[E_0] = [E] + [ES] \tag{2.3}$$

then

$$[ES] = \frac{[S][E_0]}{K_S + [S]}$$

and $$v = \frac{k_2[E_0][S]}{K_S + [S]}.$$

This latter equation is the Michaelis–Menten equation. Strictly speaking an equation similar to equation (2.3) should be written for the substrate, i.e. $[S_0] = [S] + [ES] + [P]$; since the concentration of substrate is usually much larger than that of the enzyme and the velocity v is the initial velocity at $t = 0$ ($[P] = 0$), the omission of this equation has little effect on the final equation. From equation (2.2) it can be seen that the dimensions of K_S are those of concentration, and will be expressed as mol l^{-1}, or M. At low substrate concentrations, only a small amount of the enzyme will be in the form of the enzyme–substrate complex, but as the substrate concentration is increased a point is reached when all the enzyme molecules are in the form of the complex. Any further increase in the substrate concentration will therefore not lead to an increase in the velocity of the reaction. The enzyme is then saturated with substrate.

The assumption that the enzyme–substrate complex was in equilibrium with the reactants was criticised by Briggs & Haldane (1925). They introduced the concept of a steady state in the enzymic reaction and showed that the equilibrium concept of Michaelis and Menten was not essential. The steady-state treatment is based on the hypothesis that the concentration of certain reaction intermediates such as the enzyme–substrate complex does not change rapidly during the course of the reaction. If [ES] is the concentration of the intermediate then $d[S]/dt \gg d[ES]/dt \approx 0$. This assumption is true except for the initial part of the reaction before the steady state is reached and also towards the end of the reaction when the condition $[S] \gg [E]$ is no longer true. Briggs & Haldane argued that at all other times the net rate of change of the ES complex was zero. The kinetic equations for equation (2.1) are:

$$\frac{d[ES]}{dt} = k_1[E][S] - k_{-1}[ES] - k_2[ES], \tag{2.4}$$

$$\frac{d[P]}{dt} = k_2[ES], \tag{2.5}$$

and $$\frac{d[E]}{dt} = -k_1[E][S] + k_{-1}[ES] + k_2[ES].$$

We can also write two conservation equations

$$[E_0] = [E] + [ES] \tag{2.6}$$

and $$[S_0] = [S] + [ES] + [P].$$

In the latter case, with $[S_0] \gg [E_0]$, at $t = 0$ ($[P] = 0$), the conservation equation for [S] approximates to $[S_0] \approx [S]$. If the assumption is made that $[S] \gg [E]$, there will be established a steady-state situation in which the concentration of the ES complex will change very slowly relative to [S] and [P] i.e. d[ES]/d$t \approx 0$. Equations (2.4) and (2.6) can be combined to give

$$k_1([E_0] - [ES])\,[S] = (k_{-1} + k_2)\,[ES];$$

hence

$$[ES] = \frac{k_1[E_0]\,[S]}{(k_2 + k_{-1}) + k_1[S]} \tag{2.7}$$

and so the rate of the reaction, using equations (2.5) and (2.7), becomes

$$v = \frac{d[P]}{dt} = \frac{k_1 k_2 [E_0]\,[S]}{k_2 + k_{-1} + k_1[S]}$$

$$\text{and } v = \frac{k_2[E_0]\,[S]}{K_m + [S]} \tag{2.8}$$

where $K_m = (k_2 + k_{-1})/k_1$. Equation (2.8) is therefore the same as the Michaelis–Menten equation except that K_m replaces K_S. Notice that $K_m \approx K_S$ if $k_{-1} \gg k_2$. At high substrate concentrations where $[S] \gg K_m$ equation (2.8) becomes $v = k_2[E_0]$. The reaction therefore reaches a maximum value $V = k_2[E_0]$ and hence equation (2.8) can be written as

$$v = \frac{V[S]}{K_m + [S]}. \tag{2.9}$$

V and K_m are two important kinetic parameters that are very useful for describing the properties of an enzyme-catalysed reaction. Equation (2.9) is then the steady-state rate equation for a homogeneous reaction involving a recycling catalyst. It describes a rectangular hyperbola, with asymptotes at $[S] = -K_m$ and $v = V$. The centre of this conic section is at the point $(-K_m, V)$. The portion of the curve relevant to enzyme kinetics is that section of the hyperbola having positive values of [S]. At high values of [S], $v = V$, and under these conditions the reaction velocity is constant and independent of [S] and hence is an example of zero order kinetics. When $[S] > 100\,K_m$ the deviation from zero order kinetics is less than 1%; even when $[S] > 10\,K_m$ the deviation from zero order kinetics is only 9%. If $[S] \ll K_m$ then the equation becomes

$$v = \frac{V[S]}{K_m}$$

and the reaction obeys first order kinetics with V/K_m as the rate constant. When [S] is between 0.1 K_m and 10 K_m the reaction is intermediate between zero and first order. The two kinetic parameters V and K_m define the shape

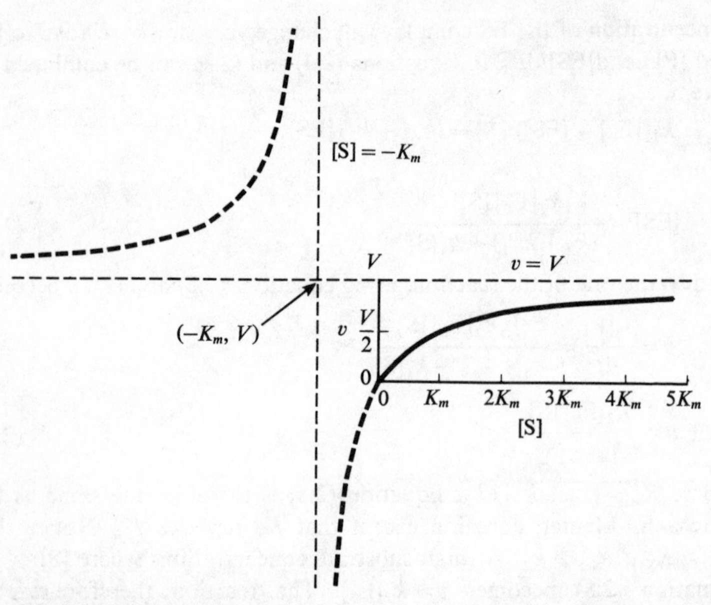

Fig. 2.2. Schematic plot of the Michaelis–Menten equation, equation (2.9), showing the rectangular hyperbolic nature of the equation. Only the solid line, i.e. positive values of [S], is relevant to enzyme kinetic studies. The asymptotes to the curve are at $v = V$ and $[S] = -K_m$ and the centre of the conic section is at the point $(-K_m, V)$.

of the curve. The properties of this curve may be seen from fig. 2.2 or from table 2.1. It will be seen that $v = V/2$ when $[S] = K_m$. This provides a simple definition of K_m; it is that concentration of substrate which gives half the maximum velocity. The more firmly an enzyme binds its substrate, the smaller will be the value of K_m. Since K_m appears in the denominator of equation (2.9), the smaller the value of K_m, the bigger v will be, provided that [S] is not approaching the level required to saturate

TABLE 2.1. *Properties of the Michaelis–Menten equation* $v = V[S]/(K_m + [S])$

[S]	v
1000 K_m	0.999 V
100 K_m	0.99 V
10 K_m	0.91 V
3 K_m	0.75 V
K_m	0.50 V
0.33 K_m	0.25 V
0.10 K_m	0.091 V
0.01 K_m	0.01 V

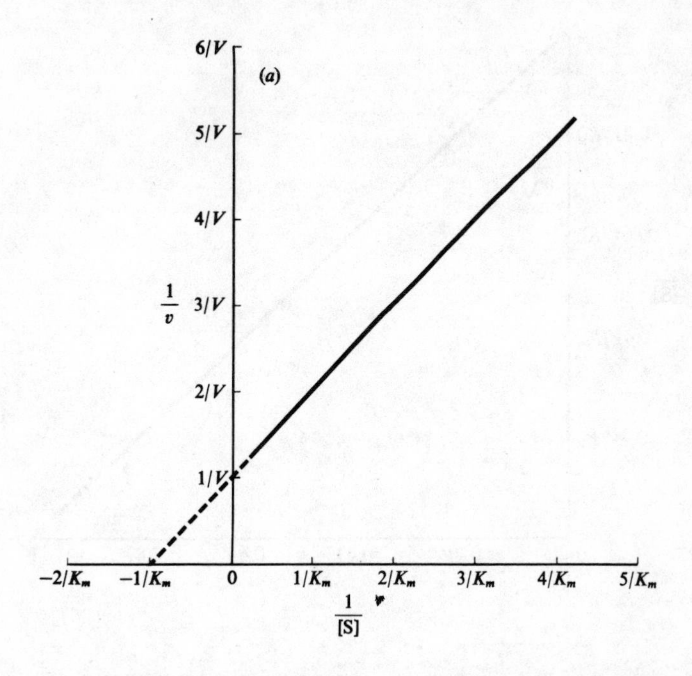

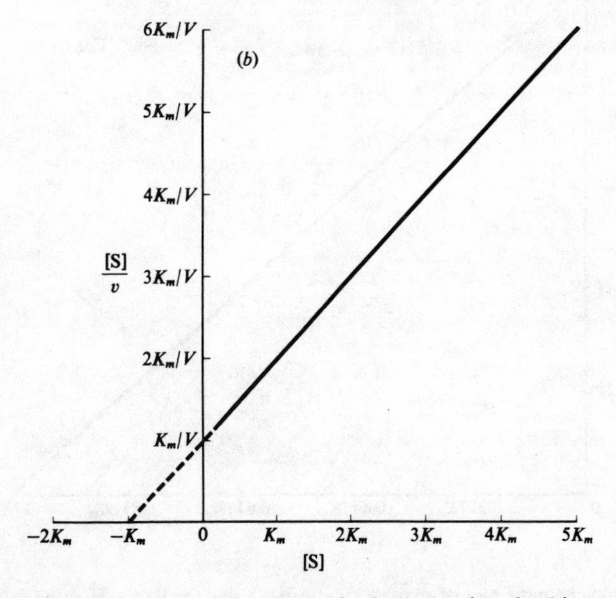

Fig. 2.3. (*a*) Schematic plot of enzyme kinetic data using the Lineweaver–Burk equation, equation (2.10). (*b*) Schematic plot of enzyme kinetic data using the Hanes equation.

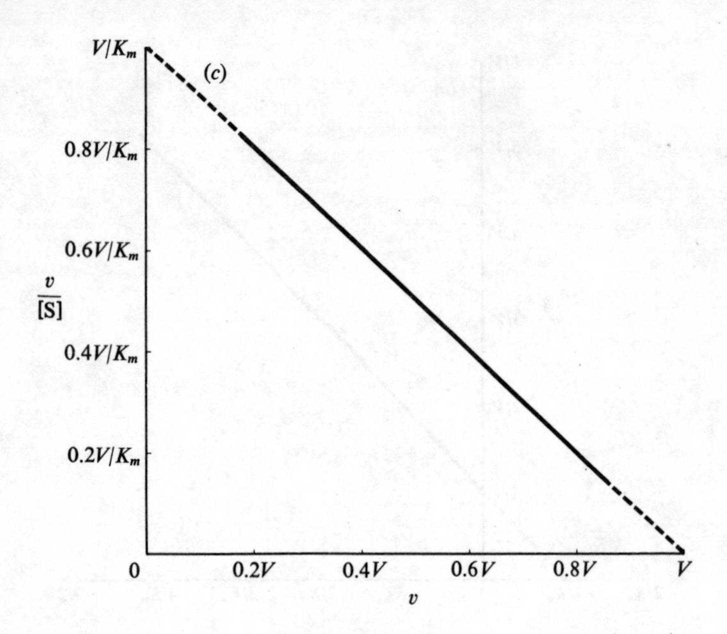

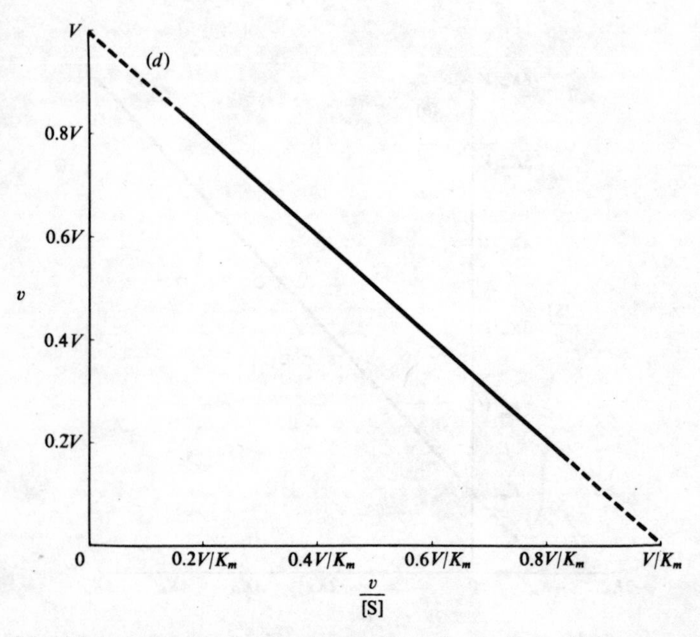

Fig. 2.3. (*c*) Schematic plot of enzyme kinetic data using the Eadie–Hofstee equation, equation (2.11). (*d*) Schematic plot of enzyme kinetic data using the Woolf equation, equation (2.12).

the enzyme. This has led to the concept of 'better binding–better catalysis', but since K_m is not a simple dissociation constant, it is only partly true.

The equation of a rectangular hyperbola is not a convenient form for evaluating the kinetic constants V and K_m. The most common procedure is that due to Lineweaver & Burk (1934). If equation (2.9) is inverted it becomes

$$\frac{1}{v} = \frac{K_m}{V}\frac{1}{[S]} + \frac{1}{V} \tag{2.10}$$

and hence a graph of $1/v$ against $1/[S]$ gives a straight line with a slope of K_m/V and ordinal intercept $1/V$ (fig. 2.3*a*). A simple least squares linear regression analysis of data in this form is definitely not recommended as inversion distorts the error span (Wilkinson, 1961). Data can be handled in this form, however, if they are appropriately weighted (appendix II). Several other linear plots may be employed for determining V and K_m. For example, multiplying equation (2.10) by [S] gives

$$\frac{[S]}{v} = \frac{K_m}{V} + \frac{[S]}{V}$$

(Hanes, 1932), so that a plot of $[S]/v$ against [S] gives a straight line with a slope of $1/V$ and ordinal intercept K_m/V (fig. 2.3*b*). Again, equation (2.9) can be arranged into the form proposed by Eadie (1952) and Hofstee (1952):

$$\frac{v}{[S]} = \frac{V}{K_m} - \frac{v}{K_m}. \tag{2.11}$$

A plot of $v/[S]$ against v is a straight line of slope $-1/K_m$ and ordinal intercept V/K_m (fig. 2.3*c*). Another version is that due to Woolf (1932):

$$v = V - \frac{vK_m}{[S]}, \tag{2.12}$$

so that a plot of v against $v/[S]$ is a straight line of slope $-K_m$ and ordinal intercept V (fig. 2.3*d*). Equations (2.11) and (2.12) are not recommended as the velocity term, v, which is subject to much greater error than [S], appears on both sides of the equations. A recommended range of substrate concentrations for the evaluation of V and K_m is from 0.2 K_m to 5 K_m (Cleland, 1967). Dowd & Riggs (1965) have compared the reliability of three of the linear transformations of the Michaelis–Menten equation (equations (2.10), (2.11) and (2.12)) and stress that the Lineweaver–Burk plot is most unreliable when using unweighted data and fitting lines by eye. The most reliable method, statistically, is to use a computer program to fit the data directly to the rectangular hyperbola, equation (2.9), by an iterative procedure and to use one of the linear plots to provide the initial starting values of K_m and V for the program (appendix II).

All the methods outlined above for the treatment of enzyme kinetic data, involve the determination of the initial velocity of the enzyme-catalysed reaction for a series of differing initial concentrations of substrate. These data are then plotted in accordance with one of the linear transforms of the Michaelis–Menten equation and the parameters K_m and V are obtained. This procedure is of course equivalent to the differential method outlined in section 1.3.1. An alternative procedure, which, at least in theory, could be used to obtain the values of K_m and V, is to fit the data from a single time course experiment to a suitable integrated rate equation. The Michaelis–Menten equation, equation (2.9), is a simple differential equation describing the rate of loss of substrate or rate of formation of product and therefore can readily be integrated by separation of the variables:

$$v = \frac{d[P]}{dt} = \frac{-d[S]}{dt} = \frac{V[S]}{K_m + [S]}$$

hence

$$-\int_{t=0}^{t} V dt = \int_{[S_0]}^{[S_t]} \left(\frac{K_m}{[S]} + 1 \right) d[S]$$

and $$Vt = K_m \ln([S_0]/[S_t]) + ([S_0] - [S_t]) \qquad (2.13)$$

where

$$[S_0] = [S_t] + [P_t].$$

Equation (2.13) can be rearranged to give

$$\frac{([S_0] - [S_t])}{t} = V - K_m \frac{\ln([S_0]/[S_t])}{t} \qquad (2.14)$$

so that a plot of $([S_0] - [S_t])/t$ against $(1/t)\ln([S_0]/[S_t])$ is a straight line of ordinal intercept V, abscissal intercept V/K_m and slope $-K_m$ (fig. 2.4). The problem associated with using equation (2.14) is that no allowance is made for any change in the initial-velocity mechanism shown above during the progress of the reaction. The reaction may be inhibited by products, approach an equilibrium position or there may be changes in the experimental conditions, such as pH, during the reaction. If the

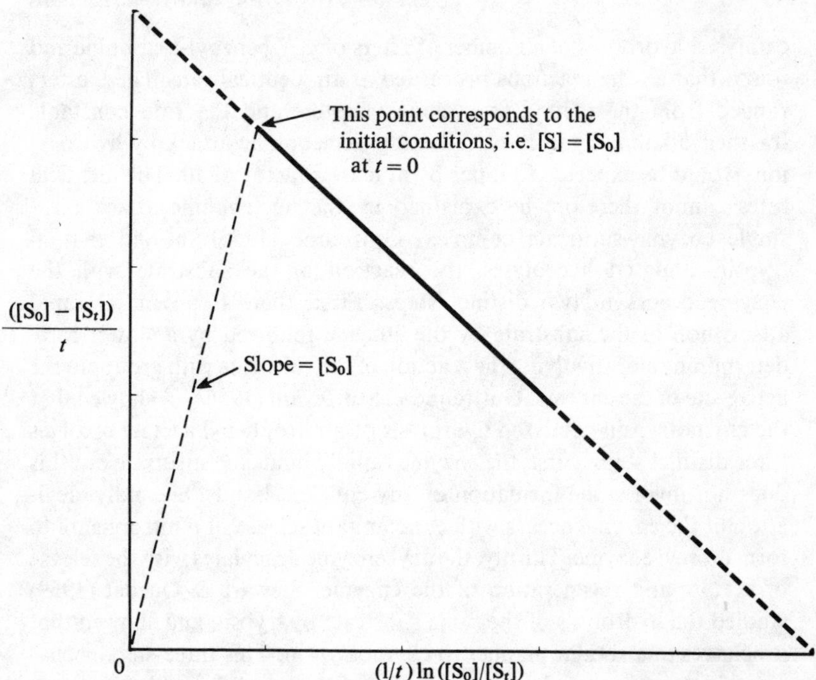

Fig. 2.4. Schematic plot of enzyme kinetic data using the integrated Michaelis–Menten equation, equation (2.14). The results from a complete product/time plot are plotted as $([S_0] - [S_t])/t$ against $(1/t)\ln([S_0]/[S_t])$.

enzymic reaction obeys Michaelis–Menten kinetics throughout the complete time course of the experiment, then it is apparent from equation (2.14) and fig. 2.4 that the results obtained for differing initial substrate concentrations will all lie along the same line. If the position of the line changes with initial substrate concentration then the behaviour is not described by equation (2.9) and possible explanations for these deviations are mentioned above. Corresponding integrated rate equations can be obtained that describe the analysis of a complete time course experiment in the presence of an inhibitor (equations (3.31), (3.32) and (3.33)) or when the product of the reaction inhibits the forward reaction (equation (3.35)).

2.3. Double-intermediate mechanism

Is the simple two-step mechanism proposed in equation (2.1) capable of explaining all the kinetic results obtained for single substrate–single enzyme reactions? Schwert & Eisenberg (1949) studied the trypsin-catalysed hydrolysis of a number of esters of α-*N*-benzoyl-L-arginine and found that all the reactions proceeded at an identical rate. These esters ranged from the methyl to cyclohexyl esters and the rate constants for their alkaline hydrolysis, reflecting nucleophilic attack by hydroxyl ion, would be expected to differ by at least a factor of 10. The identical rates cannot therefore be explained by the nucleophilic attack on a single enzyme–substrate complex. Gutfreund (1955) showed that in trypsin-catalysed hydrolyses, the reaction of the substrate with the enzyme occurs in two distinct steps. First, there is an initial rapid adsorption of the substrate by the enzyme followed by a slower rate-determining step involving the reaction of the substrate with groups in the active site of the enzyme. Gutfreund & Sturtevant (1956*a*, *b*) showed that the chymotrypsin-catalysed hydrolysis of *p*-nitrophenyl acetate involves three distinct steps. First, the enzyme rapidly binds the substrate but this does not involve the formation of any covalent bonds. Secondly, acylation of the enzyme occurs with concomitant release of *p*-nitrophenol to form an acyl enzyme. Thirdly, the acyl enzyme deacylates with the release of acetate and regeneration of the enzyme. Stewart & Ouellet (1959) studied the hydrolysis of the same substrate by trypsin and showed that it behaves in a similar manner to chymotrypsin. This three-step mechanism for trypsin and chymotrypsin, as well as a number of other serine proteinases, has been confirmed by many workers. The reaction mechanism can be described by equation (2.15)

$$E + S \underset{k_{-1}}{\overset{k_1}{\rightleftharpoons}} ES \xrightarrow[\downarrow P_1]{k_2} ES' \xrightarrow{k_3} E + P_2. \qquad (2.15)$$

For the chymotrypsin-catalysed hydrolysis of *p*-nitrophenyl acetate, the first step, involving the formation of a non-covalent enzyme–substrate complex (ES), has a rate constant $k_1 > 10^6$ l mol^{-1} s^{-1}. The acylation step, involving the formation of the covalently linked acyl enzyme (ES′) and the release of *p*-nitrophenol (P_1), has a rate constant k_2 of ~3 s^{-1}. Deacylation of the acyl enzyme takes place at a rate determined by k_3 (~0.025 s^{-1}) with the liberation of the acetate group (P_2) and regeneration of the enzyme. This mechanism can now explain the results obtained by Schwert & Eisenberg. In the trypsin-catalysed hydrolysis of a series of esters of α-*N*-benzoyl-L-arginine, a common acyl enzyme, α-*N*-benzoyl-L-arginyl-trypsin, is formed and it is the deacylation of this common intermediate that is observed experimentally. Obviously in this case k_3 must be the rate-limiting step for the reaction.

Does this three-step mechanism have any effect on the Michaelis–Menten equation, equation (2.9), for the steady-state region of the reaction? The reaction can be described by the differential equations (2.16), (2.17), (2.18), (2.19) and (2.20) and the conservation equations (2.21) and (2.22).

$$\frac{\mathrm{d}[\mathrm{E}]}{\mathrm{d}t} = -k_1[\mathrm{E}][\mathrm{S}] + k_{-1}[\mathrm{ES}] + k_3[\mathrm{ES'}] \tag{2.16}$$

$$\frac{\mathrm{d}[\mathrm{ES}]}{\mathrm{d}t} = k_1[\mathrm{E}][\mathrm{S}] - [\mathrm{ES}](k_{-1} + k_2) \tag{2.17}$$

$$\frac{\mathrm{d}[\mathrm{ES'}]}{\mathrm{d}t} = k_2[\mathrm{ES}] - k_3[\mathrm{ES'}] \tag{2.18}$$

$$\frac{\mathrm{d}[\mathrm{P_1}]}{\mathrm{d}t} = k_2[\mathrm{ES}] \tag{2.19}$$

$$\frac{\mathrm{d}[\mathrm{P_2}]}{\mathrm{d}t} = k_3[\mathrm{ES'}] \tag{2.20}$$

$$[\mathrm{S_0}] = [\mathrm{S}] + [\mathrm{ES}] + [\mathrm{ES'}] + [\mathrm{P_2}] \tag{2.21}$$

which at $t = 0$ and with $[\mathrm{S_0}] \gg [\mathrm{E_0}]$ approximates to

$$[\mathrm{S_0}] = [\mathrm{S}]$$

$$[\mathrm{E_0}] = [\mathrm{E}] + [\mathrm{ES}] + [\mathrm{ES'}]. \tag{2.22}$$

These could be solved in exactly the same way as the two-step mechanism by the steady-state approach of Briggs & Haldane, setting $\mathrm{d}[\mathrm{E}]/\mathrm{d}t = \mathrm{d}[\mathrm{ES}]/\mathrm{d}t = \mathrm{d}[\mathrm{ES'}]/\mathrm{d}t = 0$. It is convenient at this point, however, to introduce a diagrammatic method for deriving steady-state rate equations which was formulated by King & Altman (1956). First, it is necessary to decide on the mechanism of the reaction and assign rate constants to the various steps. The different possible forms of the enzyme are then arranged into a convenient geometrical pattern with arrows to indicate the possible interconversions of enzyme species. Each arrow is labelled with the appropriate rate constant and the concentration of any substrate or product that may be consumed in that step. For reversible steps, there will of course be two arrows, each with its own rate constant and associated concentration factors. The three-step mechanism of equation (2.15), in which both the acylation and deacylation steps are irreversible, can be represented by the scheme:

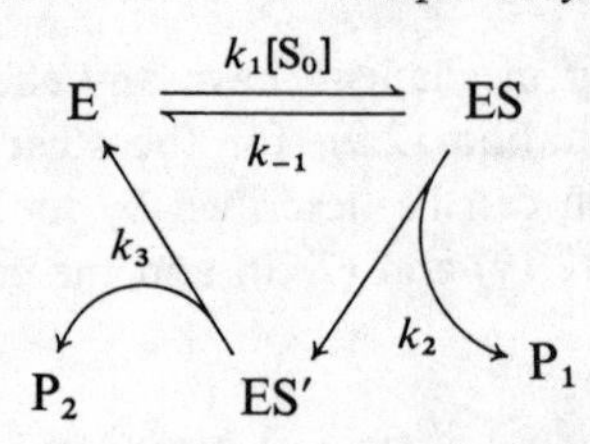

The next step is to write down all the possible patterns (made up of one less line than there are enzyme species) with which all the enzyme forms can be connected. Not more than one arrow can emanate from each enzyme species and hence closed loops are not allowed. For this reaction, there are the following vector diagrams:

Enzyme species	*Vector diagrams*	*Kinetic terms*
E	E ← k_3, ↙ k_2 ; E ← k_{-1}, ← k_3	$k_2 k_3 + k_{-1} k_3$
ES	$\xrightarrow{k_1[S]}$ ES, ↖ k_3	$k_1 k_3[S]$
ES′	$\xrightarrow{k_1[S]}$, ↙ k_2 ES′	$k_1 k_2[S]$

Once all the patterns have been found and written down, the difficult part is over. For each enzyme species, EX_i, $i = 1, n$, there is an equation of the form

$$\frac{[EX_i]}{[E_0]} = \frac{\text{sum of the terms in the vector diagrams for the species } EX_i}{\text{sum of all the terms in all the vector diagrams for all the species } EX_i}$$

$$i = 1, n \qquad\qquad i = 1, n$$

where

$$[E_0] = [EX_1] + [EX_2] + \cdots + [EX_n] = \sum_{i=1}^{n} [EX_i].$$

For this example, therefore,

$$\frac{[E]}{[E_0]} = \frac{k_2 k_3 + k_{-1} k_3}{\Sigma}$$

$$\frac{[ES]}{[E_0]} = \frac{k_1 k_3[S]}{\Sigma} \tag{2.23}$$

$$\frac{[ES']}{[E_0]} = \frac{k_1 k_2[S]}{\Sigma} \tag{2.24}$$

where

$$\Sigma = k_2 k_3 + k_{-1} k_3 + k_1 k_3[S] + k_1 k_2[S]$$

and $[E_0] = [E] + [ES] + [ES']$.

In the scheme shown in equation (2.15), the reaction can be followed by monitoring either the formation of product P_1 or the formation of P_2.

For P_1,

$$\frac{d[P_1]}{dt} = k_2[ES],$$

from equation (2.23)

$$[ES] = \frac{k_1 k_3 [S] [E_0]}{k_3(k_2 + k_{-1}) + k_1[S] (k_2 + k_3)}$$

and hence

$$\frac{d[P_1]}{dt} = \frac{k_1 k_2 k_3[S] [E_0]}{k_3(k_2 + k_{-1}) + k_1[S] (k_2 + k_3)}.$$

Similarly for P_2,

$$\frac{d[P_2]}{dt} = k_3[ES'],$$

but from equation (2.24)

$$[ES'] = \frac{k_1 k_2[S] [E_0]}{k_3(k_{-1} + k_2) + k_1[S] (k_2 + k_3)}$$

and hence

$$\frac{d[P_2]}{dt} = \frac{d[P_1]}{dt} = v = \frac{k_1 k_2 k_3[S] [E_0]}{k_3(k_{-1} + k_2) + k_1[S] (k_2 + k_3)} \tag{2.25}$$

or

$$v = \frac{V[S]}{K_m + [S]}$$

where

$$V = \frac{k_2 k_3[E_0]}{(k_2 + k_3)} = k_{cat}[E_0] \quad \text{and} \quad K_m = \frac{k_3(k_{-1} + k_2)}{(k_2 + k_3)k_1}.$$

Note that the observed catalytic breakdown constant, k_{cat}, is no longer a single rate constant as in the two-step mechanism, but a combination of rate constants. The three-step mechanism of equation (2.15) gives an equation identical in form to that derived for the two-step mechanism, but the definition of the kinetic parameters, V and K_m, in terms of the individual rate constants for the reaction, are different. It is therefore not possible, from purely steady-state kinetic studies, to ascertain the number of enzyme intermediates in a reaction pathway. Note, from equation (2.25), that the initial velocity, v, of the reaction does not depend on the product that is monitored. This is generally true of most enzyme reactions, the exceptions being a few more complicated multi-substrate/multi-product reactions in which a reactant or product occurs more than once in the reaction pathway, i.e. there is no longer a 1:1 stoichiometry between reactants and products. The hydrolysis of *p*-nitrophenyl acetate by chymotrypsin can therefore be followed equally well by either ultraviolet absorption spectroscopy, monitoring the liberation of *p*-nitrophenol (P_1) or by pH-stat techniques, monitoring the formation of the acetate group (P_2).

For more complex reactions, it is possible to calculate the number of vector diagrams that have to be drawn for the King–Altman method. It is necessary to assume initially that all the steps are reversible. If there are m reversible steps interconnecting the n different enzyme species then there will be a total of $m!/(n-1)!(m-n+1)!$ vector diagrams. If the reaction mechanism contains cycles (i.e. more than one pathway connecting enzyme species), then it is necessary to subtract a certain number of vector diagrams depending on the number of cycles and the number of steps in each cycle. For every cycle of r steps ($2 < r \leqslant n-1$) there are $(m-r)!/(n-1-r)!(m-n+1)!$ combinations which involve this cycle and these must be subtracted from the total. A simple example will illustrate this more clearly. In the following mechanism, a modifier, M, can bind to the ES complex and provide two alternative breakdown paths for the complex.

$$\mathrm{E+S} \underset{k_{-1}}{\overset{k_1}{\rightleftharpoons}} \mathrm{ES} \underset{k_{-2}}{\overset{k_2}{\rightleftharpoons}} \mathrm{E+P}$$

$$\mathrm{ES+M} \underset{k_{-3}}{\overset{k_3}{\rightleftharpoons}} \mathrm{ESM} \underset{k_{-4}}{\overset{k_4}{\rightleftharpoons}} \mathrm{E+M+P}.$$

The geometric pattern that describes this mechanism is

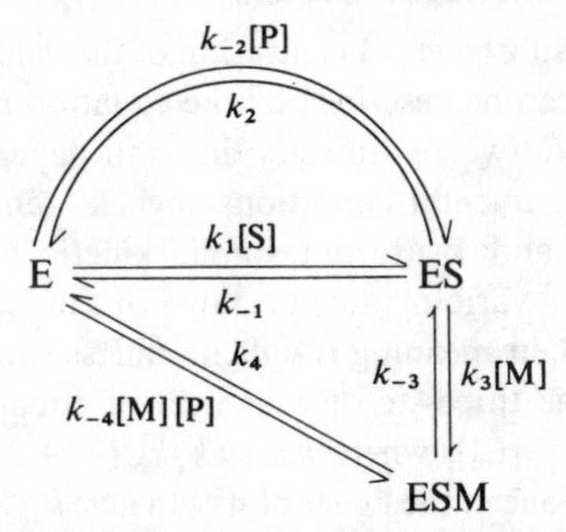

There are four reversible steps ($m = 4$) and three enzyme species ($n = 3$). The total number of patterns is $4!/(3-1)!(4-3+1)! = 6$. There is however one cycle, E ⇌ ES, consisting of two steps ($r = 2$) so that the number of vector diagrams that have to be subtracted is given by $(4-2)!/(3-1-2)!(4-3+1)! = 1$ $(0! = 1)$. The total number of vector diagrams is therefore five for each species. The five patterns are

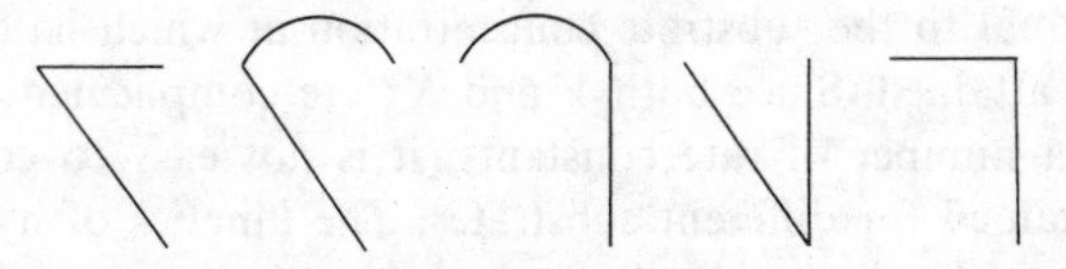

so that for E, for example, there would be

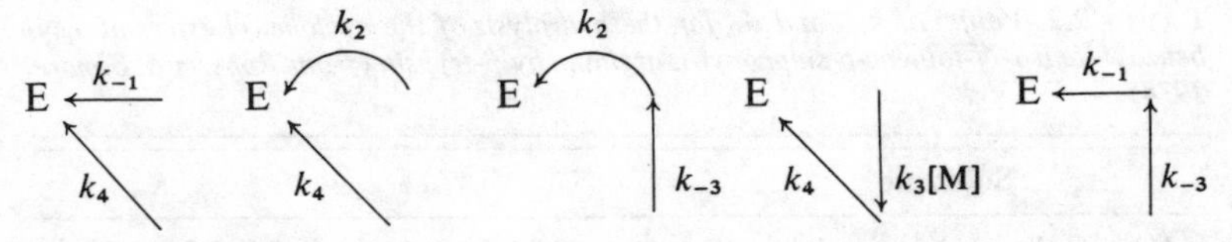

The King–Altman method is therefore very useful for deriving steady-state rate equations. For more complicated reaction mechanisms, however, the actual task of writing down all the vector diagrams for each enzyme species and the summation of all these terms becomes very tedious. Hurst (1969) has written a computer program that will evaluate all the combinations of rate constants and concentration terms for each enzyme species so that the task of producing steady-state rate equations for complicated mechanisms has been considerably reduced.

2.4. Comparison of enzyme parameters

The variation with substrate concentration of the velocity of an enzyme-catalysed reaction can be described by the equation $v = V[S]/([S] + K_m)$ where V and K_m are two parameters that will depend on the enzyme, substrate, and experimental conditions such as temperature, pH and ionic strength. K_m and V are not simple kinetic parameters but are combinations of the various rate constants in the reaction mechanism. The evaluation of their meaning is difficult unless some of the terms are predominant. In the three-step mechanism of equation (2.15), for instance, since $V = k_{cat}[E_0]$, where $k_{cat} = k_2k_3/(k_2 + k_3)$, then the determination of V from one of the linear plots can give k_{cat} (knowing $[E_0]$) but it will not be possible to obtain k_2 or k_3 unless (*a*) k_2 or k_3 is determined by a different method or (*b*) it can be shown that one of the rate constants k_2 or k_3 is rate-determining. If $k_2 \gg k_3$, then deacylation is the rate-determining step and hence $k_{cat} \approx k_3$. Similarly, if $k_3 \gg k_2$ then acylation is the rate-determining step and $k_{cat} \approx k_2$. Originally, Michaelis & Menten equated K_m with K_S, the enzyme–substrate dissociation constant, by assuming k_2 was small in relation to k_{-1}. For the three-step mechanism, K_m is not related directly to the affinity of the enzyme for the substrate but is an operationally determined parameter for a particular reaction equal to the substrate concentration at which half maximum velocity is attained. Since both V and K_m are complicated parameters involving a number of rate constants, it is not easy to compare the results obtained for different substrates. The kinetics of hydrolysis of two substrates by β-trypsin will illustrate the dilemma (table 2.2).

TABLE 2.2. *Values of k_{cat} and K_m for the hydrolysis of the cyclohexyl esters of α-N-benzoyl- and α-N-toluene-p-sulphonyl-L-arginine by β-trypsin (from Roberts & Elmore, 1974)*

Substrate	k_{cat}	K_m
α-*N*-benzoyl-L-arginine cyclohexyl ester	$24.5 \pm 0.4\ s^{-1}$	$1.5 \pm 0.1 \times 10^{-6}$ M
α-*N*-toluene-*p*-sulphonyl-L-arginine cyclohexyl ester	$99 \pm 1\ s^{-1}$	$5.8 \pm 0.1 \times 10^{-6}$ M

What criteria can be used to compare these two substrates? Is the α-*N*-toluene-*p*-sulphonyl derivative a better substrate because of its higher k_{cat} or is the α-*N*-benzoyl derivative a better substrate because of its lower K_m? From the definition of K_m, it can be seen that if $k_2 \gg k_3$ (which is true in this case because deacylation can be shown to be rate-determining) and if k_1 and k_{-1} are essentially the same for the two substrates, then an increase in k_3 will cause a similar increase in K_m. For these reasons, Brot & Bender (1969) argued that neither V nor K_m alone are suitable parameters for comparing the kinetic results obtained for different substrates. They proposed that the ratio k_{cat}/K_m, which they called the *specificity constant*, gives a better indication of the effectiveness of the substrate.

The ratio

$$\frac{k_{cat}}{K_m} = \frac{k_2 k_3 k_1(k_2+k_3)}{(k_2+k_3)k_3(k_{-1}+k_2)}$$

$$= \frac{k_2 k_1}{(k_{-1}+k_2)}$$

$$= \frac{k_2}{K'_m}$$

where K'_m is the Michaelis constant for the acylation step. In some cases, K'_m is not very different from the K_I^D (the inhibition constant) for the D-stereoisomer of the substrate, so that K'_m is similar to K_S, the dissociation constant for the enzyme–substrate complex. An approximate value of k_2 could therefore be obtained from the substitution of K_I^D for K'_m. If K'_m or K_S do not vary appreciably with pH, then the variation of k_{cat}/K_m with pH indicates that k_2 is changing with pH. If the concept of better binding–better catalysis is accepted, then the higher the value of the specificity constant, the better is the substrate. From these results for the two substrates mentioned earlier, the specificity constants are virtually the same, 1.7×10^7. The large value indicates that they are highly specific substrates for β-trypsin.

2.5. Steady-state kinetics in the presence of added nucleophiles

The aim of kinetic investigations is to obtain as much information as possible about the reaction steps of the mechanism being studied using the experimental techniques that are available. Steady-state methods of studying three-step mechanisms yield the two kinetic parameters V and K_m, both of which are functions of a number of rate constants.

$$V = k_{cat}[E_0] \quad \text{where} \quad k_{cat} = \frac{k_2 k_3}{(k_2 + k_3)}$$

and

$$K_m = \frac{(k_{-1} + k_2)k_3}{k_1(k_2 + k_3)}.$$

Only when $k_2 \gg k_3$ or $k_3 \gg k_2$, will a knowledge of V allow the determination of a single rate constant, k_3 or k_2.

An interesting experimental method which is applicable to those enzymes that form an acyl intermediate and catalyse reactions by the three-step mechanism (equation (2.15)) allows the determination of both k_2 and k_3 (Berezin, Kazanskaya & Klyosov, 1971; Hinberg & Laidler, 1972*a,b*). Enzymes for which this method is suitable are serine proteinases, for example trypsin and chymotrypsin, elastase, thrombin, acetyl cholinesterase, some phosphotransferases and possibly other enzymes in which a water molecule takes part in the overall mechanism. For these enzymes, the second stage in the mechanism is the nucleophilic attack by a water molecule on the acyl enzyme to give free enzyme and the product P_2. In this method, an additional nucleophile is added to the reaction mixture to compete with the water molecule and give rise to an alternative product P_3. Berezin *et al.* (1971) studied the hydrolysis of ester substrates of α-chymotrypsin in the presence of 1,4-butanediol as the added nucleophile.

The new mechanism can be described by

$$E + S \underset{k_{-1}}{\overset{k_1}{\rightleftharpoons}} ES \xrightarrow[\downarrow P_1]{k_2} ES' \xrightarrow{k_3} E + P_2$$

$$ES' \xrightarrow{k_4[N]} E + P_3$$

where [N] is the concentration of the added nucleophile. The rate constant k_3 is actually a composite rate constant, $k_3 = k_3'$ $[H_2O]$, where $[H_2O]$ is the concentration of water (~55 M). The mechanism does not propose that there is a binding site for either water molecules or the nucleophile.

We can derive the steady-state rate equations for this system using the King–Altman method:

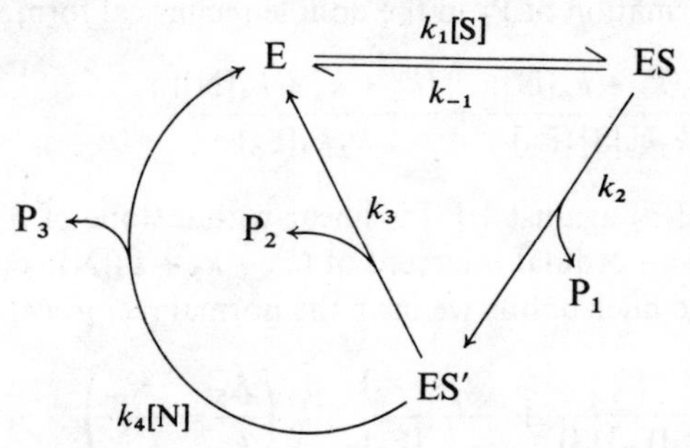

$$[E]/[E_0] = \{k_3(k_{-1}+k_2) + k_4[N](k_{-1}+k_2)\}/\Sigma$$
$$[ES]/[E_0] = \{k_1[S](k_3+k_4[N])\}/\Sigma$$
$$[ES']/[E_0] = k_1 k_2[S]/\Sigma$$

where $\Sigma = k_3(k_{-1}+k_2) + k_4[N](k_{-1}+k_2) + k_1[S](k_2+k_3+k_4[N])$. For the three products P_i, $i = 1, 3$, we can write velocity equations

$$v_i = \frac{k_{\text{cat}_i}[E_0][S]}{K_m + [S]}$$

where

$$k_{\text{cat}_1} = \frac{k_2(k_3+k_4[N])}{(k_2+k_3+k_4[N])}$$

$$k_{\text{cat}_2} = \frac{k_2 k_3}{(k_2+k_3+k_4[N])}$$

$$k_{\text{cat}_3} = \frac{k_2 k_4[N]}{(k_2+k_3+k_4[N])}$$

and

$$K_m = \frac{(k_{-1}+k_2)(k_3+k_4[N])}{k_1(k_2+k_3+k_4[N])}$$
$$= K_S\frac{(k_3+k_4[N])}{(k_2+k_3+k_4[N])}.$$

One method of following the hydrolysis of the ester substrate is to monitor the release of the acidic product P_2 by means of a pH-stat. We are therefore interested only in the equation that relates to the velocity of formation of product P_2 since the new product P_3 is an ester and its presence would not be detected by pH-stat techniques. The equation for the velocity of formation of P_2 in the double reciprocal form is

$$\frac{1}{v_2} = \frac{K_S(k_3 + k_4[\mathrm{N}])}{k_2 k_3 [\mathrm{S}][\mathrm{E_0}]} + \frac{(k_2 + k_3 + k_4[\mathrm{N}])}{k_2 k_3 [\mathrm{E_0}]}.$$

So a plot of $1/v_2$ against $1/[\mathrm{S}]$ is linear with a slope of $K_S(k_3 + k_4[\mathrm{N}])/k_2 k_3[\mathrm{E_0}]$ and an ordinal intercept of $(k_2 + k_3 + k_4[\mathrm{N}])/k_2 k_3[\mathrm{E_0}]$. In the absence of the nucleophile we have the normal reciprocal equation

$$\frac{1}{v_2} = \frac{K_S}{k_2[\mathrm{E_0}][\mathrm{S}]} + \frac{(k_2 + k_3)}{k_2 k_3 [\mathrm{E_0}]} \quad \left(\frac{K_S}{k_2} = \frac{K_m}{k_{cat}}\right). \tag{2.26}$$

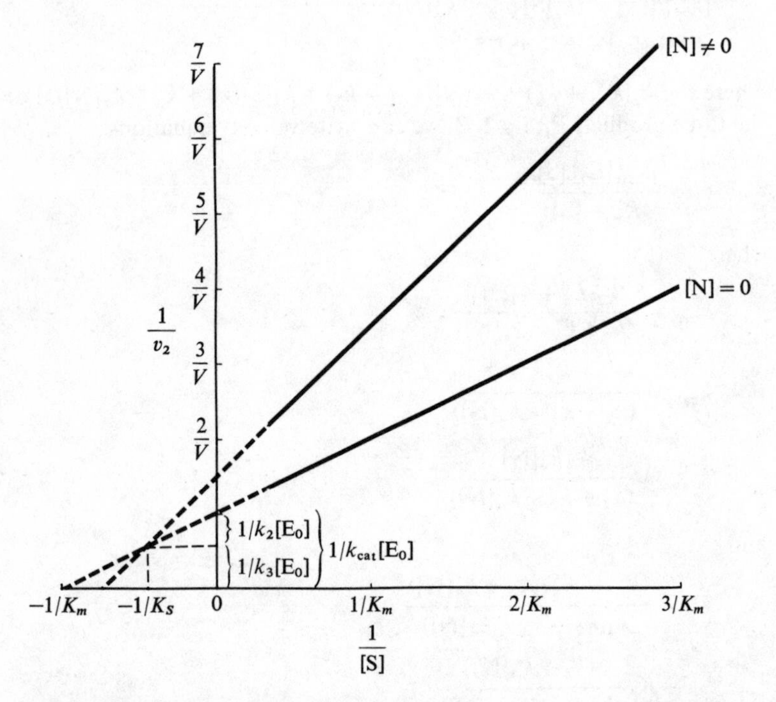

Fig. 2.5. Schematic plot of enzyme kinetic data obtained from reactions carried out in the presence of added nucleophiles. The results are plotted in the reciprocal form of $1/v_2$ against $1/[\mathrm{S}]$. The kinetic parameters are: $k_1 = 10^6\ \mathrm{M^{-1}\ s^{-1}}$, $k_2 = 10^3\ \mathrm{s^{-1}}$, $k_3 = 10^3\ \mathrm{s^{-1}}$, $k_4 = 5 \times 10^3\ \mathrm{s^{-1}}$, $k_{-1} = 10^3\ \mathrm{s^{-1}}$, $[\mathrm{E_0}] = 3 \times 10^{-8}$ M, $[\mathrm{N}] = 0$ and 0.2 M, $[\mathrm{S_0}] = 3 \times 10^{-3}$ M.

These two lines having different slopes will intersect at a point given by

$$\frac{K_S}{k_2[E_0][S]} + \frac{(k_2+k_3)}{k_2k_3[E_0]} = \frac{K_S(k_3+k_4[N])}{k_2k_3[S][E_0]} + \frac{(k_2+k_3+k_4[N])}{k_2k_3[E_0]},$$

i.e. when $1/[S] = -1/K_S$. Replacing this value in equation (2.26) we get $1/v_2 = 1/k_3[E_0]$. Hence from the ordinal intercepts in the presence and absence of the nucleophile and the intercept of the two lines we can obtain k_2, k_3, k_4 and K_S. A plot of $1/v_2$ against $1/[S]$ is shown in fig. 2.5. There are two main restrictions to the use of this method for evaluating k_2 and k_3. First, in the interests of precision, the ratio of k_2/k_3 should be in the range $10 > k_2/k_3 > 0.1$. Secondly, the nucleophile should not be bound by any of the enzyme species as this will alter the mechanism and the lines may not intersect. This method is, however, quite useful in that under certain conditions it allows the determination of the individual rate constants k_2 and k_3 without the need to resort to fast reaction techniques.

3 Enzyme inhibition

3.1. Introduction

The addition of a foreign substance to an organism can lead to the inactivation of a single enzyme in a major metabolic pathway. This inactivation may result in fatality, as in the case of cyanide poisoning. The effect of cyanide poisoning is to inactivate cytochrome oxidase, an important enzyme in the aerobic oxidation process. Failure to supply an antidote, in this case a compound for which cyanide has a high affinity, can lead to death in a few minutes. The discovery that many enzymes can be inhibited by the addition of small amounts of foreign substances, has led to the development of civil and military laboratories whose main purpose is the study of metabolic inhibitors. Modern insecticides are effective because they inactivate only essential enzymes in insect metabolism and not those in higher animals. Similarly, the 'nerve gases', such as phosphorofluoridate esters, are fairly specific inhibitors for the enzyme acetyl cholinesterase which is important in the control of muscle contraction. Thus the effect of a 'nerve gas' is to cause muscular spasm, vomiting, choking and cardiac arrest. In the cell, inhibition of certain key reactions plays a vital role in the fine control of metabolic pathways. Early stages in a multi-enzyme pathway may be finely controlled by the concentration level of the final product(s). Such control is called *feedback control* and is a common feature of many metabolic pathways (chapter 8). The mechanism can be likened to the use of negative feedback to control the gain of an electronic amplifier.

3.2. Types of inhibition

Inhibition of an enzyme can be either reversible or irreversible. When inhibition is reversible, the activity of the enzyme can be regained by removal of the inhibitor by physical or chemical means. There is a chemical equilibrium between the enzyme and the inhibitor which can be displaced in favour of the enzyme by the removal of the inhibitor by some means, e.g. dialysis, gel filtration. The removal of the inhibitor restores the enzyme activity to its original value. Irreversible inhibition, however, is characterised by complete loss of enzyme activity after a period of time even in the presence of low concentrations of the inhibitor (provided it is in excess of the enzyme). The enzyme activity cannot be regained by any physical treatment, e.g. dialysis, but it may be possible to regenerate the enzyme by suitable chemical treatment. The inhibition of acetyl cholinesterase by the 'nerve gases' is an example of irreversible inhibition. The 'nerve gas' acylates a serine hydroxyl group in the active site of the enzyme and since this involves the formation of a covalent bond, the acyl group cannot be removed by any physical treatment. The activity of the enzyme can, however, be restored by the displacement of the acyl group by the use of a powerful nucleophile, such as hydroxylamine. Irreversible inhibition is usually quantified in terms of the velocity of inhibition whereas reversible inhibition is expressed in terms of an equilibrium constant, K_I, for the enzyme and inhibitor. We shall only be concerned with the effect of reversible inhibitors but the reader who is interested in the design and mode of action of irreversible inhibitors is referred to a book by Baker (1967).

3.2.1. Competitive inhibition. This has sometimes been called 'inhibition by affinity' since the inhibitor and substrate both *compete* for the same binding site in the enzyme. The inhibitor is usually structurally related to the substrate (or more effectively to the substrate transition state (Lienhard, 1973)) so that it is bound at the site of the enzyme which determines specificity. The inhibitor, therefore, does not form a complex with any enzyme species in which the binding site is already occupied by the substrate. The simplest mechanism can be described by the following equations:

$$\mathrm{E}+\mathrm{S} \underset{k_{-1}}{\overset{k_1}{\rightleftharpoons}} \mathrm{ES} \xrightarrow{k_2} \mathrm{E}+\mathrm{P}$$

$$\mathrm{E}+\mathrm{I} \underset{k_{-3}}{\overset{k_3}{\rightleftharpoons}} \mathrm{EI}$$

$$K_I = \frac{[\mathrm{E}][\mathrm{I}]}{[\mathrm{EI}]} = \frac{k_{-3}}{k_3}.$$

Using the King–Altman method, we have

$$\mathrm{E} \underset{k_{-1}}{\overset{k_1[\mathrm{S}]}{\rightleftharpoons}} \mathrm{ES} \xrightarrow{k_2} \mathrm{E}$$

$$\mathrm{E} \underset{k_{-3}}{\overset{k_3[\mathrm{I}]}{\rightleftharpoons}} \mathrm{EI}$$

Enzyme species	*Vector diagrams*	*Kinetic terms*
E		$k_{-3}k_{-1} + k_{-3}k_2$
ES		$k_{-3}k_1[\mathrm{S}]$
EI		$k_{-1}k_3[\mathrm{I}] + k_2k_3[\mathrm{I}]$

but
$$\frac{[\mathrm{ES}]}{[\mathrm{E_0}]} = \frac{k_1k_{-3}[\mathrm{S}]}{k_{-3}(k_2+k_{-1}) + k_3[\mathrm{I}](k_2+k_{-1}) + k_1k_{-3}[\mathrm{S}]}$$

therefore

$$v = \frac{k_1k_2k_{-3}[\mathrm{E_0}][\mathrm{S}]}{(k_2+k_{-1})(k_{-3}+k_3[\mathrm{I}]) + k_1k_{-3}[\mathrm{S}]}$$

$$v = \frac{V[\mathrm{S}]}{K_m\left(1+\dfrac{[\mathrm{I}]}{K_I}\right) + [\mathrm{S}]} \tag{3.1}$$

where $V = k_2[\mathrm{E_0}]$ and $K_m = (k_2+k_{-1})/k_1$. Inversion of this equation (3.1) to a form similar to the Lineweaver–Burk equation gives

$$\frac{1}{v} = \frac{1}{V} + \frac{K_m}{V}\left(1+\frac{[\mathrm{I}]}{K_I}\right)\frac{1}{[\mathrm{S}]}. \tag{3.2}$$

A plot of $1/v$ against $1/[\mathrm{S}]$, therefore, has the same ordinal intercept $(1/V)$ in the presence or absence of inhibitor, as shown in fig. 3.1(*a*).

The criterion for competitive inhibition is that the intercept on the ordinal axis is unchanged on the addition of the inhibitor. The effect of the competitive inhibitor, therefore, is to produce an apparent increase in K_m by the factor $(1 + [I]/K_I)$, which, for this scheme, increases without limit as [I] increases. A replot of the slope, $K_m(1 + [I]/K_I)/V$, of the Lineweaver–Burk plot against [I] is a straight line (fig. 3.2, line 2) and hence this form of inhibition expressed by equation (3.2) is known as *linear competitive inhibition.*

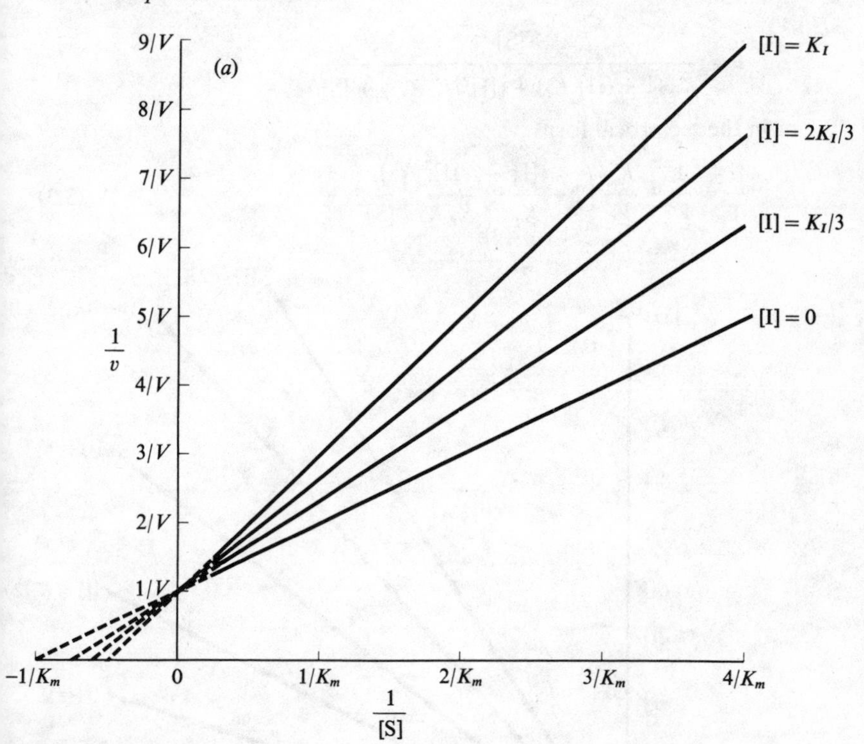

Fig. 3.1. (*a*) Schematic Lineweaver–Burk plot of linear competitive inhibition.

If the combination of a molecule of inhibitor, I, at the active site of the enzyme allows the reaction of a second molecule of the inhibitor so that two molecules of inhibitor contribute to the exclusion of the substrate, S, *parabolic competitive inhibition* is obtained:

$$E + S \underset{k_{-1}}{\overset{k_1}{\rightleftharpoons}} ES \xrightarrow{k_2} E + P$$

$$\mathrm{E} + \mathrm{I} \underset{k_{-3}}{\overset{k_3}{\rightleftharpoons}} \mathrm{EI} \qquad K_I = \frac{k_{-3}}{k_3} = \frac{[\mathrm{E}][\mathrm{I}]}{[\mathrm{EI}]}$$

$$\mathrm{EI} + \mathrm{I} \underset{k_{-4}}{\overset{k_4}{\rightleftharpoons}} \mathrm{EI_2} \qquad K_i = \frac{k_{-4}}{k_4} = \frac{[\mathrm{EI}][\mathrm{I}]}{[\mathrm{EI_2}]}$$

EI and EI_2 are dead-end complexes and again saturation with substrate will eliminate the inhibition. The initial velocity, v, is given by the equation

$$v = \frac{V[\mathrm{S}]}{K_m\{1 + ([\mathrm{I}]/K_I) + ([\mathrm{I}]^2/K_i\,K_I)\} + [\mathrm{S}]}$$

or in the reciprocal form

$$\frac{1}{v} = \frac{1}{V} + \frac{K_m}{V}\left(1 + \frac{[\mathrm{I}]}{K_I} + \frac{[\mathrm{I}]^2}{K_i\,K_I}\right)\frac{1}{[\mathrm{S}]}. \tag{3.3}$$

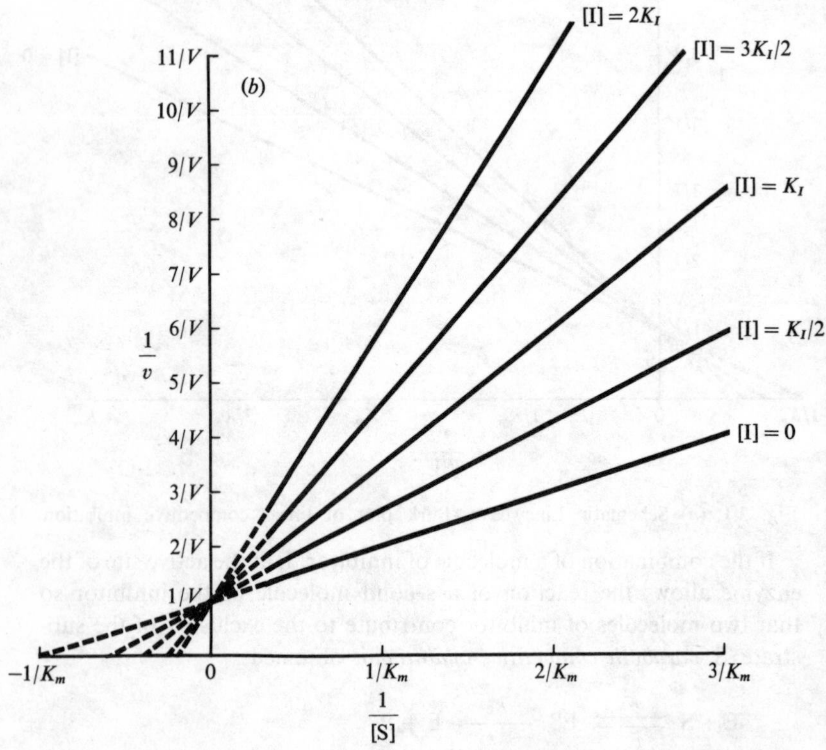

Fig. 3.1. (*b*) Schematic Lineweaver–Burk plot of parabolic competitive inhibition in which $K_i = 2K_I$.

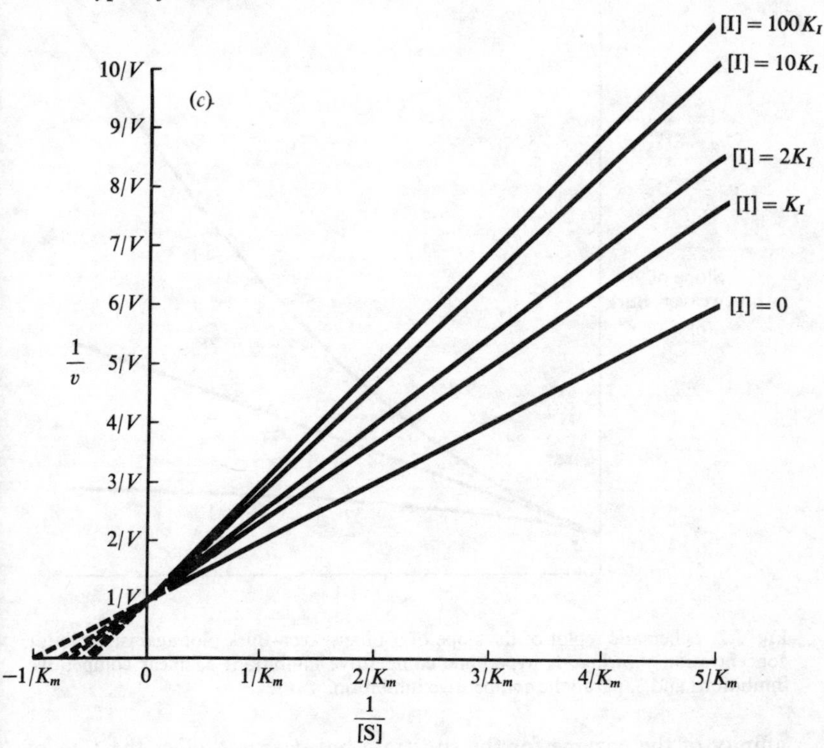

Fig. 3.1. (*c*) Schematic Lineweaver–Burk plot of hyperbolic competitive inhibition in which $K_i = 2K_I$.

A plot of $1/v$ against $1/[S]$ is a straight line which, for different concentrations of the inhibitor, yields a family of lines with a common intercept $1/V$ (fig. 3.1*b*). The criterion for competitive inhibition is therefore satisfied by the unchanged ordinal intercept, $1/V$. The slopes of the Lineweaver–Burk plot when replotted against inhibitor concentration, now vary in a parabolic manner (fig. 3.2, line 3).

In the classical definition of competitive inhibition, it was assumed that the inhibitor and substrate were structually related and that they both competed for the same site on the enzyme. This idea leads to equation (3.1) describing linear competitive inhibition in which the apparent affinity of the enzyme for the substrate, as measured by the observed K_m, is reduced by the factor $(1 + [I]/K_I)$. Since, in this case, the two ligands are competing for the same site, the formation of the complex EIS is not allowed. If, however, a compound binds to a site on the enzyme other than the substrate binding site, and in so doing causes a reduction in the

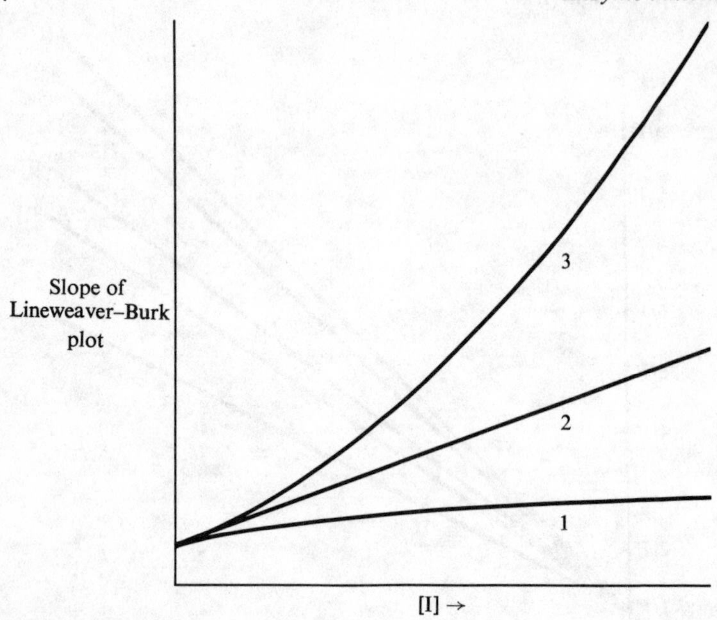

Fig. 3.2. Schematic replot of the slope of a Lineweaver–Burk plot against inhibitor concentration showing: 1, hyperbolic competitive inhibition; 2, linear competitive inhibition; and 3, parabolic competitive inhibition.

affinity of the enzyme for the substrate but *does not affect* the rate of catalytic breakdown of enzyme–substrate complex(es) to give product(s), then this will also show competitive inhibition. The mechanism can be described by the following reaction scheme:

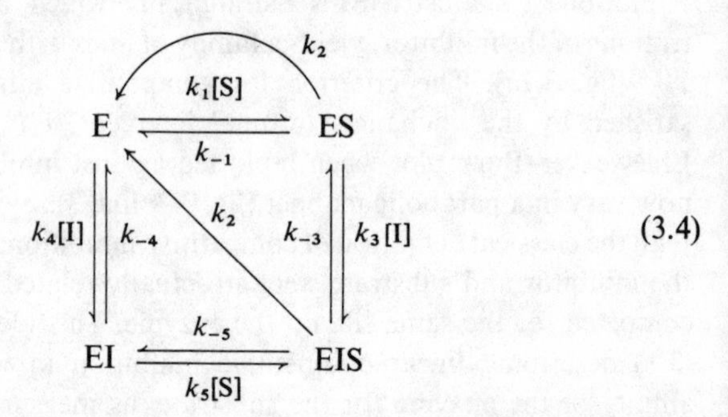

(3.4)

A rigorous treatment of this mechanism by either the steady-state method or the King–Altman procedure leads to a rather complicated rate equation which is difficult to simplify and not very useful for the analysis of experimental results. A satisfactory approximate solution can be obtained by treating the vertical reactions, involving only binding and dissociation processes, as rapid and in equilibrium and applying the steady-state treatment to those reactions involved in the overall catalytic process.

The equilibrium equations for the two vertical reactions are

$$K_I = \frac{k_{-4}}{k_4} = \frac{[\mathrm{E}][\mathrm{I}]}{[\mathrm{EI}]} \tag{3.5}$$

and $$K_i = \frac{k_{-3}}{k_3} = \frac{[\mathrm{ES}][\mathrm{I}]}{[\mathrm{EIS}]}. \tag{3.6}$$

The steady-state equations for ES and EIS are

$$\frac{\mathrm{d}[\mathrm{ES}]}{\mathrm{d}t} = k_1[\mathrm{E}][\mathrm{S}] - (k_{-1} + k_2)[\mathrm{ES}] = 0 \tag{3.7}$$

and $$\frac{\mathrm{d}[\mathrm{EIS}]}{\mathrm{d}t} = k_5[\mathrm{EI}][\mathrm{S}] - (k_{-5} + k_2)[\mathrm{EIS}] = 0. \tag{3.8}$$

Since

$$v = k_2([\mathrm{ES}] + [\mathrm{EIS}])$$

$$\frac{v}{[\mathrm{E_0}]} = \frac{k_2([\mathrm{ES}] + [\mathrm{EIS}])}{[\mathrm{E}] + [\mathrm{EI}] + [\mathrm{ES}] + [\mathrm{EIS}]}.$$

Using equations (3.5), (3.6) and (3.7), it follows that

$$v = \frac{k_2\{1 + ([\mathrm{I}]/K_i)\}[\mathrm{E_0}][\mathrm{S}]}{K_m\{1 + ([\mathrm{I}]/K_I)\} + [\mathrm{S}]\{1 + ([\mathrm{I}]/K_i)\}}$$

or $$v = \frac{V[\mathrm{S}]}{K_m\dfrac{\{1 + ([\mathrm{I}]/K_I)\}}{\{1 + ([\mathrm{I}]/K_i)\}} + [\mathrm{S}]} \tag{3.9}$$

where $V = k_2[\mathrm{E_0}]$ and $K_m = (k_{-1} + k_2)/k_1$. In the reciprocal form equation (3.9) becomes

$$\frac{1}{v} = \frac{K_m}{V}\frac{\{1 + ([\mathrm{I}]/K_I)\}}{\{1 + ([\mathrm{I}]/K_i)\}}\frac{1}{[\mathrm{S}]} + \frac{1}{V}. \tag{3.10}$$

Notice that the criterion for competitive inhibition is again satisfied since a plot of $1/v$ against $1/[\mathrm{S}]$, for a series of different inhibitor concentrations, yields a family of lines with a common ordinal intercept $1/V$ (fig. 3.1*c*). In this case, however, a replot of the slopes of the Lineweaver–Burk plot against [I] is hyperbolic and consequently the inhibition does not increase indefinitely with inhibitor concentration (fig. 3.2, line 1). This type of inhibition is called *hyperbolic competitive inhibition* or sometimes, partially competitive inhibition. When the inhibitor concentration is very large, equation (3.9) reduces to

$$v = \frac{V[\mathrm{S}]}{(K_m K_i/K_I) + [\mathrm{S}]}$$

From equations (3.5), (3.6), (3.7) and (3.8) it can be shown that, if $k_2 \ll k_{-1}$ and $k_2 \ll k_{-5}$, then

$$K_m' = \frac{K_m K_i}{K_I}$$

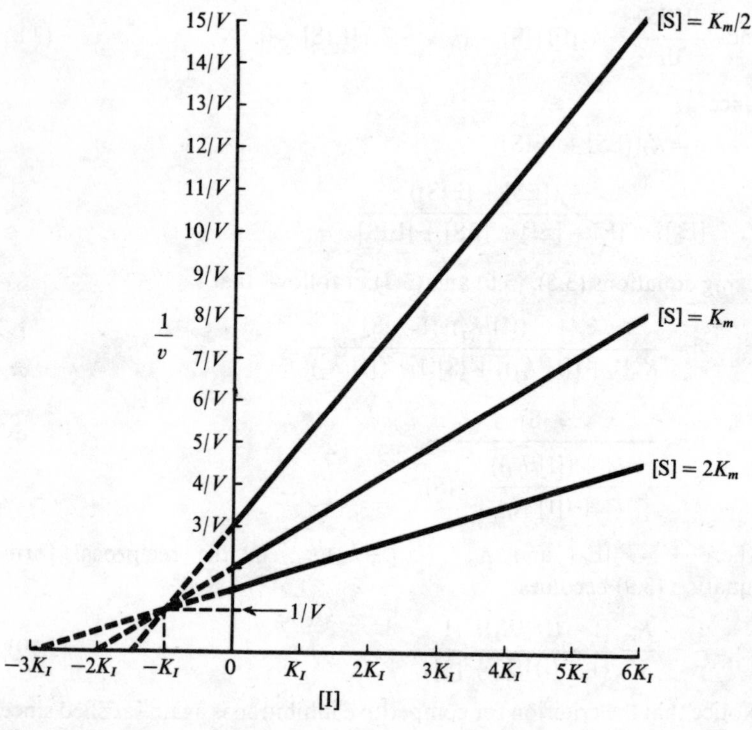

Fig. 3.3. Dixon plot of $1/v$ against [I] for linear competitive inhibition.

where

$$K'_m = \frac{(k_{-5} + k_2)}{k_5}$$

and hence

$$v = \frac{V[\mathrm{S}]}{K'_m + [\mathrm{S}]}.$$

The effect of the inhibitor is to produce a new enzyme, EI, with a different affinity, K'_m, for the substrate. The inhibition, therefore, reaches a maximum when all the enzyme is in the EI form.

Dixon (1953) devised a graphical method for determining K_I. From equation (3.2), a plot of $1/v$ against [I], at a fixed initial substrate concentration [S], is linear with a slope $K_m/VK_I[\mathrm{S}]$ and an ordinal intercept $(1/V + K_m/[\mathrm{S}]V)$. The slope and intercept vary with substrate concentration in different ways and hence, for two substrate concentrations $[\mathrm{S}_1]$ and $[\mathrm{S}_2]$, the lines will intersect at a point given by

$$\frac{1}{v} = \frac{1}{V} + \frac{K_m}{V}\left(1 + \frac{[\mathrm{I}]}{K_I}\right)\frac{1}{[\mathrm{S}_1]} = \frac{1}{V} + \frac{K_m}{V}\left(1 + \frac{[\mathrm{I}]}{K_I}\right)\frac{1}{[\mathrm{S}_2]},$$

i.e. when $[\mathrm{I}] = -K_I$ (since $[\mathrm{S}_1] \neq [\mathrm{S}_2]$) (fig. 3.3). This method is only applicable to linear competitive inhibition since both parabolic and hyperbolic competitive inhibition give rise to curved plots (fig. 3.4).

Hunter & Downs (1945) described the inhibition of arginase in terms of the fractional activity $\alpha = v_I/v$ where v_I is the velocity in the presence of the inhibitor and v is the velocity in the absence of inhibitor. The fractional inhibition is then $1 - \alpha$. In the absence of the inhibitor,

$$v = \frac{V[\mathrm{S}]}{K_m + [\mathrm{S}]}.$$

Utilising this and equation (3.1) we obtain

$$\frac{v}{v_I} = \frac{K_m\{1 + ([\mathrm{I}]/K_I)\} + [\mathrm{S}]}{K_m + [\mathrm{S}]},$$

therefore

$$\frac{1}{\alpha} = 1 + \frac{K_m[\mathrm{I}]}{K_I(K_m + [\mathrm{S}])}$$

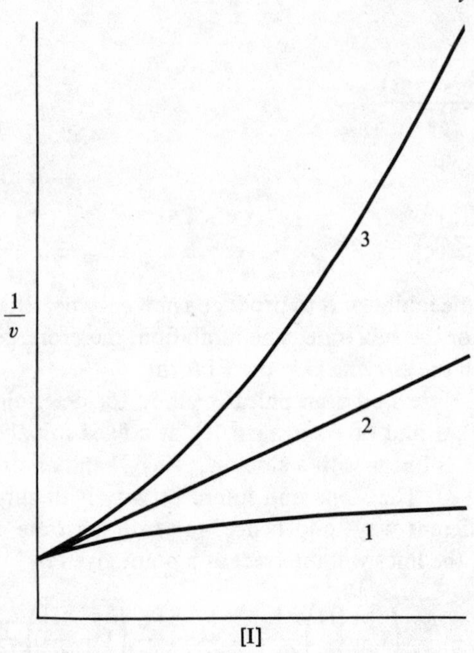

Fig. 3.4. Dixon plot of $1/v$ against [I] for: 1, hyperbolic competitive inhibition; 2, linear competitive inhibition, and 3, parabolic competitive inhibition.

and

$$K_I = \frac{[\mathrm{I}]K_m}{(K_m + [\mathrm{S}])\left(\dfrac{1}{\alpha} - 1\right)}$$

or alternatively

$$\frac{[\mathrm{I}]\alpha}{(1-\alpha)} = K_I + \frac{K_I}{K_m}[\mathrm{S}],$$

so that a plot of $[\mathrm{I}]\alpha/(1-\alpha)$ against [S] is a straight line whose intercept on the ordinal axis is K_I and whose slope is K_I/K_m. Both parameters K_I and K_m can therefore be determined.

3.2.2. Non-competitive inhibition. This mode of inhibition has been referred to as 'inhibition without affinity' since the inhibitor is not bound at the substrate binding site but at some other binding site on the enzyme.

The inhibitor can bind with both free enzyme, E, and also the enzyme–substrate complex, ES, to form EI and EIS complexes. The presence of the inhibitor does not prevent the enzyme from binding the substrate and vice versa. The reaction mechanism is similar to the mechanism for hyperbolic competitive inhibition given in scheme (3.4) except in this case it is assumed that the presence of bound inhibitor or substrate does not affect the binding of the other (i.e. $k_3 = k_4$, $k_{-3} = k_{-4}$, $k_1 = k_5$ and $k_{-1} = k_{-5}$). There are two possible mechanisms:

(1) the ternary complex EIS is a dead-end complex and does not break down to yield products;

(2) the EIS complex breaks down at a slower rate than the ES complex.

For the first case, the system can be described by the following mechanism:

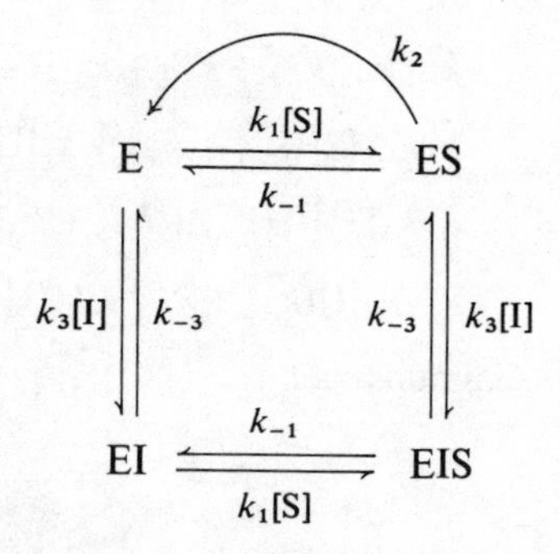

If $k_2 \ll k_{-1}$, it can be shown by the King–Altman method that

$$v = \frac{V[\mathrm{S}]}{\{1 + ([\mathrm{I}]/K_I)\}(K_S + [\mathrm{S}])} \tag{3.11}$$

where

$$K_S = k_{-1}/k_1, \tag{3.12}$$

$$K_I = k_{-3}/k_3 \tag{3.13}$$

and

$$V = k_2[\mathrm{E_0}].$$

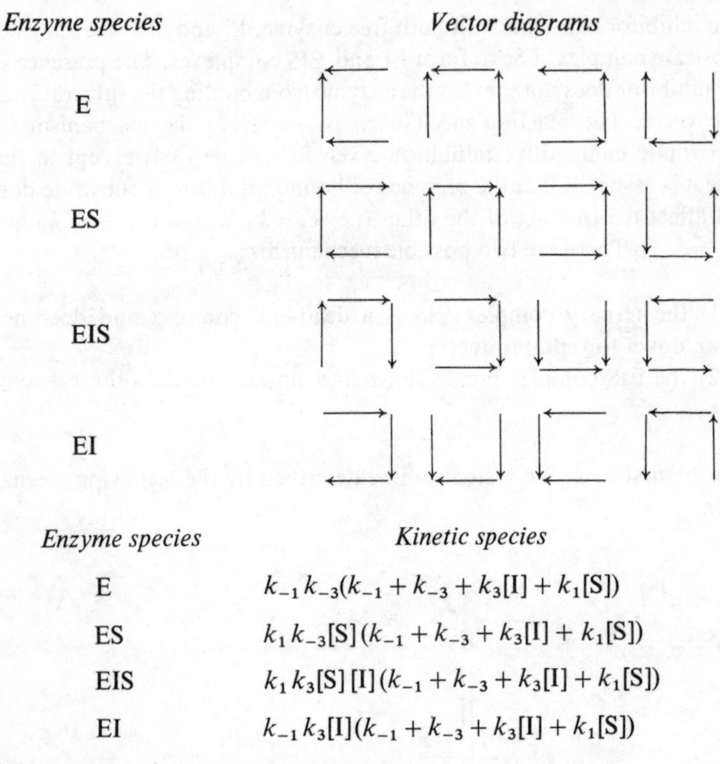

Enzyme species	*Kinetic species*
E	$k_{-1}k_{-3}(k_{-1}+k_{-3}+k_3[\mathrm{I}]+k_1[\mathrm{S}])$
ES	$k_1k_{-3}[\mathrm{S}](k_{-1}+k_{-3}+k_3[\mathrm{I}]+k_1[\mathrm{S}])$
EIS	$k_1k_3[\mathrm{S}][\mathrm{I}](k_{-1}+k_{-3}+k_3[\mathrm{I}]+k_1[\mathrm{S}])$
EI	$k_{-1}k_3[\mathrm{I}](k_{-1}+k_{-3}+k_3[\mathrm{I}]+k_1[\mathrm{S}])$

Notice that six vector diagrams such as

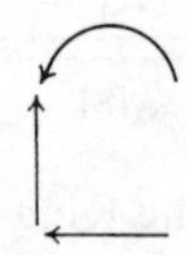

have been omitted since it is assumed that $k_2 \ll k_{-1}$ and consequently corresponding kinetic terms may be neglected. This procedure is justified by the fact that the four reactions used in the King–Altman pattern are simply binding and dissociation processes, involving the two ligands, I and S, and are therefore rapid in comparison with the breakdown of the ES complex to give product(s), which involves a chemical change. This is often a convenient way of obtaining steady-state rate equations for systems that are too complex to treat in a completely rigorous manner. The velocity of the reaction is given by

$$v = k_2[\mathrm{ES}]$$

$$\frac{v}{[E_0]} = \frac{k_2[ES]}{[E]+[ES]+[EIS]+[EI]}$$

$$v = \frac{k_2 k_1 k_{-3}[S](k_{-1}+k_{-3}+k_3[I]+k_1[S])[E_0]}{\{(k_{-1}+k_{-3}+k_3[I]+k_1[S])(k_{-1}k_{-3}+k_1k_{-3}[S] + k_1k_3[S][I]+k_{-1}k_3[I])\}}$$

Using equations (3.12) and (3.13), it follows that

$$v = \frac{k_2[E_0][S]}{(K_S+[S])\{1+([I]/K_I)\}} = \frac{V[S]}{(K_S+[S])\{1+([I]/K_I)\}}.$$

It is important to notice that if any of the conditions, $k_3 = k_4$, $k_{-3} = k_{-4}$, $k_1 = k_5$, $k_{-1} = k_{-5}$ and $k_2 \ll k_{-1}$, do not hold, equations are obtained which are much more complicated than equation (3.11). Notice that if an enzyme obeys equation (3.11) then the observed K_m is in fact the dissociation constant, K_S, for the ES complex. If the condition $k_2 \ll k_{-1}$ does not hold, then k_2 makes an important contribution to K_m ($K_m =$

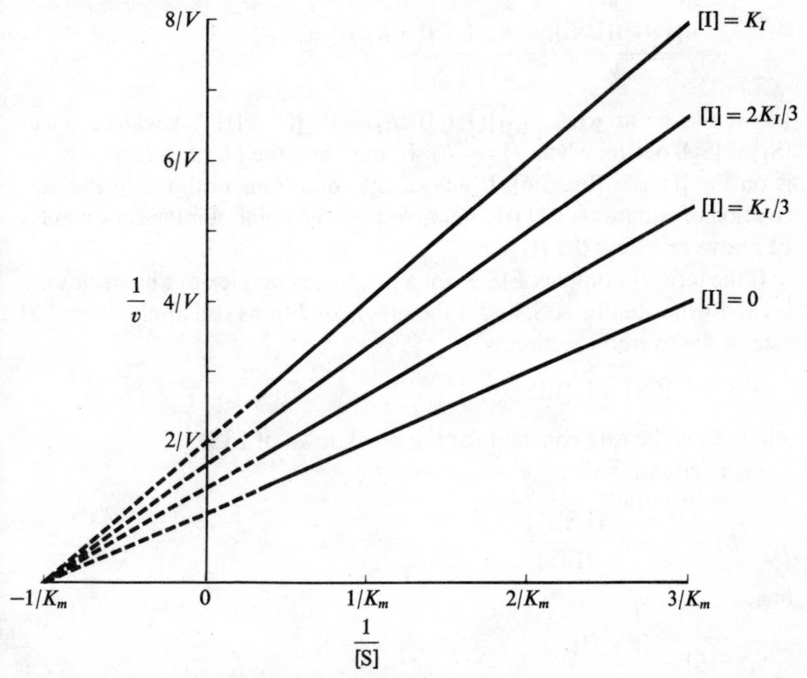

Fig. 3.5. Schematic reciprocal plot of $1/v$ against $1/[S]$ for non-competitive inhibition.

$(k_{-2}+k_{-1})/k_1$) and an inhibitor affecting the catalytic step, involving k_2, must also affect K_m as well as V. In this case the inhibition will be mixed. Inversion of equation (3.11) gives

$$\frac{1}{v}=\frac{\{1+([\mathrm{I}]/K_I)\}}{V}+\frac{K_S\{1+([\mathrm{I}]/K_I)\}}{V[\mathrm{S}]}.$$

A plot of $1/v$ against $1/[\mathrm{S}]$ has a slope equal to $K_S(1+[\mathrm{I}]/K_I)/V$, an ordinal intercept of $(1+[\mathrm{I}]/K_I)/V$ and an abscissal intercept of $-1/K_S$ (fig. 3.5). In contrast to competitive inhibition, the maximum velocity is $V_I=V/(1+[\mathrm{I}]/K_I)$ in the presence of the inhibitor and no matter how high a substrate concentration is used, the maximum velocity in the absence of the inhibitor will not be reached. The $K_m(\approx K_S)$ for the system remains unchanged. The K_I can be determined from a Dixon plot of $1/v$ against [I] for a series of substrate concentrations. The point of intersection of any two lines is given by

$$\frac{1}{v}=\frac{\{1+([\mathrm{I}]/K_I)\}}{V}+\frac{K_m\{1+([\mathrm{I}]/K_I)\}}{V[\mathrm{S}_1]}$$

$$=\frac{\{1+([\mathrm{I}]/K_I)\}}{V}+\frac{K_m\{1+([\mathrm{I}]/K_I)\}}{V[\mathrm{S}_2]},$$

i.e. at the point where $[\mathrm{S}_1](1+[\mathrm{I}]/K_I)=[\mathrm{S}_2](1+[\mathrm{I}]/K_I)$ which, since $[\mathrm{S}_1]\neq[\mathrm{S}_2]$, occurs when $[\mathrm{I}]=-K_I$. In this case, the point of intersection is on the [I] axis (fig. 3.6). If any of the conditions outlined in the derivation of equation (3.11) do not hold then the point of intersection may be above or below the [I] axis.

If the ternary complex EIS is not a dead-end complex but has a slower breakdown rate than ES, and if the other conditions still apply, then the rate of the overall reaction is

$$v=k_2[\mathrm{ES}]+k_6[\mathrm{EIS}]$$

where k_6 is the rate constant for the breakdown of EIS.

From equation (3.13),

$$K_I=\frac{k_{-3}}{k_3}=\frac{[\mathrm{ES}][\mathrm{I}]}{[\mathrm{EIS}]}$$

hence

$$[\mathrm{EIS}]=\frac{[\mathrm{ES}][\mathrm{I}]}{K_I}$$

and $v = k_2[\mathrm{ES}]\left(1 + \dfrac{k_6[\mathrm{I}]}{k_2 K_I}\right)$

so $\quad v = \dfrac{V[\mathrm{S}]\{1 + (k_6[\mathrm{I}]/k_2 K_I)\}}{(K_m + [\mathrm{S}])\{1 + ([\mathrm{I}]/K_I)\}}$

or $\quad \dfrac{1}{v} = \dfrac{K_m\{1 + ([\mathrm{I}]/K_I)\}}{V\{1 + (k_6[\mathrm{I}]/k_2 K_I)\}}\,\dfrac{1}{[\mathrm{S}]} + \dfrac{\{1 + ([\mathrm{I}]/K_I)\}}{V\{1 + (k_6[\mathrm{I}]/k_2 K_I)\}}.$

A plot of $1/v$ against $1/[\mathrm{S}]$ has an intercept of $(1 + [\mathrm{I}]/K_I)/V(1 + k_6[\mathrm{I}]/k_2 K_I)$. The intercept on the $1/[\mathrm{S}]$ axis is still $-1/K_m$ so that the breakdown of EIS again has no effect on the K_m of the system. This type of behaviour is known as *partially non-competitive inhibition* (fig. 3.7).

3.2.3. Uncompetitive inhibition. Another possible form of inhibition arises if the inhibitor combines only with the ES complex, i.e. there is no binding site on the enzyme until a molecule of substrate has bound to the enzyme. In this respect it is similar to ordered sequential reactions with

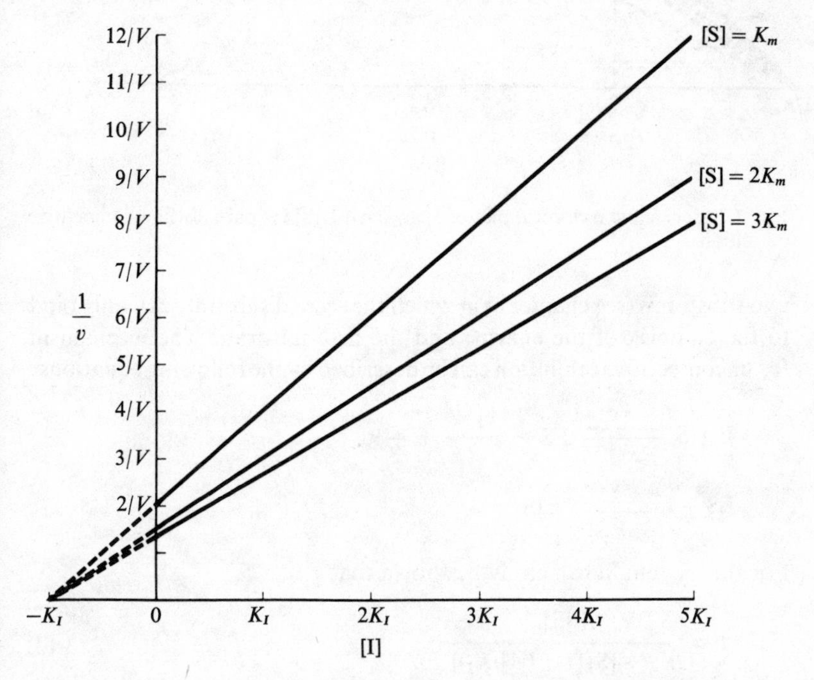

Fig. 3.6. Dixon plot of $1/v$ against [I] for non-competitive inhibition.

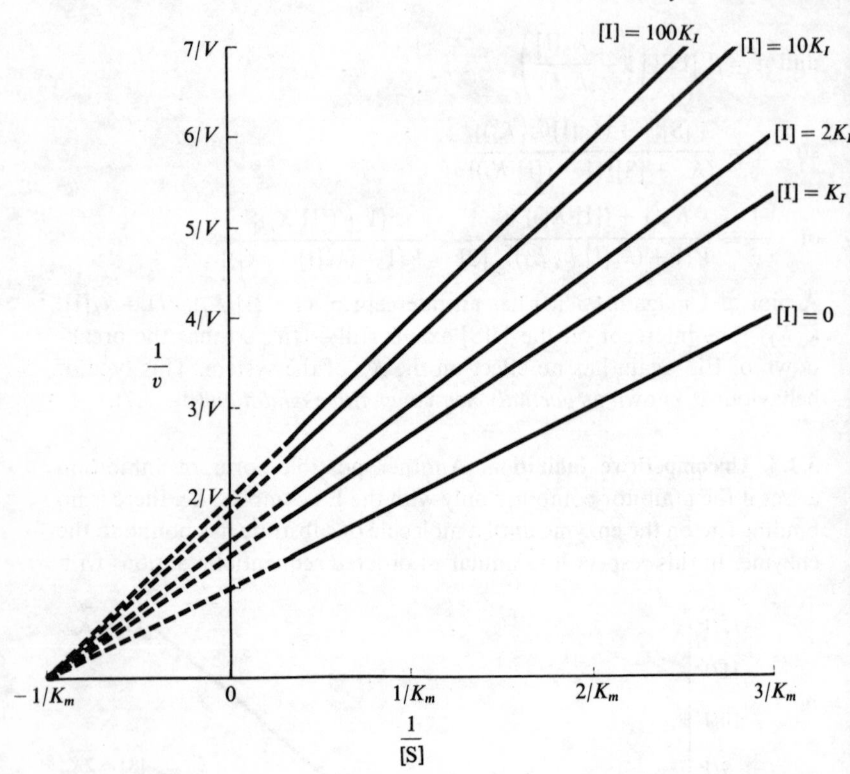

Fig. 3.7. Schematic reciprocal plot of $1/v$ against $1/[S]$ for partially non-competitive inhibition.

two substrates (see chapter 6) in which the second substrate can only bind to the complex of the enzyme and the first substrate. The mechanism for uncompetitive inhibition can be described by the following equations:

$$\mathrm{E} + \mathrm{S} \underset{k_{-1}}{\overset{k_1}{\rightleftharpoons}} \mathrm{ES} \xrightarrow{k_2} \mathrm{E} + \mathrm{P}$$

$$\mathrm{ES} + \mathrm{I} \underset{k_{-3}}{\overset{k_3}{\rightleftharpoons}} \mathrm{EIS}.$$

For this system, it can easily be shown that

$$v = \frac{V[\mathrm{S}]}{K_m + [\mathrm{S}]\{1 + ([\mathrm{I}]/K_I)\}} \tag{3.14}$$

$$\text{or} \quad \frac{1}{v} = \frac{\{1 + ([\mathrm{I}]/K_I)\}}{V} + \frac{K_m}{V[\mathrm{S}]}.$$

In this case both V and K_m are reduced by the factor $(1 + [\mathrm{I}]/K_I)$, so that a plot of $1/v$ against $1/[\mathrm{S}]$ for a series of inhibitor concentrations gives a number of parallel lines (fig. 3.8). In the case of uncompetitive inhibition, K_I cannot be obtained from a Dixon plot of $1/v$ against [I] as the lines do not intersect.

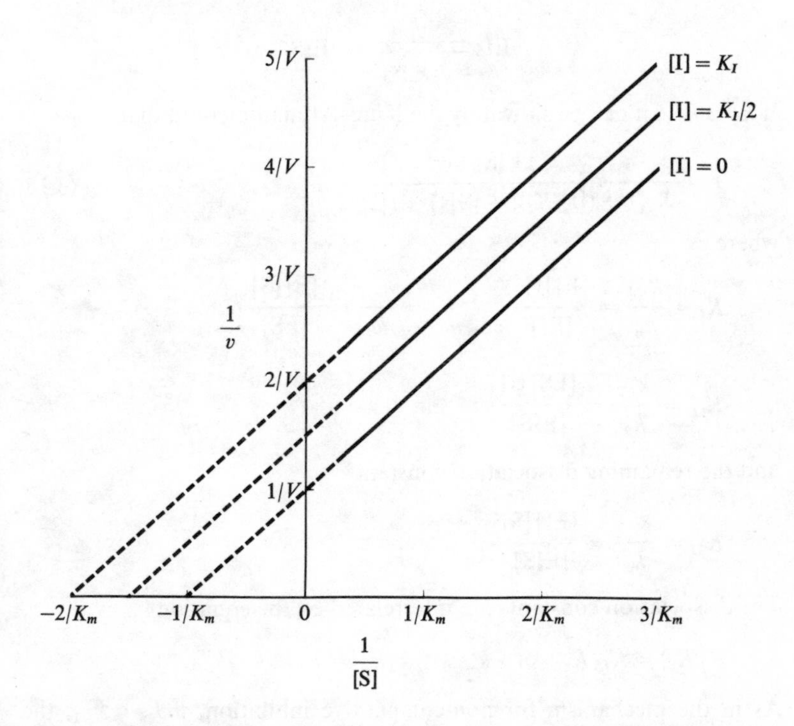

Fig. 3.8. Schematic reciprocal plot of $1/v$ against $1/[\mathrm{S}]$ for uncompetitive inhibition.

3.2.4. Mixed inhibition. In the mechanism for non-competitive inhibition it was assumed that the presence of inhibitor or substrate did not affect the binding of the other species, i.e. K_S is the dissociation constant for S bound as ES or EIS and K_I is the dissociation constant for I bound as EI or EIS. If these conditions are not observed then the system can be described by the following mechanism:

$$\begin{array}{ccc} \mathrm{E} & \underset{k_{-1}}{\overset{k_1[\mathrm{S}]}{\rightleftharpoons}} & \mathrm{ES} \xrightarrow{k_5} \mathrm{E} \\ k_4[\mathrm{I}] \downarrow\uparrow k_{-4} & & k_{-2} \uparrow\downarrow k_2[\mathrm{I}] \\ \mathrm{EI} & \underset{k_3[\mathrm{S}]}{\overset{k_{-3}}{\leftrightharpoons}} & \mathrm{EIS} \end{array}$$

If $k_5 \ll k_{-1}$, it can be shown by the King–Altman method that

$$v = \frac{V[\mathrm{S}]}{K_S\{1 + ([\mathrm{I}]/K_I)\} + [\mathrm{S}]\{1 + ([\mathrm{I}]/K_{IS})\}} \tag{3.15}$$

where

$$K_I = \frac{k_{-4}}{k_4} = \frac{[\mathrm{E}][\mathrm{I}]}{[\mathrm{EI}]}, \qquad K_S = \frac{k_{-1}}{k_1} = \frac{[\mathrm{E}][\mathrm{S}]}{[\mathrm{ES}]},$$

$$K_{IS} = \frac{k_{-2}}{k_2} = \frac{[\mathrm{ES}][\mathrm{I}]}{[\mathrm{EIS}]}$$

and the remaining dissociation constant

$$K_{SI} = \frac{k_{-3}}{k_3} = \frac{[\mathrm{EI}][\mathrm{S}]}{[\mathrm{EIS}]}.$$

The dissociation constants are interrelated by the equation

$$K_I K_{SI} = K_{IS} K_S \quad \text{or} \quad k_{-1}k_{-2}k_3k_4 = k_1k_2k_{-3}k_{-4}.$$

As in the mechanism for non-competitive inhibition, if $k_5 \ll k_{-1}$, the vector diagrams containing k_5 can be neglected.

Enzyme species	*Kinetic terms*
E	$k_{-1}k_{-2}k_3[\mathrm{S}] + k_{-1}k_{-2}k_{-4} + k_{-1}k_{-3}k_{-4} + k_2k_{-3}k_{-4}[\mathrm{I}]$
ES	$[\mathrm{S}](k_1k_{-2}k_3[\mathrm{S}] + k_1k_{-2}k_{-4} + k_1k_{-3}k_{-4} + k_{-2}k_3k_4[\mathrm{I}])$

EIS $\quad [S][I](k_1k_2k_3[S] + k_1k_2k_{-4} + k_{-1}k_3k_4 + k_2k_3k_4[I])$

EI $\quad [I](k_1k_2k_{-3}[S] + k_{-1}k_{-2}k_4 + k_{-1}k_{-3}k_4 + k_2k_{-3}k_4[I])$

Factorising out $k_1k_2k_3$ and using the various equilibrium relationships, the kinetic terms become

E $\quad A \cdot K_S K_{IS}$

ES $\quad A \cdot [S] K_{IS}$

EIS $\quad A \cdot [S][I]$

EI $\quad A \cdot [I] K_{SI}$

where $A = ([S] + k_{-4}/k_3 + k_{-4}k_{-3}/k_{-2}k_3 + k_4[I]/k_1)$.

The velocity of the reaction is given by

$$v = k_5[ES]$$

hence

$$v = \frac{k_5[S][E_0]K_{IS}}{K_S K_{IS} + [S]K_{IS} + [S][I] + [I]K_{SI}}$$

or
$$v = \frac{V[S]}{K_S\{1 + ([I]/K_I)\} + [S]\{1 + ([I]/K_{IS})\}}.$$

Inversion of equation (3.15) gives

$$\frac{1}{v} = \frac{K_S\{1 + ([I]/K_I)\}}{V}\frac{1}{[S]} + \frac{\{1 + ([I]/K_{IS})\}}{V}.$$

A plot of $1/v$ against $1/[S]$ has a slope equal to $K_S\{1 + ([I]/K_I)\}/V$, an ordinal intercept of $\{1 + ([I]/K_{IS})\}/V$ and an abscissal intercept of

$$-\frac{\{1 + ([I]/K_{IS})\}}{K_S\{1 + ([I]/K_I)\}}.$$

The intersection of Lineweaver–Burk plots with different inhibitor concentrations occurs at the point

$$\frac{1}{v} = \frac{(K_{IS} - K_I)}{K_{IS}V} \quad \text{and} \quad 1/[S] = \frac{-K_I}{K_S K_{IS}}$$

If $K_{IS} > K_I$, the point of intersection will lie in the upper left-hand quadrant (fig. 3.9) whereas if $K_{IS} < K_I$, the point of intersection will lie

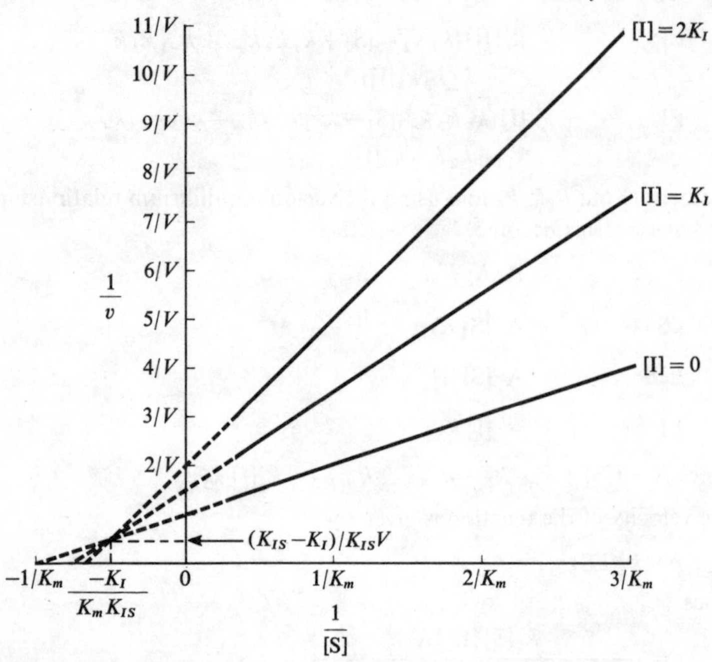

Fig. 3.9. Schematic reciprocal plot of $1/v$ against $1/[S]$ for mixed inhibition where $K_{IS} = 2K_I$.

in the lower left-hand quadrant. It should be noted that this mechanism is often referred to as a form of non-competitive inhibition. The term non-competitive inhibition should be restricted to the specific case where $K_I = K_{IS}$ and consequently the family of Lineweaver–Burk plots, at different inhibitor concentrations, all intersect at a common point on the abscissal axis.

An alternative way of distinguishing the various types of inhibition is from the variation in the degree of inhibition with change in substrate concentration. The degree of inhibition, i, is related to the velocity of the inhibited reaction, v_i and that of the uninhibited reaction, v_0 by the relationship

$$i = \frac{v_0 - v_i}{v_0}.$$

For classical competitive inhibition, therefore

$$i = \frac{K_m[\mathrm{I}]}{K_I(K_m\{1 + ([\mathrm{I}]/K_I)\} + [\mathrm{S}])}$$

and hence $i \to 0$ as [S] increases. For non-competitive inhibition,

$$i = \frac{[\mathrm{I}]}{K_I + [\mathrm{I}]}$$

and is therefore independent of [S]. For uncompetitive inhibition,

$$i = \frac{[\mathrm{S}][\mathrm{I}]}{K_I(K_m + [\mathrm{S}]\{1 + ([\mathrm{I}]/K_I)\})}$$

which rises from zero when $[\mathrm{S}] = 0$ to a limiting value of $[\mathrm{I}]/(K_I + [\mathrm{I}])$ at high values of substrate concentration. For mixed inhibition the change in the degree of inhibition will depend on the magnitudes of K_I and K_{IS}. If $K_{IS} > K_I$, the inhibition contains both competitive and non-competitive character and hence the degree of inhibition is reduced by an increase in substrate concentration. The inhibition will not, however, be completely overcome by saturation with substrate unless $K_{IS} \gg K_I$, in which case the mechanism reduces to that of competitive inhibition (equation (3.1)). Similarly, if $K_I > K_{IS}$, the inhibition will contain both non-competitive and uncompetitive character and hence the degree of inhibition will be increased by an increase in substrate concentration. If $K_I \gg K_{IS}$, then the mechanism reduces to that of pure uncompetitive inhibition (equation (3.14)). The model for mixed inhibition can be further extended to allow for the breakdown of the EIS complex ($\mathrm{EIS} \xrightarrow{k_6} \mathrm{EI} + \mathrm{P}$). This leads to the following equation:

$$v = \frac{V[\mathrm{S}]\{1 + (k_6[\mathrm{I}]/k_5 K_{IS})\}}{K_S\{1 + ([\mathrm{I}]/K_I)\} + [\mathrm{S}]\{1 + ([\mathrm{I}]/K_{IS})\}}.$$

There is now only a limited effect due to the inhibitor since at high concentration of the inhibitor the equation becomes

$$v = \frac{V'[\mathrm{S}]}{K_S' + [\mathrm{S}]}$$

where $V' = k_6[\mathrm{E}_0]$ and $K_S' = K_S K_{IS}/K_I$.

3.3. Reversibility and product inhibition

In chapter 2, kinetic equations for the two- and three-step mechanisms were derived for cases (equations (2.1) and (2.15)) involving the irreversible conversion of the enzyme complex(es) into products. It is useful at this point to derive the equations for the fully reversible system shown below in equation (3.16), as it will introduce concepts that are found in the more complicated multi-substrate reactions. The system can be described by the following set of equations under steady-state conditions:

$$\mathrm{E}+\mathrm{S} \underset{k_{-1}}{\overset{k_1}{\rightleftharpoons}} \mathrm{ES} \underset{k_{-2}}{\overset{k_2}{\rightleftharpoons}} \mathrm{EP} \underset{k_{-3}}{\overset{k_3}{\rightleftharpoons}} \mathrm{E}+\mathrm{P} \tag{3.16}$$

$$[\mathrm{E_0}] = [\mathrm{E}] + [\mathrm{ES}] + [\mathrm{EP}] \tag{3.17}$$

$$\frac{\mathrm{d[ES]}}{\mathrm{d}t} = k_1[\mathrm{E}][\mathrm{S}] - (k_{-1}+k_2)[\mathrm{ES}] + k_{-2}[\mathrm{EP}] = 0 \tag{3.18}$$

$$\frac{\mathrm{d[EP]}}{\mathrm{d}t} = k_2[\mathrm{ES}] - (k_{-2}+k_3)[\mathrm{EP}] + k_{-3}[\mathrm{E}][\mathrm{P}] = 0 \tag{3.19}$$

$$\frac{\mathrm{d[P]}}{\mathrm{d}t} = k_3[\mathrm{EP}] - k_{-3}[\mathrm{E}][\mathrm{P}]. \tag{3.20}$$

The final rate equation could be derived by the King–Altman method (chapter 2). Alternatively, we could solve the three simultaneous equations (3.17)–(3.19) by elimination, but to make a change, we shall solve them by using determinants. In order to make the procedure more obvious, equations (3.17)–(3.19) are rewritten as

$$[\mathrm{E}] + [\mathrm{ES}] + [\mathrm{EP}] - [\mathrm{E_0}] = 0 \tag{3.21}$$

$$k_1[\mathrm{S}][\mathrm{E}] - (k_{-1}+k_2)[\mathrm{ES}] + k_{-2}[\mathrm{EP}] + 0 = 0 \tag{3.22}$$

$$k_{-3}[\mathrm{P}][\mathrm{E}] + k_2[\mathrm{ES}] - (k_{-2}+k_3)[\mathrm{EP}] + 0 = 0 \tag{3.23}$$

The solutions for [E], [ES] and [EP] are given by

$$\frac{[\mathrm{E}]}{\Delta_E} = -\frac{[\mathrm{ES}]}{\Delta_{ES}} = \frac{[\mathrm{EP}]}{\Delta_{EP}} = -\frac{1}{\Delta}$$

where Δ_E, Δ_{ES} and Δ_{EP} refer to the determinants for E, ES and EP:

$$\Delta_E = \begin{vmatrix} 1 & 1 & -[\mathrm{E_0}] \\ -(k_{-1}+k_2) & k_{-2} & 0 \\ k_2 & -(k_{-2}+k_3) & 0 \end{vmatrix}$$

$$\Delta_{ES} = \begin{vmatrix} 1 & 1 & -[\mathrm{E_0}] \\ k_1[\mathrm{S}] & k_{-2} & 0 \\ k_{-3}[\mathrm{P}] & -(k_{-2}+k_3) & 0 \end{vmatrix}$$

$$\Delta_{EP} = \begin{vmatrix} 1 & 1 & -[\mathrm{E_0}] \\ k_1[\mathrm{S}] & -(k_{-1}+k_2) & 0 \\ k_{-3}[\mathrm{P}] & k_2 & 0 \end{vmatrix}$$

$$\Delta = \begin{vmatrix} 1 & 1 & 1 \\ k_1[\mathrm{S}] & -(k_{-1}+k_2) & k_{-2} \\ k_{-3}[\mathrm{P}] & k_2 & -(k_{-2}+k_3) \end{vmatrix}$$

In order to solve equation (3.20), we require the solutions for [E] and [EP], and so

$$[\mathrm{EP}] = \frac{-1(-[\mathrm{E_0}])(k_1 k_2[\mathrm{S}] + k_{-1}k_{-3}[\mathrm{P}] + k_2 k_{-3}[\mathrm{P}])}{\Delta}$$

$$[\mathrm{E}] = \frac{-1(-[\mathrm{E_0}])(k_{-1}k_{-2} + k_{-1}k_3 + k_2 k_3)}{\Delta}$$

where $\Delta = \{(k_{-1}k_{-2} + k_{-1}k_3 + k_2k_3) + k_1[\mathrm{S}](k_{-2} + k_3 + k_2) + k_{-3}[\mathrm{P}]\,(k_{-1} + k_2 + k_{-2})\}$.

Hence the velocity of the reaction is given by

$$v = \frac{(k_1 k_2 k_3[\mathrm{S}] - k_{-1}k_{-2}k_{-3}[\mathrm{P}])\,[\mathrm{E_0}]}{\{(k_{-1}k_{-2} + k_{-1}k_3 + k_2k_3) + k_1[\mathrm{S}](k_{-2}+k_3+k_2) + k_{-3}[\mathrm{P}](k_{-1}+k_2+k_{-2})\}} \quad (3.24)$$

Equation (3.24) can be written in the form

$$v = \frac{(\mathit{numerator}\ 1)\,[\mathrm{S}] - (\mathit{numerator}\ 2)\,[\mathrm{P}]}{\mathit{constant} + (\mathit{coef}\ \mathrm{S})\,[\mathrm{S}] + (\mathit{coef}\ \mathrm{P})\,[\mathrm{P}]} \quad (3.25)$$

which is relevant to the later discussion of two substrate reactions. Note that *numerator* 1 contains all the rate constants in the forward direction whilst *numerator* 2 contains all the reverse rate constants. We can define the maximum velocity in the forward direction, V_F, as

$$V_F = \frac{\mathit{numerator}\ 1}{\mathit{coef}\ \mathrm{S}} \quad ([\mathrm{P}] = 0 \text{ and } (\mathit{coef}\ \mathrm{S})\,[\mathrm{S}] \gg \mathit{constant})$$

and the maximum velocity in the reverse direction

$$V_R = \frac{\mathit{numerator}\ 2}{\mathit{coef}\ \mathrm{P}} \quad ([\mathrm{S}] = 0 \text{ and } (\mathit{coef}\ \mathrm{P})\,[\mathrm{P}] \gg \mathit{constant}).$$

The Michaelis constants for this scheme are defined as

$$K_S = \frac{constant}{coef\,\mathrm{S}} \quad \text{and} \quad K_P = \frac{constant}{coef\,\mathrm{P}}.$$

Using rate constants,

$$V_F = \frac{k_2 k_3[\mathrm{E_o}]}{(k_{-2} + k_3 + k_2)},$$

$$V_R = \frac{k_{-1} k_{-2}[\mathrm{E_o}]}{(k_{-1} + k_2 + k_{-2})},$$

$$K_S = \frac{k_{-1} k_{-2} + k_{-1} k_3 + k_2 k_3}{k_1(k_2 + k_{-2} + k_3)},$$

$$K_P = \frac{k_{-1} k_{-2} + k_{-1} k_3 + k_2 k_3}{k_{-3}(k_{-1} + k_2 + k_{-2})}.$$

Since all the reactions steps are reversible, the system will eventually reach an equilibrium and v will be zero. At this point the concentrations of [S] and [P] will be $[\mathrm{S}]_{\mathrm{eq}}$ and $[\mathrm{P}]_{\mathrm{eq}}$. Hence

$$K_{\mathrm{eq}} = \frac{[\mathrm{P}]_{\mathrm{eq}}}{[\mathrm{S}]_{\mathrm{eq}}} = \frac{(numerator\ 1)}{(numerator\ 2)} = \frac{V_F K_P}{V_R K_S}. \tag{3.26}$$

This latter relationship is known as a Haldane equation. There are more complex Haldane equations for more complex reactions. If we multiply numerator and denominator in equation (3.25) by the factor (*constant*)/(*coef* P) (*coef* S) we obtain

$$v = \frac{V_F K_P[\mathrm{S}] - V_R K_S[\mathrm{P}]}{K_S K_P + K_P[\mathrm{S}] + K_S[\mathrm{P}]} \tag{3.27}$$

which can be written as

$$v = \frac{V_F[\mathrm{S}]}{K_S\{1 + ([\mathrm{P}]/K_P)\} + [\mathrm{S}]} - \frac{V_R[\mathrm{P}]}{K_P\{1 + ([\mathrm{S}]/K_S)\} + [\mathrm{P}]}$$

where v_F, the velocity of the forward reaction, is given by

$$v_F = \frac{V_F[\mathrm{S}]}{K_S\{1 + ([\mathrm{P}]/K_P)\} + [\mathrm{S}]} \tag{3.28}$$

and v_R, the velocity of the reverse reaction, is given by

$$v_R = \frac{V_R[\text{P}]}{K_P\{1 + ([\text{S}]/K_S)\} + [\text{P}]}. \tag{3.29}$$

Equations (3.28) and (3.29) show that the product is always a potential competitive inhibitor for the forward reaction and the substrate is a competitive inhibitor for the reverse reaction. This relationship does not necessarily hold for reactions involving more than one substrate and product (chapter 6). If the product is only weakly bound by the enzyme, the equilibrium position lies very much to the right in equation (3.16) and the reaction is almost irreversible.

We can further simplify equation (3.27) by making use of the Haldane relationship. If we multiply equation (3.25) by the factor *numerator* 2/(*coef* S) (*coef* P), and utilise the relationship, *numerator* 2 = *numerator* 1/K_{eq}, we obtain:

$$v = \frac{V_F V_R[\text{S}] - (V_F V_R[\text{P}]/K_{\text{eq}})}{V_R K_S + V_R[\text{S}] + (V_F[\text{P}]/K_{\text{eq}})}$$

and hence from equation (3.26)

$$v = \frac{V_F\{[\text{S}] - ([\text{P}]/K_{\text{eq}})\}}{K_S + [\text{S}] + (K_S/K_P)[\text{P}]}. \tag{3.30}$$

For initial velocity studies where [P] = 0, equation (3.30) simplifies to the Michaelis–Menten equation derived earlier. For more complex mechanisms with more than one substrate and product, product inhibition studies are extremely useful for determining the mechanism of the reaction (see chapter 6).

3.4. Integrated rate equations

In chapter 2, it was shown that the Michaelis–Menten equation, equation (2.9), could be readily integrated to give an equation (2.14) into which the results for a single time course experiment could be fitted directly. The corresponding rate equations for competitive, non-competitive and uncompetitive inhibition can be integrated in a similar manner.

For the competitive case, equation (3.1),

$$v = -\frac{\text{d}[\text{S}]}{\text{d}t} = \frac{V[\text{S}]}{K_m\{1 + ([\text{I}]/K_I)\} + [\text{S}]}$$

hence

$$-\int_0^t V\mathrm{d}t = \int_{[\mathrm{S}_0]}^{[\mathrm{S}_t]} \left\{\frac{K_m}{[\mathrm{S}]}\left(1+\frac{[\mathrm{I}]}{K_I}\right)+1\right\}\mathrm{d}[\mathrm{S}]$$

$$-Vt = K_m\left(1+\frac{[\mathrm{I}]}{K_I}\right)\ln\frac{[\mathrm{S}_t]}{[\mathrm{S}_0]}+([\mathrm{S}_t]-[\mathrm{S}_0])$$

or

$$\frac{([\mathrm{S}_0]-[\mathrm{S}_t])}{t} = V - K_m\left(1+\frac{[\mathrm{I}]}{K_I}\right)\frac{1}{t}\ln\frac{[\mathrm{S}_0]}{[\mathrm{S}_t]}. \tag{3.31}$$

A plot of $([\mathrm{S}_0]-[\mathrm{S}_t])/t$ against $(1/t)\ln([\mathrm{S}_0]/[\mathrm{S}_t])$ for a series of inhibitor concentrations yields a family of straight lines with slopes of $-K_m\{1+([\mathrm{I}]/K_I)\}$, abscissal intercepts of $V/K_\mathrm{m}\{1+([\mathrm{I}]/K_I)\}$ and a common ordinal intercept V (fig. 3.10).

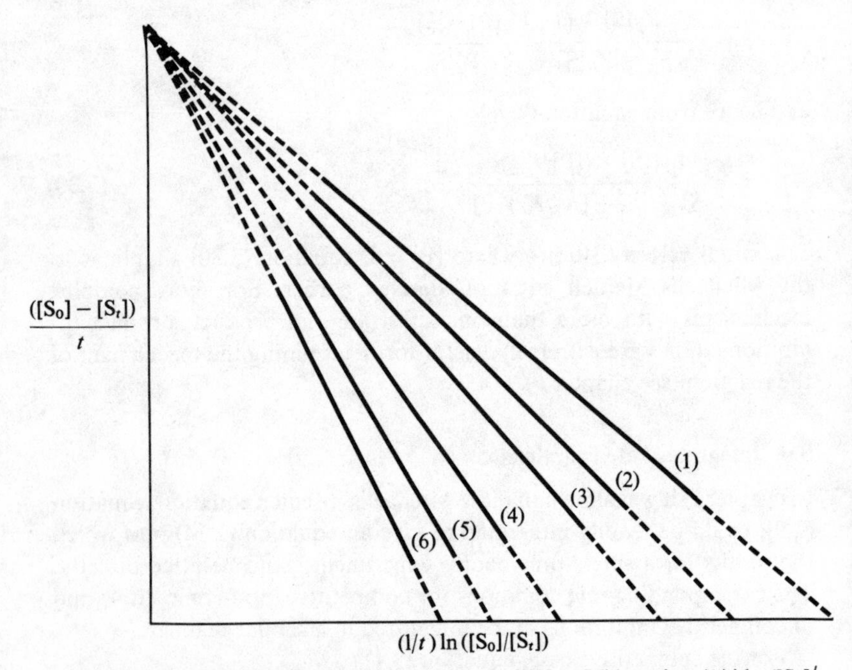

Fig. 3.10. Schematic plot of enzyme kinetic data as $([\mathrm{S}_0]-[\mathrm{S}_t])/t$ against $(1/t)\ln([\mathrm{S}_0]/[\mathrm{S}_t])$ for a reaction subject to competitive inhibition (equation (3.31)). Examples shown are: (1), $[\mathrm{I}]=0$; (2), $[\mathrm{I}]=K_I/6$; (3), $[\mathrm{I}]=K_I/3$; (4), $[\mathrm{I}]=2K_I/3$; (5), $[\mathrm{I}]=K_I$; and (6), $[\mathrm{I}]=4K_I/3$. $[\mathrm{S}_0]$ in each case equals $3K_m$.

For the non-competitive case, equation (3.11),

$$v = \frac{-\mathrm{d}[\mathrm{S}]}{\mathrm{d}t} = \frac{V[\mathrm{S}]}{\{1 + ([\mathrm{I}]/K_I)\}(K_m + [\mathrm{S}])}$$

(remembering $K_m = K_S$) and hence

$$\frac{([\mathrm{S}_0] - [\mathrm{S}_t])}{t} = \frac{V}{\{1 + ([\mathrm{I}]/K_I)\}} - \frac{K_m}{t} \ln \frac{[\mathrm{S}_0]}{[\mathrm{S}_t]}. \tag{3.32}$$

A plot of $([\mathrm{S}_0] - [\mathrm{S}_t])/t$ against $(1/t)\ln([\mathrm{S}_0]/[\mathrm{S}_t])$ for a series of inhibitor concentrations yields a family of parallel straight lines with a common slope $-K_m$, an ordinal intercept $V/\{1 + ([\mathrm{I}]/K_I)\}$ and an abscissal intercept $V/K_m\{1 + ([\mathrm{I}]/K_I)\}$ (fig. 3.11).

For the uncompetitive case, equation (3.14),

$$v = -\frac{\mathrm{d}[\mathrm{S}]}{\mathrm{d}t} = \frac{V[\mathrm{S}]}{K_m + [\mathrm{S}]\{1 + ([\mathrm{I}]/K_I)\}}$$

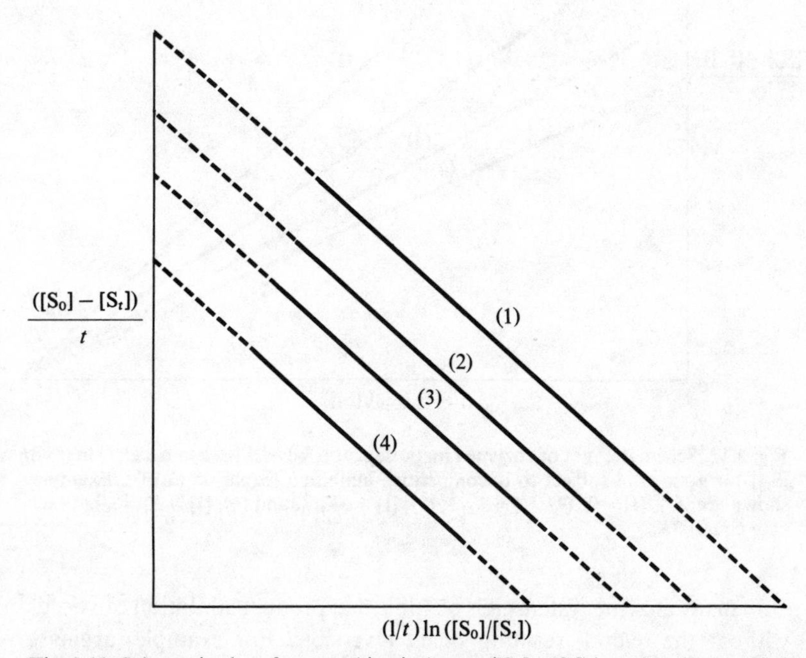

Fig. 3.11. Schematic plot of enzyme kinetic data as $([\mathrm{S}_0] - [\mathrm{S}_t])/t$ against $(1/t)\ln([\mathrm{S}_0]/[\mathrm{S}_t])$ for a reaction subject to non-competitive inhibition (equation (3.32)). Examples shown are: (1), $[\mathrm{I}] = 0$; (2), $[\mathrm{I}] = K_I/3$; (3), $[\mathrm{I}] = 2K_I/3$; and (4), $[\mathrm{I}] = K_I$. $[\mathrm{S}_0]$ in each case equals $3K_m$.

and hence the integrated rate equation is

$$\frac{([S_0]-[S_t])}{t} = \frac{V}{\{1+([I]/K_I)\}} - \frac{K_m}{\{1+([I]/K_I)\}}\frac{1}{t}\ln\frac{[S_0]}{[S_t]}. \quad (3.33)$$

A plot of $([S_0]-[S_t])/t$ against $(1/t)\ln([S_0]/[S_t])$ for a series of inhibitor concentrations has an ordinal intercept $V/\{1+([I]/K_I)\}$, a slope of $-K_m/\{1+([I]/K_I)\}$ and a common abscissal intercept V/K_m (fig. 3.12).

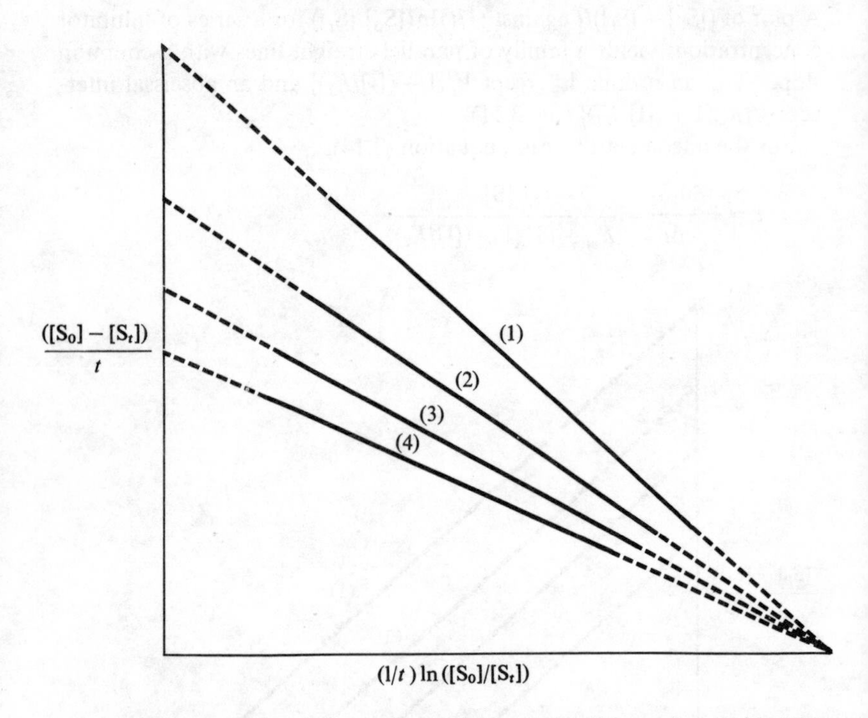

Fig. 3.12. Schematic plot of enzyme kinetic data as $([S_0]-[S_t])/t$ against $(1/t)\ln([S_0]/[S_t])$ for a reaction subject to uncompetitive inhibition (equation (3.33)). Examples shown are: (1), $[I]=0$; (2), $[I]=K_I/3$; (3), $[I]=2K_I/3$; and (4), $[I]=K_I$. $[S_0]$ in each case equals $3K_m$.

In many enzyme systems it is possible that product inhibition can occur without the overall reaction being reversible. For example, arginase which catalyses the hydrolysis of arginine to ornithine and urea, is strongly inhibited in a competitive manner by the product, ornithine. The K_I for ornithine is 3 mM, whilst the K_m for arginine is 5 mM (Kuchel,

Nichol & Jeffrey, 1975). A similar example is the hydrolysis of acetyl-L-tyrosine hydroxamide by α-chymotrypsin to give hydroxylamine and acetyl-L-tyrosine. In this case α-chymotrypsin is inhibited by acetyl-L-tyrosine (Foster & Niemann, 1953). These and other hydrolytic reactions are in fact two-substrate reactions with a water molecule acting as the second substrate. The concentration of water (normally 55 M but not necessarily this high in the cell or in the enzyme) is sufficiently high to ensure that the reaction is virtually irreversible. This does not, however, prevent the product from binding to the enzyme to form an EP complex, but it does prevent the further reaction of the EP complex in the reverse direction. For these systems, equation (3.1) for competitive inhibition can be written as

$$v = \frac{-d[S]}{dt} = \frac{V[S]}{K_m\{1 + ([P]/K_P)\} + [S]} \tag{3.34}$$

where $[P] = [S_0] - [S]$ and K_P is the dissociation constant for the EP complex. Hence from equation (3.34)

$$-\int_0^t V dt = \int_{[S_0]}^{[S_t]} \left\{ \frac{K_m}{[S]}\left(1 + \frac{[S_0]}{K_P}\right) + \left(1 - \frac{K_m}{K_P}\right) \right\} d[S]$$

and therefore

$$-Vt = K_m\left(1 + \frac{[S_0]}{K_P}\right) \ln \frac{[S_t]}{[S_0]} + \left(1 - \frac{K_m}{K_P}\right)([S_t] - [S_0])$$

or

$$\frac{([S_0] - [S_t])}{t} = \frac{VK_P}{(K_P - K_m)} - \frac{K_m(K_P + [S_0])}{(K_P - K_m)} \frac{1}{t} \ln \frac{[S_0]}{[S_t]} \tag{3.35}$$

so that a plot of $([S_0] - [S_t])/t$ against $(1/t)\ln([S_0]/[S_t])$ has an ordinal intercept of $VK_P/(K_P - K_m)$, a slope of $-K_m(K_P + [S_0])/(K_P - K_m)$ and an abscissal intercept of $VK_P/K_m(K_P + [S_0])$ (fig. 3.13). Since the term $(K_P - K_m)$ appears both in the terms for the ordinal intercept and slope, the position of the line will depend on the value of K_P in relation to K_m. If $K_P < K_m$, as is the case for arginase, then the slope is positive and the ordinal intercept is negative, whereas if $K_P > K_m$ the opposite is the case. If K_P is very large, equation (3.35) reduces to the integrated equation (2.14) for the simple uninhibited Michaelis–Menten equation. If K_P

is very small, i.e. extremely tight binding of the product, then equation (3.35) reduces to

$$\frac{([S_0]-[S_t])}{t} = -\frac{VK_P}{K_m} + \frac{[S_0]}{t}\ln\frac{[S_0]}{[S_t]} \tag{3.36}$$

and hence in the limit as $K_P \to 0$, the ordinal intercept tends to zero and the slope of the line tends to $[S_0]$. For values of K_P that lie between these two limits, the plot of $([S_0]-[S_t])/t$ against $(1/t)\ln([S_0]/[S_t])$ will lie between the two limiting cases (fig. 3.13). For a particular enzyme reaction with a fixed value of K_P, plots of $([S_0]-[S_t])/t$ against $(1/t)\ln([S_0]/[S_t])$ for differing initial concentrations of substrate $[S_0]$, all have a common ordinal intercept $VK_P/(K_P-K_m)$. Lines drawn through the origin with slopes equal to the initial substrate concentration intersect

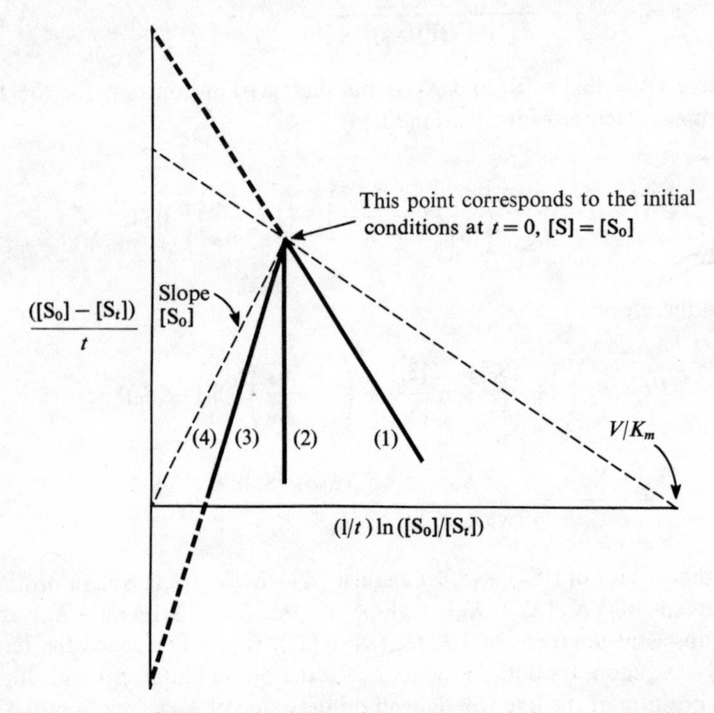

Fig. 3.13. A plot of enzyme kinetic data from a time depletion curve in which the product of the reaction is a competitive inhibitor. The data are plotted as $([S_0]-[S_t])/t$ against $(1/t)\ln([S_0]/[S_t])$. The three examples are: (1), K_P, the inhibition constant for the product, is greater than the K_m for the substrate; (2), $K_P=K_m$; (3), $K_P<K_m$; and (4), $K_P=0$. A line (-----) joining the initial condition points (i.e. at $t=0$) corresponds to the uninhibited example shown in fig. 2.4.

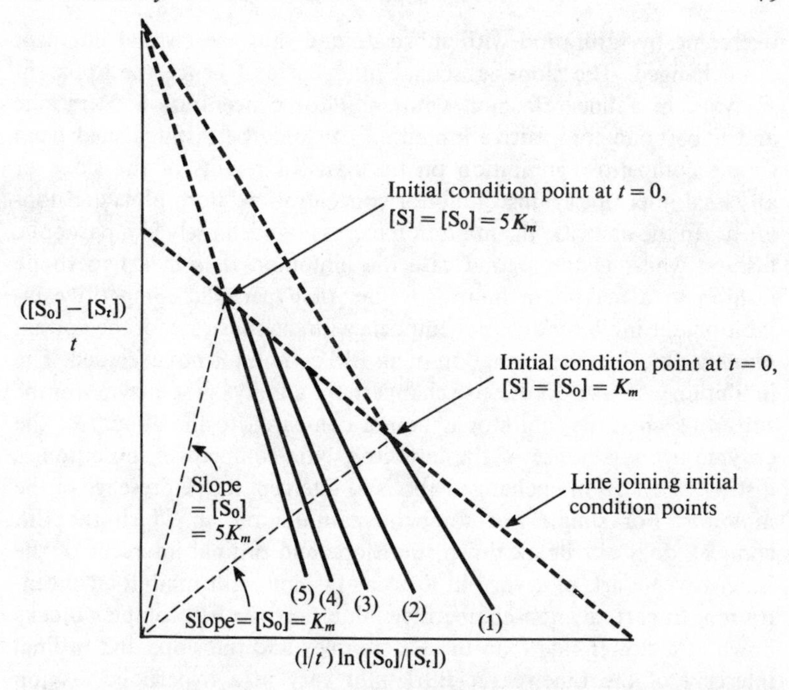

Fig. 3.14. A plot of enzyme kinetic data (solid lines) as $([S_0] - [S_t])/t$ against $(1/t)\ln([S_0]/[S_t])$ for different initial substrate concentrations for a reaction in which the product acts as a competitive inhibitor ($K_P = 3K_m$) (equation (3.35)). The examples shown are: (1), $[S_0] = K_m$; (2), $[S_0] = 2K_m$; (3), $[S_0] = 3K_m$; (4), $[S_0] = 4K_m$; and (5), $[S_0] = 5K_m$. The line joining the initial condition points is equivalent to the uninhibited reaction.

the corresponding plot of $([S_0] - [S_t])/t$ against $(1/t)\ln([S_0]/[S_t])$ at a point corresponding to the initial conditions. A line drawn through all these initial condition points, is equivalent to the line for the hypothetical uninhibited reaction with ordinal intercept V, slope $-K_m$ and abscissal intercept V/K_m (fig. 3.14).

3.5. Summary

The various types of inhibition can be separated primarily by the effect of the inhibitor on the ordinal intercept, abscissal intercept or slope of a Lineweaver–Burk plot. On this basis the four classes of inhibition are competitive, non-competitive, uncompetitive and mixed inhibition. In classical competitive inhibition the inhibitor binds only to the active site and thus competes directly with the substrate. The inhibition is overcome by saturation with substrate and thus the ordinal intercept is unchanged. The slope, abscissal intercept and hence the apparent K_m vary in a linear fashion with inhibitor concentration. Parabolic and hyperbolic competitive inhibition can only be distinguished from simple competitive inhibition on the basis of replots of the slope or abscissal intercept against inhibitor concentration. Both plots are non-linear. In the first case the inhibition increases indefinitely in a parabolic fashion whilst in the second case the inhibition rises in a hyperbolic fashion to a maximum limiting value. In hyperbolic competitive inhibition, the inhibitor does not compete with the substrate for the normal binding site; thus the formation of an EIS complex is not excluded. The inhibition in this case is due to a change in the affinity of the enzyme for the substrate when the inhibitor is bound compared to the affinity of the enzyme in the absence of the inhibitor. Non-competitive inhibition is distinguished by an unchanged abscissal intercept in the presence of the inhibitor. For simple non-competitive inhibition, in which the EIS complex does not break down, the slope and ordinal intercept of the Lineweaver–Burk plot vary in a linear fashion with inhibitor concentration. In partially non-competitive inhibition, the EIS complex breaks down at a slower rate than the ES complex and the slope and ordinal intercept of the Lineweaver–Burk plot vary in a hyperbolic fashion with inhibitor concentration; the maximum inhibition depends on the ratio of the rate constants for the breakdown of the ES and EIS complexes. In uncompetitive inhibition the slope of the Lineweaver–Burk plot is unchanged in the presence of the inhibitor and this readily distinguishes it from other forms of inhibition. Mixed inhibition is distinguished from the other forms of inhibition by the variation of all the parameters of the Lineweaver–Burk plot. The various inhibition equations are summarised in Table 3.1.

It should be noted that all the mechanisms that have been described in this chapter are the simplest schemes involving only the single-intermediate or two-step mechanism of equation (2.1). Many enzymes are known to catalyse reactions by a three-step mechanism such as equation (2.15) and hence, in these cases, there is the added possibility of the inhibitor binding to the second intermediate, ES′. In this case, the behaviour of a particular inhibitor may be competitive, non-competitive, uncompetitive or mixed depending on the particular substrate being studied (Krupka & Laidler, 1961; Kaplan & Laidler, 1967). The ratio of the rate constants k_2 and k_3 (equation (2.15)) is now important. The behaviour of an inhibitor with a substrate in which $k_2 \gg k_3$ will in general

TABLE 3.1. *Summary of inhibition relationships for Lineweaver–Burk plots for the single-intermediate mechanism*

Mechanism	Slope	Ordinal intercept	Abscissal intercept
Competitive	$\frac{K_m\{1+([I]/K_I)\}}{V}$	$\frac{1}{V}$	$\frac{-1}{K_m\{1+([I]/K_I)\}}$
Parabolic competitive	$\frac{K_m\{1+([I]/K_I)+([I]^2/K_i K_I)\}}{V}$	$\frac{1}{V}$	$\frac{-1}{K_m\{1+([I]/K_I)+([I]^2/K_i K_I)\}}$
Hyperbolic competitive	$\frac{K_m\{1+([I]/K_I)\}}{V\{1+([I]/K_{IS})\}}$	$\frac{1}{V}$	$\frac{-\{1+([I]/K_{IS})\}}{K_m\{1+([I]/K_I)\}}$
Non-competitive	$\frac{K_m\{1+([I]/K_I)\}}{V}$	$\frac{\{1+([I]/K_I)\}}{V}$	$\frac{-1}{K_m}$
Partially or hyperbolic non-competitive	$\frac{K_m\{1+([I]/K_I)\}}{V\{1+(\alpha[I]/K_I)\}}$	$\frac{\{1+([I]/K_I)\}}{V\{1+(\alpha[I]/K_I)\}}$	$\frac{-1}{K_m}$
	(α = ratio of the breakdown rate constants)		
Uncompetitive	$\frac{K_m}{V}$	$\frac{\{1+([I]/K_I)\}}{V}$	$\frac{-\{1+([I]/K_I)\}}{K_m}$
Mixed	$\frac{K_m\{1+([I]/K_I)\}}{V}$	$\frac{\{1+([I]/K_{IS})\}}{V}$	$\frac{-\{1+([I]/K_{IS})\}}{K_m\{1+([I]/K_I)\}}$

be different from that observed with a substrate in which $k_3 \gg k_2$. Finally, examples of activation caused by a modifier, M, can also be treated by the equations derived in this chapter. Thus, activation can be caused by the breakdown of the EMS complex at a faster rate than the ES complex, e.g. $k_6 \gg k_2$ in the partially non-competitive scheme, or by a reduction in the apparent K_m, e.g. $K_I > K_i$ in the hyperbolic competitive scheme.

4 The effect of pH on the rate of enzyme-catalysed reactions

4.1. Qualitative considerations

To describe completely the changes that occur in an enzyme due to a change in pH is an impossible task. The enzyme contains many amino acids, e.g. lysine, arginine, aspartic acid, which have ionisable groups in their side chains. These ionisable groups are in environments of different polarities in the enzyme, so that the pK_a value for any particular ionisable group may differ considerably from that for a small peptide in free solution containing the amino acid. This makes the assignment of experimentally determined pK_a values to a particular group in the enzyme quite a difficult and sometimes impossible task without some other supporting evidence. The initial step in the investigation is to determine the range of pH values over which changes are reversible. Tests for reversibility can be carried out quite simply by determining the activity of the enzyme at some standard pH after exposing it for different lengths of time to some other pH at which its stability is unknown. Irreversible changes due to pH extremes are interesting in themselves but are of no concern in a kinetic study of enzyme-catalysed processes. Since the effect of a change in pH on the kinetic parameters, V and K_m, can be quite large, it is unwise to assume that a substrate concentration that is high enough to saturate the enzyme and give zero order kinetics at one pH will necessarily do so at some other pH. It is essential, therefore, that the determination of the kinetic parameters, V and K_m, is carried out at each pH. Some of the possible effects that are caused by a change in pH are listed below:

(1) a change in the ionisation of groups involved in catalysis;
(2) a change in the ionisation of groups involved in binding the substrate;
(3) a change in the ionisation of groups in the substrate;
(4) a change in the ionisation of other groups in the enzyme.

Changes in the ionisation of groups involved in the catalytic mechanism may completely disrupt the mechanism. Changes in the groups involved in the binding site may reduce the affinity of the enzyme for the substrate and hence lead to a reduction in catalysis. Changes in the state of ionisation of groups in the substrate may also alter its affinity for the enzyme. Changes in the ionisation of other groups in the enzyme may cause minor or even major changes in the three-dimensional structure of the enzyme causing disruption of the active site. For example, the important groups in trypsin that are involved in the catalytic mechanism include the carboxyl group of Asp-102, the imidazole ring of His-57 and the hydroxyl group of Ser-195 (the numbering is that of chymotrypsinogen as proposed by Hartley, Brown, Kauffman & Smillie (1965)). These three groups form a charge-relay system (Blow, Birktoft & Hartley, 1969) which allows the transfer of a proton to and from the 'buried' Asp-102 carboxyl group and accounts for the unusual reactivity of the hydroxyl group of Ser-195. Changes in the ionisation of any of these groups would lead to an impairment of catalysis. Trypsin normally hydrolyses peptide bonds involving the carbonyl group of L-lysine or L-arginine and the –NH– group of some other amino acid. The side chain of the substrate contains either a terminal amino group (L-lysine) or a guanidino group (L-arginine). This side chain is known to be bound in a hydrophobic cleft in the enzyme by the formation of an ionic bond with a carboxyl group of Asp-189 situated at the base of the cleft. The effect of lowering the pH to about 3–4, a point where the carboxyl group of Asp-189 is no longer ionised or only partly ionised, can not be determined by a change in catalysis towards the normal peptide substrates since at these low pH values trypsin has negligible activity. Some synthetic aryl ester substrates, however, are hydrolysed by trypsin at these low pH values and could be used to study changes in the ionisation of this and other acidic groups. Another way of determining the pK_a values of groups involved in the specificity binding site of an enzyme is to carry out binding studies using inhibitors which are substrate analogues but which are not converted into products. This enables groups involved in binding to be differentiated from those that control catalysis. Two other groups that are known to be important in trypsin are the carboxyl group of Asp-194 and the N-terminal amino group of Ile-16. From X-ray studies, it has been suggested that these two groups form an ionic bond which maintains the stable conformation of the active site. Although neither group is directly involved in the binding of the substrate or in the catalytic mechanism, any change in the state of their ionisation leads to a change in the structure of the active site and is reflected in changes in the kinetic parameters, K_m and V. Although trypsin has three important carboxyl groups, an imidazole group, a seryl hydroxyl group and an α-amino group, all of which have been shown to be important in the catalytic mechanism, the kinetics of hydrolysis of a number of substrates have been shown to depend upon only two groups with pK_a values of approximately 6–7 and 9.5–10.5 (Roberts & Elmore, 1974). The pK_a values vary with the substrate since the nature of the substrate alters the polarity of the active site and consequently alters the observed pK_a values for the ionising groups. The pK_a in the region 6–7 can be assigned to Asp-102 or the imidazole group of His-57, whilst that in the range 9.5–10.5 can be assigned to the N-terminal amino group of Ile-16.

4.2. Derivation of pH-dependent rate equations

4.2.1. Ionisation of the free enzyme. Let us assume that an enzyme has two ionisable groups that control its activity and that the substrate only binds to one of the three ionisable forms so that the other two forms are inactive. This can be described by the following system:

$$\begin{array}{lcl} & E^- & \\ K_{2e} & \upharpoonleft\downharpoonright & \\ & EH + S & \underset{k_{-1}}{\overset{k_1}{\rightleftharpoons}} \; EHS \; \xrightarrow{k_2} \; EH + P \\ K_{1e} & \upharpoonleft\downharpoonright & \\ & EH_2^+ & \end{array}$$

Under steady-state conditions, the following equilibria exist,

$$K_S = \frac{[EH]\,[S]}{[EHS]}, \qquad K_{2e} = \frac{[E^-]\,[H^+]}{[EH]}, \qquad K_{1e} = \frac{[EH]\,[H^+]}{[EH_2^+]} \tag{4.1}$$

and the conservation equation for the enzyme species must be obeyed, i.e.

$$[E_0] = [E^-] + [EH] + [EHS] + [EH_2^+].$$

Using the relationships in (4.1),

$$[E_0] = [EHS]\left\{1 + \frac{K_S}{[S]}\left(1 + \frac{K_{2e}}{[H^+]} + \frac{[H^+]}{K_{1e}}\right)\right\}$$

and since $v = k_2[EHS]$

$$v = \frac{k_2[E_0][S]}{[S] + K_S\{1 + (K_{2e}/[H^+]) + ([H^+]/K_{1e})\}} \tag{4.2}$$

$$v = \frac{V[S]}{[S] + K_m}$$

where $V = k_2[E_0]$

$$\text{and} \quad K_m = K_S\left(1 + \frac{K_{2e}}{[H^+]} + \frac{[H^+]}{K_{1e}}\right). \tag{4.3}$$

At any pH, the term $(1 + K_{2e}/[H^+] + [H^+]/K_{1e})$ is greater than 1, so that K_m is always greater than K_S. Notice also that at high substrate concentrations such that $[S] \gg K_m$, there will be no inhibition and therefore the maximum rate V is unaffected by pH. At low pH values where $[H^+] \gg K_{2e}$, equation (4.2) reduces to

$$v = \frac{V[S]}{[S] + K_S\{1 + ([H^+]/K_{1e})\}}$$

which has the same form as that for competitive inhibition (equation (3.1)); thus, hydrogen ions in this pH region behave as competitive inhibitors. For this particular reaction scheme, only K_m varies with pH. Since $K_{1e} > K_{2e}$, the function $K_m = K_S(1 + K_{2e}/[H^+] + [H^+]/K_{1e})$ will pass through a minimum as $[H^+]$ is varied. The position of this minimum can be determined by differentiating the function with respect to $[H^+]$,

$$\frac{dK_m}{d[H^+]} = K_S\left(\frac{1}{K_{1e}} - \frac{K_{2e}}{[H^+]^2}\right),$$

which is zero when $[H^+] = \sqrt{(K_{1e}K_{2e})}$. Hence, the optimum pH for the reaction is $(pK_{1e} + pK_{2e})/2$.

At low pH where $[H^+] \gg K_{1e} \gg K_{2e}$, equation (4.3) becomes

$$K_m = \frac{K_S[H^+]}{K_{1e}}.$$

Taking logarithms:

$$\log K_m = \log K_S + \log [H^+] - \log K_{1e}$$

$$\text{or} \quad pK_m = pK_S + pH - pK_{1e}.$$

A plot of pK_m against pH is therefore a straight line of slope +1. Similarly, at high pH where $K_{1e} \gg K_{2e} \gg [\mathrm{H}^+]$

$$K_m = \frac{K_S K_{2e}}{[\mathrm{H}^+]}$$

and so pK_m = pK_S + pK_{2e} − pH. So a plot of pK_m against pH is linear with a slope of −1. If K_{1e} and K_{2e} are widely separated, there will be an intermediate range of pH where $K_{1e} \gg [\mathrm{H}^+] \gg K_{2e}$ so that $K_m \approx K_S$ and hence a plot of pK_m against pH will be horizontal. These changes of pK_m with pH are shown in fig. 4.1. The intersections of the straight

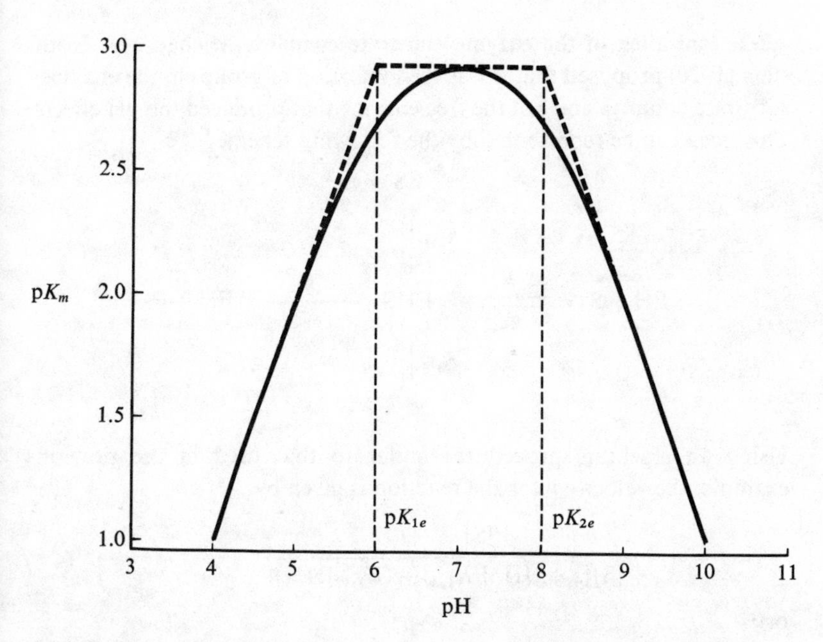

Fig. 4.1. Diagrammatic plot of pK_m against pH for the reaction scheme described by equation (4.2). The enzyme parameters are $K_m = 10^{-3}$, pK_{1e} = 6.0 and pK_{2e} = 8.0.

line sections occur at pH values equal to pK_{1e} and pK_{2e}. An alternative way of determining K_{1e} and K_{2e} is to plot K_m against $[\mathrm{H}^+]$ for K_{1e}, and K_m against $1/[\mathrm{H}^+]$ for K_{2e}. If $[\mathrm{H}^+] \gg K_{2e}$, equation (4.3) reduces to

$$K_m = \frac{K_S[\mathrm{H}^+]}{K_{1e}} + K_S$$

so a plot of K_m against $[H^+]$ is linear with a slope of K_S/K_{1e} and an intercept of K_S. If $K_{1e} \gg [H^+]$, equation (4.3) becomes

$$K_m = \frac{K_S K_{2e}}{[H^+]} + K_S$$

and hence a plot of K_m against $1/[H^+]$ is linear with a slope of $K_S K_{2e}$ and an intercept K_S. It should be noted that the plot of pK_m against pH is only linear in the two pH regions where $[H^+]/K_{1e} \gg 1 \gg K_{2e}/[H^+]$ and $K_{2e}/[H^+] \gg 1 \gg [H^+]/K_{1e}$. The plot of K_m against $[H^+]$ or $1/[H^+]$ is linear over a range of pH that includes the linear and curved portion of the pK_m verus pH plot.

4.2.2. Ionisation of the enzyme–substrate complex. Michaelis & Rothstein (1920) proposed that it was the ionisation of groups in the enzyme–substrate complex and not the free enzyme that produced the pH effects. This idea can be represented by the following scheme:

$$\begin{array}{ccccc} & & ES^- & & \\ & & K_{2es} \Big\Updownarrow & & \\ EH + S & \underset{k_{-1}}{\overset{k_1}{\rightleftharpoons}} & EHS & \xrightarrow{k_2} & EH + P \\ & & K_{1es} \Big\Updownarrow & & \\ & & EH_2S^+ & & \end{array}$$

Using an algebraic procedure similar to that used in the previous example, the velocity v for the reaction is given by

$$v = \frac{V[S]}{K_S + [S]\{1 + ([H^+]/K_{1es}) + (K_{2es}/[H^+])\}} \tag{4.4}$$

or

$$v = \frac{V'[S]}{K_m + [S]}$$

$$K_{2es} = \frac{[ES^-][H^+]}{[EHS]} \quad \text{and} \quad K_{1es} = \frac{[EHS][H^+]}{[EH_2S^+]}$$

where

$$V' = \frac{V}{\{1 + ([H^+]/K_{1es}) + (K_{2es}/[H^+])\}} \tag{4.5}$$

and

$$K_m = \frac{K_S}{\{1 + ([H^+]/K_{1es} + K_{2es}/[H^+])\}},$$

so that in this particular scheme both V' and K_m are pH-dependent. At low pH values where $[H^+] \gg K_{2es}$, equation (4.4) reduces to

$$v = \frac{V[S]}{K_S + [S]\{1 + ([H^+]/K_{1es})\}}$$

and hence by analogy with equation (3.14) the H^+ ion is an uncompetitive inhibitor. Equation (4.4) also predicts that at sufficiently low substrate concentrations where $K_S \gg [S](1 + [H^+]/K_{1es} + K_{2es}/[H^+])$ there will be no variation of the velocity of the reaction with pH. At low pH where $[H^+] \gg K_{1es} \gg K_{2es}$

$$V' = \frac{VK_{1es}}{[H^+]} \quad \text{and} \quad K_m = \frac{K_S K_{1es}}{[H^+]}$$

so that $\log V' = \log V - pK_{1es} + pH$

and $pK_m = pK_S + pK_{1es} - pH.$ (4.6)

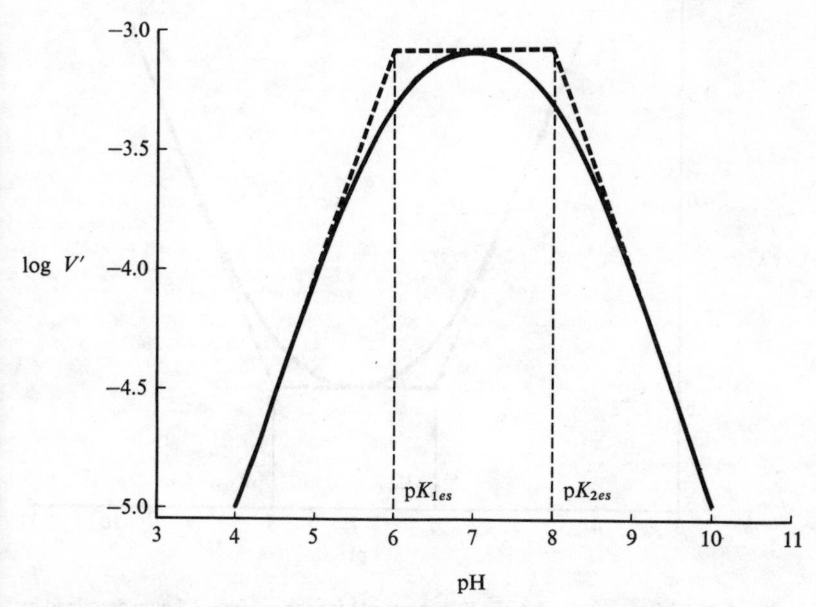

Fig. 4.2. Diagrammatic plot of $\log V'$ against pH for the reaction scheme described by equation (4.4). The enzyme parameters are $V = 10^{-3}$, $pK_{1es} = 6.0$ and $pK_{2es} = 8.0$.

At high pH, where $K_{1es} \gg K_{2es} \gg [H^+]$

$$V' = \frac{V[H^+]}{K_{2es}} \quad \text{and} \quad K_m = \frac{K_S[H^+]}{K_{2es}}$$

so that

$$\log V' = \log V + pK_{2es} - pH$$

$$\text{and} \quad pK_m = pK_S - pK_{2es} + pH. \tag{4.7}$$

At the optimum pH (again given by $pH = (pK_{1es} + pK_{2es})/2$), both V' and K_m are independent of pH. A plot of $\log V'$ against pH (fig. 4.2) is linear with a slope of +1 at low pH, horizontal near the optimum pH region and linear with a slope of –1 at high pH. If $[H^+] \gg K_{2es}$ then from equation (4.5)

$$V' = \frac{V}{1 + ([H^+]/K_{1es})}$$

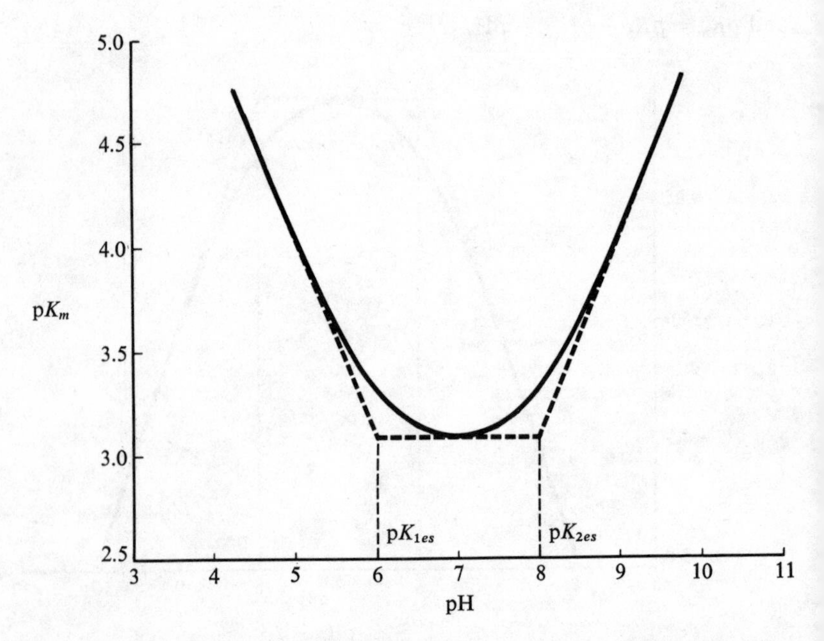

Fig. 4.3. Diagrammatic plot of pK_m against pH for the reaction scheme described by equation (4.4). The enzyme parameters are $K_m = 10^{-3}$, $pK_{1es} = 6.0$ and $pK_{2es} = 8.0$.

which can be inverted to give

$$\frac{1}{V'} = \frac{1}{V} + \frac{[H^+]}{VK_{1es}}$$

so that a plot of $1/V'$ against $[H^+]$ will be linear with a slope of $1/VK_{1es}$ and intercept $1/V$.

If $K_{1es} \gg [H^+]$, then

$$V' = \frac{V}{\{1 + (K_{2es}/[H^+])\}}$$

and $$\frac{1}{V'} = \frac{1}{V} + \frac{K_{2es}}{V[H^+]}$$

so that a plot of $1/V'$ against $1/[H^+]$ is linear with a slope of K_{2es}/V and intercept $1/V$. From equations (4.6) and (4.7) the variation of pK_m with pH is shown in fig. 4.3. Again the plot of pK_m against pH is linear over a smaller range of pH than the corresponding K_m against $[H^+]$ or $1/[H^+]$ plot. In this scheme the observed K_m is higher at the optimum pH than at other pH values.

4.2.3. Ionisation of both the free enzyme and ES complex. The previous two schemes can be combined (Von Euler, Josephson & Myrbäck, 1924; Waley, 1953) to give

$$\begin{array}{ccccc} E^- & & ES^- & & \\ K_{2e} \updownarrows & & K_{2es} \updownarrows & & \\ EH + S & \underset{k_{-1}}{\overset{k_1}{\rightleftharpoons}} & EHS & \xrightarrow{k_2} & EH + P \\ K_{1e} \updownarrows & & K_{1es} \updownarrows & & \\ EH_2^+ & & EH_2S^+ & & \end{array}$$

Using the same method as before, the velocity v is given by

$$v = \frac{k_2[E_0][S]}{\langle K_S\{1 + ([H^+]/K_{1e}) + (K_{2e}/[H^+])\} + [S]\{1 + ([H^+]/K_{1es}) + (K_{2es}/[H^+])\}\rangle} \qquad (4.8)$$

or $$v = \frac{V'[S]}{K_m + [S]}$$

where

$$V' = \frac{V}{\{1 + ([H^+]/K_{1es}) + (K_{2es}/[H^+])\}}$$

and $$K_m = \frac{K_S\{1 + ([H^+]/K_{1e}) + (K_{2e}/[H^+])\}}{\{1 + ([H^+]/K_{1es}) + (K_{2es}/[H^+])\}}.$$

If the pK_a values of the two groups in the enzyme are altered upon binding the substrate, such that $K_{1e} \neq K_{1es}$ and $K_{2e} \neq K_{2es}$ then both V' and K_m will be pH-dependent. The variation of V' with pH will be similar to that shown in fig. 4.2. The overall change in K_m with pH will not be too large if the perturbations in pK_{1e} and pK_{2e} upon binding the substrate are small. There are a total of eight possible changes that could happen, each one giving rise to a different relationship between pK_m and pH:

(1) $pK_{1e} > pK_{1es}$, $pK_{2e} = pK_{2es}$
(2) $pK_{1e} = pK_{1es}$, $pK_{2e} < pK_{2es}$
(3) $pK_{1e} > pK_{1es}$, $pK_{2e} < pK_{2es}$
(4) $pK_{1e} < pK_{1es}$, $pK_{2e} = pK_{2es}$
(5) $pK_{1e} > pK_{1es}$, $pK_{2e} > pK_{2es}$
(6) $pK_{1e} = pK_{1es}$, $pK_{2e} > pK_{2es}$
(7) $pK_{1e} < pK_{1es}$, $pK_{2e} > pK_{2es}$
(8) $pK_{1e} < pK_{1es}$, $pK_{2e} < pK_{2es}$

Possible curves for the first four cases are shown in fig. 4.4 for an enzyme with two groups of $pK_{1e} = 6.0$ and $pK_{2e} = 9.0$ and a possible variation of ± 0.5 pK unit upon binding the substrate. If one assumes that the pK_a values of the ionisable groups are unchanged by the binding of the substrate then $K_{1e} = K_{1es}$ and $K_{2e} = K_{2es}$ and equation (4.8) simplifies to

$$v = \frac{V[S]}{(K_S + [S])\{1 + ([H^+]/K_{1e}) + (K_{2e}/[H^+])\}},$$

so that at low pH where $[H^+] \gg K_{2e}$, hydrogen ions will inhibit in a non-competitive manner (see equation (3.11)):

$$v = \frac{V[S]}{(K_S + [S])\{1 + ([H^+]/K_{1e})\}}.$$

Under these circumstances only the maximum velocity, V', will vary with pH (fig. 4.2).

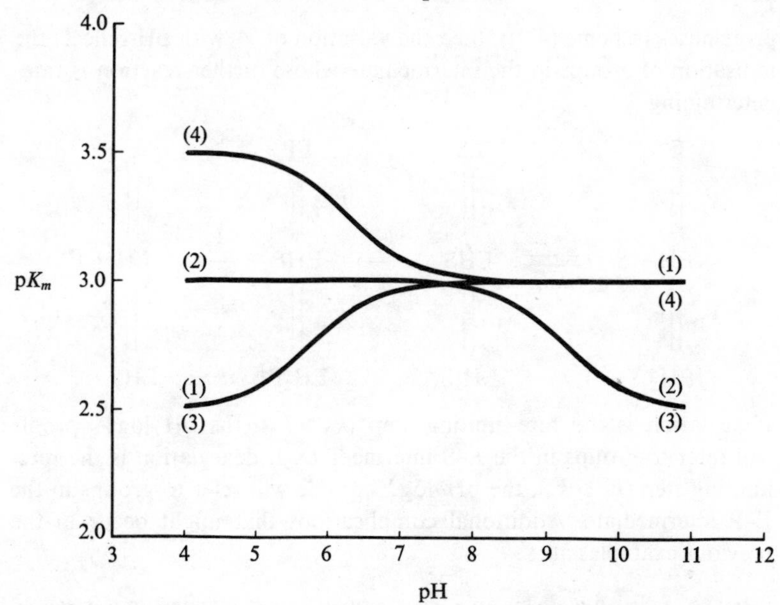

Fig. 4.4. Diagrammatic plot of pK_m against pH for the reaction scheme described by equation (4.8). The examples are: (1) $pK_{1e} = 6.0$, $pK_{1es} = 5.5$, $pK_{2e} = 9.0$, $pK_{2es} = 9.0$; (2) $pK_{1e} = 6.0$, $pK_{1es} = 6.0$, $pK_{2e} = 9.0$, $pK_{2es} = 9.5$; (3) $pK_{1e} = 6.0$, $pK_{1es} = 5.5$, $pK_{2e} = 9.0$, $pK_{2es} = 9.5$; and (4) $pK_{1e} = 6.0$, $pK_{1es} = 6.5$, $pK_{2e} = 9.0$, $pK_{2es} = 9.0$.

In all these cases, it is assumed that reactions involving the transfer or removal of a proton are very much more rapid than any catalytic reaction step. In summary, if K_m varies with pH (section 4.2.1), there are ionising groups in the free enzyme which control the binding of the substrate; the K_m in this case rises on both sides of the optimum pH (fig. 4.1). If both V' and K_m vary with pH (section 4.2.2), there are groups in the enzyme–substrate complex which control both the binding of the substrate and the breakdown of the ES complex. The K_m in this case is reduced on either side of the optimum pH (fig. 4.3). Thirdly if V' varies with pH and there is little or no variation in K_m, then this can be described by the third reaction scheme in which there are ionising groups in both the free enzyme and enzyme–substrate complex whose pK_a values are relatively unaffected by the binding of the substrate (section 4.2.3).

These previous examples are relatively simple, since they take no account of the formation of other reaction intermediates. If one includes the formation of an acyl-enzyme intermediate as in the case of the serine proteinases (scheme (4.9)), then the variation of V' with pH reflects the ionisation of groups in the intermediate whose further reaction is rate-determining

$$
\begin{array}{ccccccc}
E^- & & ES^- & & EP^- & & E^- \\
{\scriptstyle K_{2e}}\upharpoonleft\downharpoonright & & {\scriptstyle K_{2es}}\upharpoonleft\downharpoonright & & {\scriptstyle K_{2ep}}\upharpoonleft\downharpoonright & & \upharpoonleft\downharpoonright \\
EH + S & \underset{k_{-1}}{\overset{k_1}{\rightleftharpoons}} & EHS & \xrightarrow{k_2} & EHP & \xrightarrow{k_3} & EH + P \\
{\scriptstyle K_{1e}}\upharpoonleft\downharpoonright & & {\scriptstyle K_{1es}}\upharpoonleft\downharpoonright & & {\scriptstyle K_{1ep}}\upharpoonleft\downharpoonright & & \upharpoonleft\downharpoonright \\
EH_2^+ & & EH_2S^+ & & EH_2P^+ & & EH_2^+
\end{array}
\qquad (4.9)
$$

If acylation is the rate-limiting step ($k_3 \gg k_2$), the pH–log V' profile will refer to groups in the E–S intermediate. If deacylation is the rate-limiting step ($k_2 \gg k_3$), the pH–log V' profile will refer to groups in the E–P intermediate. Additional complications that might occur in the previous examples are:

(1) binding of S to E^- ($S + E^- \rightleftharpoons ES^-$) and S to EH_2^+ ($S + EH_2^+ \rightleftharpoons EH_2S^+$);

(2) breakdown of ES^- and EH_2S^+, in the case of the single-intermediate mechanisms or of EP^- and EH_2P^+ in the case of the double-intermediate mechanisms, to give products;

(3) involvement of several intermediates including Michaelis complexes of products if the reaction is reversible;

(4) the presence of more than two groups that control the catalytic mechanism of the enzyme;

(5) the presence of ionisable groups in the substrate.

4.3. Ionisation of the substrate

Four possible reaction schemes that include the ionisation of the substrate are given in (*a*) to (*d*) below:

(*a*)

$$
\begin{array}{ccccc}
S + E & \underset{k_{-1}}{\overset{k_1}{\rightleftharpoons}} & ES & \xrightarrow{k_2} & E + P. \\
{\scriptstyle K}\upharpoonleft\downharpoonright & & & & \\
HS^+ & & & &
\end{array}
$$

In this case, we assume there are no ionisable groups in either E or ES.

This scheme leads to the following equation for the velocity of the reaction:

$$v = \frac{V[S_0]}{K_m\{1+([H^+]/K)\}+[S_0]}$$

where

$$K = \frac{[H^+][S]}{[HS^+]} \quad \text{and} \quad [S_0] = [S]+[HS^+];$$

hence K_m will vary with pH according to the equation

$$K'_m = K_m\left(1+\frac{[H^+]}{K}\right).$$

Note that this is the same form as equation (4.2) when $[H^+] \gg K_{2e}$ and could not be distinguished kinetically.

(*b*) is similar to (*a*) in which HS^+ rather than S binds to the enzyme.

$$\begin{array}{ccccc} HS^+ + E & \underset{k_{-1}}{\overset{k_1}{\rightleftharpoons}} & EHS^+ & \xrightarrow{k_2} & E + HP^+. \\ K \updownarrow & & & & \\ S & & & & \end{array}$$

For this scheme

$$v = \frac{V[S_0]}{K'_m+[S_0]},$$

where

$$K'_m = K_m\left(1+\frac{K}{[H^+]}\right);$$

which again could not be distinguished from equation (4.2) when $K_{1e} \gg [H^+]$.

(*c*) If both the enzyme and substrate have ionisation processes,

$$\begin{array}{ccccc} HS^+ + E & \underset{k_{-1}}{\overset{k_1}{\rightleftharpoons}} & EHS^+ & \xrightarrow{k_2} & E + HP^+ \\ K \updownarrow \quad K_e \updownarrow & & & & \\ S \quad HE^+ & & & & \end{array}$$

then $$V = \frac{V[S_0]}{K_m' + [S_0]} \tag{4.10}$$

where

$$K_m' = K_m\left(1 + \frac{K}{[H^+]}\right)\left(1 + \frac{[H^+]}{K_e}\right). \tag{4.11}$$

(*d*) is a similar scheme to (*c*) except that the unprotonated form of the substrate is bound to the protonated enzyme:

$$\begin{array}{ccccccc} HS^+ & & E & & & & \\ K \big\updownarrow & & K_e \big\updownarrow & & & & \\ S & + & EH^+ & \underset{k_{-1}}{\overset{k_1}{\rightleftharpoons}} & EHS^+ & \xrightarrow{k_2} & EH^+ + P \end{array}$$

The variation of velocity with pH for this scheme is also described by equation (4.10) where

$$K_m' = K_m\left(1 + \frac{K_e}{[H^+]}\right)\left(1 + \frac{[H^+]}{K}\right).$$

Although it is possible to determine the pK_a of the ionising group in the substrate in the absence of the enzyme, the different polarity of the binding site in the enzyme may produce a significant shift in the pK_a of the group in the substrate when bound to the enzyme. From these discussions, it is clear that the interpretation of the effect of pH on enzyme-catalysed reactions is complicated by the large number of possible ionising groups in the enzyme. In general, however, the variation of V' or K_m with pH can usually be described in terms of one of the foregoing models.

5 The effect of temperature on enzyme-catalysed reactions

5.1. Introduction

In chapter 1, the variation in the rate of a chemical reaction with temperature was described by the Arrhenius equation, equation (1.25). The frequency factor, A, in the Arrhenius equation could be equated with the term $(\mathrm{e}kT/h)\mathrm{e}^{\Delta S^{\ddagger}/R}$ in equation (1.31). A comparison of equations (1.23) and (1.30), shows that the activation energy in the Arrhenius equation, $E_a = \Delta H^{\ddagger} + RT$, where $\Delta H^{\ddagger}$ is the enthalpy of activation for the formation of the transition state complex. The overall free energy change for the conversion of reactants to products is given by $\Delta G^{\circ} = -RT \ln K$, where K is the equilibrium constant for the reaction. From equation (1.25), it can be seen that an increase in temperature will obviously lead to an increase in the rate of reaction.

How does an increase in temperature affect an enzyme-catalysed reaction? It is difficult to make direct comparisons between the effect of temperature on the enzymic and non-enzymic reactions for two reasons: firstly most biological reactions that are catalysed by enzymes usually do not proceed at a measurable rate in the absence of the enzyme so a direct comparison of the thermodynamic parameters for the reaction may not be possible; secondly the enzymic reaction pathway is usually different, involving the formation of reaction intermediates that are not formed in the uncatalysed reaction. For each of these enzyme intermediates, there will be a transition state complex with its associated thermodynamic parameters $\Delta G^{\ddagger}$, $\Delta H^{\ddagger}$ and $\Delta S^{\ddagger}$. The three-step mechanism described earlier in equation (2.15) can therefore be described more completely by the reaction mechanism in equation (5.1):

$$\mathrm{E}+\mathrm{S} \rightleftharpoons (\mathrm{E}\underset{k_{-1}}{\overset{k_1}{\text{------}}}\mathrm{S})^{\ddagger} \rightleftharpoons \mathrm{ES} \rightleftharpoons (\mathrm{E}\text{---}\underset{k_{-2}}{\overset{k_2}{\mathrm{S}}}\text{---}\mathrm{P})^{\ddagger}$$

$$\rightleftharpoons \mathrm{EP} \rightleftharpoons (\mathrm{E}\underset{k_{-3}}{\overset{k_3}{\text{------}}}\mathrm{P})^{\ddagger} \rightleftharpoons \mathrm{E}+\mathrm{P} \quad (5.1)$$

where $(E\text{ --- }S)^{\ddagger}$, $(E\text{ --- }S\text{ --- }P)^{\ddagger}$ and $(E\text{ --- }P)^{\ddagger}$ are the transition state complexes. For any particular enzyme reaction, two points are of interest: firstly, what direction will the reaction take for a given concentration of reactants and products, and secondly, what is the rate of the reaction under these conditions? A typical free energy diagram for the scheme given in equation (5.1) is shown in fig. 2.1(*b*). Note that this is only one of a number of possible energy diagrams. The free energy of the system at any point along the reaction pathway is determined by the sum of the chemical potentials of the reactants at that point, i.e. $\Sigma\mu_i$ where $\mu_i = \mu_i^0 + RT\ln c_i$, where c_i is the concentration of the ith species. Initially, when $[P] = 0$, the reaction will proceed from left to right solely because of the excess chemical potential provided by the substrate. As the substrate is used up, the free energy of $E+S$ falls whilst that of $E+P$ rises and hence eventually a position will be reached where the free energy difference between $E+S$ and $E+P$ is zero. The reaction will then be in equilibrium. Most enzyme reactions are reversible but in some cases the equilibrium lies very much to one side. In these cases, the standard free energy change for the conversion of substrate(s) into product(s), $\Delta G^\circ = -RT\ln K$, is large and negative (i.e. positive in the reverse direction) and in order for the free energy change to be negative in the reverse direction, the concentration of product(s) must be very high. This follows immediately from the equation for the free energy change in the reverse direction

$$\Delta G = \Delta G^\circ + RT\ln \frac{[\text{Substrate(s)}]}{[\text{Product(s)}]}. \tag{5.2}$$

It is apparent from equation (5.2) that for those reactions in which the molecularity for the forward reaction is lower than for the reverse reaction, the equilibrium will tend to favour the formation of products. For example, the conversion of prephenic acid to phenyl pyruvic acid, by the enzyme prephenate dehydratase, is a single-substrate reaction in the forward direction but is a three-substrate (phenyl pyruvic acid, carbon dioxide and water) reaction in the reverse direction and the equilibrium lies very much in favour of the products. Some enzyme reactions appear to be completely irreversible but in these cases it is usually found that water takes part in the reaction mechanism and since the concentration of water is 55 M (though not necessarily in the enzyme structure), the reaction cannot be reversed by any reasonable concentration of products. The serine proteinases are typical examples; thus the hydrolysis of a peptide bond by chymotrypsin or trypsin is a Crypto

Ping Pong Uni Bi mechanism (see chapter 6) with water acting as a second substrate. The overall rate of the reaction is determined by the various activation energy barriers in the mechanism. For the reaction scheme in equation (5.1) (fig. 2.1(*b*)), the binding of the substrate, S, to the enzyme, E, is usually very rapid and reversible. A reasonable value for the equilibrium constant for the formation of the ES complex would be in the range 10^4–10^6 giving a value for ΔG°, of –22 to –34 kJ mol^{-1} (at 25 °C). The rapid attainment of equilibrium (Gutfreund & Sturtevant, 1956*a*) would indicate that the free energy of activation, $\Delta G_1^\ddagger$, for the formation of the ES transition state, is low. The free energy of activation for the transition state is related to the enthalpy and entropy of activation by the equation:

$$\Delta G^\ddagger = \Delta H^\ddagger - T\Delta S^\ddagger = -RT\ln K^\ddagger = -RT\ln k + RT\ln\frac{kT}{h}, \qquad (5.3)$$

which can be compared formally with equation (5.4) relating the classical thermodynamic parameters:

$$\Delta G^\circ = \Delta H^\circ - T\Delta S^\circ = -RT\ln K. \qquad (5.4)$$

The formation of the ES complex is usually favourable in terms of the enthalpy change ($\Delta H^\ddagger$ is negative) due to the formation of ionic, hydrogen and other bonds with groups in the enzyme. It will, however, usually be unfavourable in an entropic sense ($\Delta S^\ddagger$ is negative) due to the freezing out of various degrees of freedom on formation of the transition state. As a general rule, the values of $\Delta H^\ddagger$ and $\Delta S^\ddagger$, for widely differing reactions, are such that $\Delta G^\ddagger$ does not differ markedly from one enzymic reaction to another. Thus, if in one reaction $\Delta H^\ddagger$ is large and positive, it will be compensated by a large positive $\Delta S^\ddagger$ term, and if $\Delta H^\ddagger$ is small or even negative then $\Delta S^\ddagger$ will also be negative.

The next step in the three-step mechanism is the formation of the acyl-enzyme, EP, via the transition state complex $(\mathrm{E} \cdots \mathrm{S} \cdots \mathrm{P})^\ddagger$. In this particular example, the free energy of activation for this step, $\Delta G_2^\ddagger$, is larger than any other in the reaction pathway. This means that, at this temperature, acylation is the rate-limiting step for this particular scheme. If deacylation were the rate-limiting step then $\Delta G_3^\ddagger$ would be larger than any of the other energies of activation. Note that the size of the activation energy barrier determines the steady-state concentration of the various enzyme intermediates. Thus, in the example shown in fig. 2.1(*b*), acylation is the rate-limiting step and the conversion of ES to EP is slow. The concentration of the ES complex is therefore high whereas that of the acyl-enzyme EP is low since it is rapidly converted to free enzyme and product. If deacylation were the rate-limiting step, then the conversion of EP to E + P would be slow and consequently the concentration of EP would be high relative to ES and E.

Even for the simplest, single substrate–single enzyme system, a quantitative determination of all the thermodynamic parameters is a formidable task and as yet no enzyme reaction has been completely studied. Table 5.1 summarises the measurements that would be required to determine each of the thermodynamic parameters.

TABLE 5.1. *Relationship between the rate constants and thermodynamic constants for the reaction described by equation* (5.1)

Quantity determined	Effect of temperature on:
$\Delta G_1^{\ddagger}$	k_1
$\Delta G_{-1}^{\ddagger}$	k_{-1}
ΔG_1°	k_1/k_{-1}
$\Delta G_2^{\ddagger}$	k_2
$\Delta G_{-2}^{\ddagger}$	k_{-2}
ΔG_2°	k_2/k_{-2}
$\Delta G_3^{\ddagger}$	k_3
$\Delta G_{-3}^{\ddagger}$	k_{-3}
ΔG_3°	k_3/k_{-3}

From the variation with temperature of the various rate and equilibrium constants, the other thermodynamic parameters ΔS°, ΔG°, $\Delta S^{\ddagger}$ and $\Delta G^{\ddagger}$ could be obtained using equations (5.3) and (5.4). To determine all of these parameters would involve the research worker in undertaking steady-state, pre-steady-state and relaxation kinetics (see chapter 7). Diagrams similar to fig. 2.1(*b*) can be drawn for the variation of enthalpy and entropy with reaction co-ordinate. The various transition states are restricted spatial arrangements of the reacting molecules and this leads to a loss of entropy compared to the molecules in free solution. The rate enhancements achieved by enzymes are due not only to a reduction in the enthalpy of activation for the reaction but also due to a reduction in the *loss* of entropy on formation of the transition state complex compared to the loss of entropy on formation of the transition state complex for the uncatalysed reaction (Bruice, Brown & Harris, 1971; Page & Jencks, 1971). For the three-step mechanism, the kinetic parameters that are determined using steady-state techniques are V and K_m. Both of these are complex functions involving a number of rate constants (equations (5.5) and (5.6)):

$$V = k_{cat}[E_0] \quad \text{where} \quad k_{cat} = \frac{k_2 k_3}{k_2 + k_3}, \tag{5.5}$$

$$K_m = \frac{(k_{-1} + k_2) k_3}{k_1 (k_2 + k_3)}. \tag{5.6}$$

Arrhenius plots of $\ln k_{cat}$ or $\ln K_m$ against $1/T$ can only be carried out under certain circumstances in which the expressions for k_{cat} or K_m simplify to expressions containing only a single rate constant or equilibrium constant. First, if deacylation is the rate-limiting step, then $k_2 \gg k_3$ and the expression for k_{cat} simplifies to $k_{cat} \approx k_3$. Under these conditions,

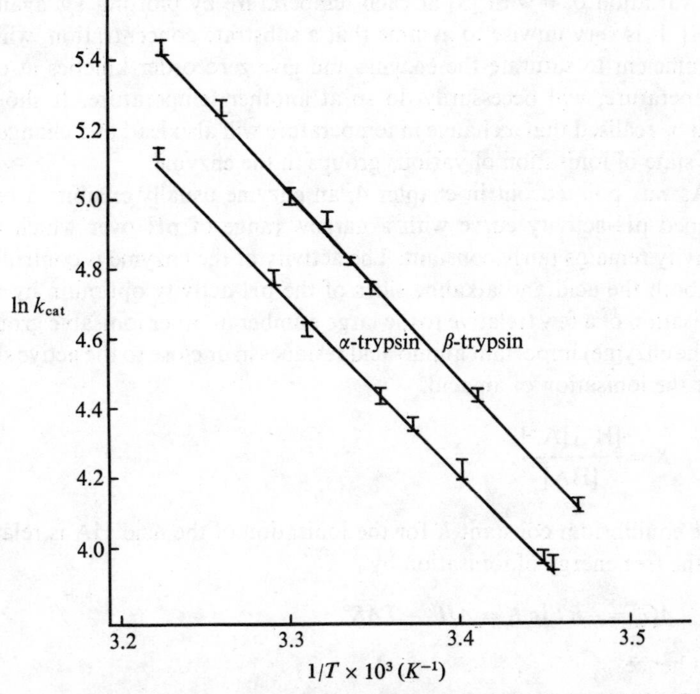

Fig. 5.1. An Arrhenius plot for the α- and β-trypsin-catalysed hydrolysis of α-*N*-toluene-*p*-sulphonyl-L-lysine methyl ester. For α-trypsin: $E_a = 41.9 \pm 2.1$ kJ mol^{-1}, $\Delta H^{*} = 39.4 \pm 2.1$ kJ mol^{-1}, $\Delta S^{*} = -76.2 \pm 8.4$ kJ mol^{-1} deg C^{-1}. For β-trypsin: $E_a = 44.4 \pm 0.8$ kJ mol^{-1}, $\Delta H^{*} = 41.9 \pm 0.8$ kJ mol^{-1}, $\Delta S^{*} = -65.3 \pm 2.5$ kJ mol^{-1} deg C^{-1}. (Adapted from Roberts & Elmore (1974). *Biochem. J.* **141**, 545–54.)

a plot of $\ln k_{cat}$ ($=\ln k_3$) against $1/T$ will provide the thermodynamic parameters for the transition state complex $(E \text{ --- } P)^{\ddagger}$. Alternatively, if acylation is the rate-limiting step, $k_3 \gg k_2$ then $k_{cat} \approx k_2$ and hence a plot of $\ln k_{cat}$ ($=\ln k_2$) against $1/T$ will give $\Delta G_2^{\ddagger}$, $\Delta S_2^{\ddagger}$ and $\Delta H_2^{\ddagger}$ for the acylation step. When acylation is the rate-limiting step, $K_m = (k_{-1} + k_2)/k_1$ and if k_2 is much less than k_{-1}, $K_m \approx k_{-1}/k_1$ which is the equilibrium constant for the *dissociation* of the enzyme–substrate complex. Only under these conditions will a plot of $\ln K_m$ against $1/T$ be meaningful and yield the parameters ΔG_1°, ΔS_1° and ΔH_1°. Fig. 5.1 shows an Arrhenius plot for the α- and β-trypsin-catalysed hydrolysis of α-*N*-toluene-*p*-sulphonyl-L-lysine methyl ester (Roberts & Elmore, 1974) for which deacylation is the rate-limiting step. It is important that the parameter V and hence k_{cat} should be determined from a complete investigation of the variation of v with [S] at each temperature by plotting $1/v$ against $1/[S]$. It is very unwise to assume that a substrate concentration, which is sufficient to saturate the enzyme and give zero order kinetics at one temperature, will necessarily do so at another temperature. It should also be realised that a change in temperature will also lead to a change in the state of ionisation of various groups in the enzyme.

As was pointed out in chapter 4, an enzyme usually exhibits a bell-shaped pH-activity curve with a narrow range of pH over which the activity remains fairly constant. The activity of the enzyme is controlled on both the acid and alkaline sides of the pH-activity optimum by the ionisation of a few (relative to the large number of other ionisable groups in the enzyme) important amino acid residues in or close to the active site. For the ionisation of an acid,

$$K = \frac{[H^+][A^-]}{[HA]}.$$

The equilibrium constant K for the ionisation of the acid HA is related to the free energy of ionisation by

$$\Delta G^{\circ} = -RT \ln K = \Delta H^{\circ} - T\Delta S^{\circ}$$

and hence

$$\ln K = \frac{\Delta S^{\circ}}{R} - \frac{\Delta H^{\circ}}{RT} \tag{5.7}$$

where ΔH° and ΔS° are respectively the enthalpy and entropy of ionisation. The change in pK for a particular group follows immediately from equation (5.7) so that for two temperatures T_1 and T_2 we have

$$\Delta pK = pK_1 - pK_2 = \frac{\Delta H^\circ (T_2 - T_1)}{2.303\, RT_1 T_2}. \tag{5.8}$$

Any change in the state of ionisation of the important residues in the active site will be reflected by a change in K_m or V. It is obviously important to determine the pH optimum at the two extremes of temperature and to choose a suitable pH for the kinetic study that ensures that the activity of the enzyme is at a maximum throughout the temperature range. The linearity of the Arrhenius plot over the whole temperature range is a good indication that there has been no change in mechanism. A change in mechanism could be due to the pH changes mentioned above or to a change in the rate-limiting step in the reaction. If the enthalpies of activation for acylation and deacylation were significantly different, then Arrhenius plots of $\ln k_2$ and $\ln k_3$ against $1/T$ would have different slopes and would cross at a particular temperature. If this temperature

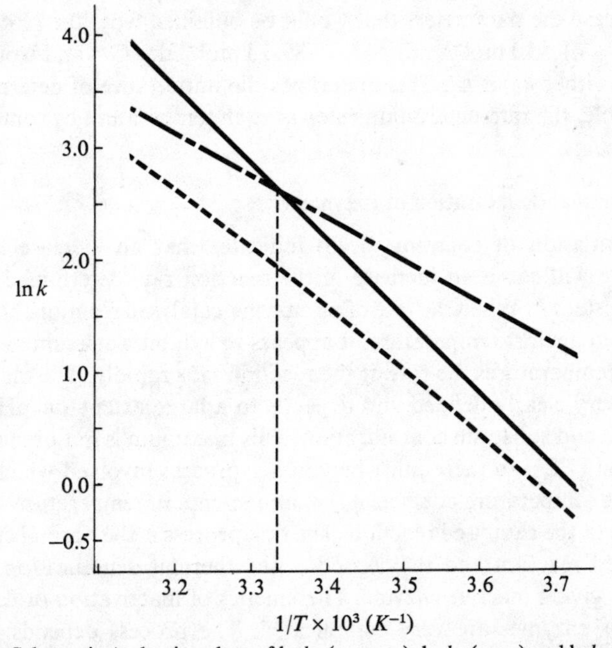

Fig. 5.2. Schematic Arrhenius plots of $\ln k_2$ (– – – –), $\ln k_3$ (——) and $\ln k_{cat}$ (- - - -) against $1/T$ showing the change in rate-limiting step at $1/T \simeq 3.33 \times 10^{-3}$. It is apparent that the plot using the steady-state kinetic parameter k_{cat}, only gives meaningful results at $1/T > 3.7 \times 10^{-3}$ where $k_{cat} \simeq k_3$ or when $1/T < 3.1 \times 10^{-3}$ where $k_{cat} \simeq k_2$.

were within the range under examination then there would be a change in the rate-limiting step. The temperature at which this would occur is given by the equation

$$T = \frac{\Delta H_2^\ddagger - \Delta H_3^\ddagger}{\Delta S_2^\ddagger - \Delta S_3^\ddagger}. \tag{5.9}$$

Fig. 5.2 shows an Arrhenius plot for $\ln k_2$ and $\ln k_3$ against $1/T$ for $\Delta H_2^\ddagger = 29.3$ kJ mol^{-1}, $\Delta S_2^\ddagger = -125.5$ J mol^{-1} deg C^{-1}, $\Delta H_3^\ddagger = 54.4$ kJ mol^{-1} and $\Delta S_3^\ddagger = -41.8$ J mol^{-1} deg C^{-1}. Obviously, if one measured the variation of the individual rate constants k_2 and k_3 with temperature and plotted the corresponding Arrhenius line, it would be immediately apparent that there had been a change in the rate-limiting step at approximately 300 K ($1/T = 3.33 \times 10^{-3}$). Since one cannot measure the individual rate constants from steady-state kinetics, but only the overall catalytic rate constant $k_{cat} = k_2k_3/(k_2 + k_3)$, a plot of $\ln k_{cat}$ against $1/T$ is only slightly curved. When one allows for the probable error for each point, the plot of $\ln k_{cat}$ against $1/T$ would probably be accepted as linear. In this case the parameters that would be obtained from the plot would be $\Delta H^\ddagger = 41.8$ kJ mol^{-1} and $\Delta S^\ddagger = -89.5$ J mol^{-1} deg C^{-1} and would not refer to either k_2 or k_3. This underlines the importance of determining, if possible, the rate-determining step at each temperature by some other experiment.

5.2. Thermal denaturation of enzymes

Differentiation of equation (1.28) indicates that an increase in temperature will cause an increase in the reaction rate determined by the rate constant k. When the rate of an enzyme-catalysed reaction, however, is plotted against temperature, it appears to exhibit a maximum and at higher temperatures the rate of the reaction falls rapidly. The maximum is not very clearly defined and depends to a large extent on pH, ionic strength and substrate concentration. This maximum is not predicted by equation (1.28) so there must be another process involved which has a negative temperature coefficient, i.e. an increase in temperature reduces the rate of the catalysed reaction. The first process is the normal enzyme-catalysed reaction and the second is the thermal denaturation of the enzyme giving inactive enzyme. The kinetics of inactivation or denaturation of enzymes are very complicated. The process depends on the disruption of a number of bonds in the protein and the enthalpy of activation for the process is high, 150–600 kJ mol^{-1}. The entropy of activation for the process is also large and positive, 200–1200 J mol^{-1}

deg C^{-1}. This large positive entropy term indicates that the protein loses its ordered structure including that of the active site. The large enthalpy and entropy terms cause the denaturation process to take place over a very narrow range of temperature. Consequently, the rate of denaturation at lower temperatures will not be significant. Most enzymes are rapidly denaturated at 70 °C and above, but some enzymes that are found in organisms growing in hot springs (*Bacillus stearothermophilus*) can withstand temperatures of 90 °C for periods of up to 1 hour (Manning & Campbell, 1961). Whilst the thermal denaturation of proteins and enzymes is an interesting phenomenon, it lies outside the scope of this book.

Some other possible explanations for discontinuities in Arrhenius plots have been suggested:

(1) A reversible structural change may take place in the enzyme; this change could alter the conformation of the active site and hence change the energy requirements for the catalytic step.

(2) The rate-limiting step in a consecutive or sequential system involving two or more enzymes may alter as the temperature is changed because of different enthalpies of activation for the individual enzyme-catalysed steps in the overall mechanism.

(3) An enzyme may undergo reversible dimerisation or polymerisation into other active forms; the reactions catalysed by the latter may have completely different entropies and enthalpies of activation.

5.3. Summary

The first step in a kinetic analysis of the effect of temperature on an enzyme-catalysed reaction is to determine the range of temperature over which the enzyme remains completely active. This can be achieved quite simply by exposing the enzyme to extremes of temperature for different lengths of time and then assaying its activity at some other temperature at which it is known to be stable. Secondly, the pH-activity optimum for the enzyme should be determined at the two extremes of temperature and a suitable pH employed for the kinetic study that will ensure that the activity of the enzyme is at an optimum at all temperatures. This will eliminate any complications due to changes in the state of ionisation of groups in the enzyme with temperature. Thirdly, the kinetic parameters, V and K_m, should be determined at each temperature. Finally, some other evidence (kinetic or otherwise) should be obtained to decide if the kinetics relate to a single rate-limiting step.

For example, deacylation appears to be rate-determining in the hydrolysis of a number of esters of a particular amino acid by trypsin or chymotrypsin. Consequently the observed thermodynamic parameters refer to that particular step in the reaction mechanism. If the observed kinetics do not refer to a single rate-limiting step or equilibrium reaction, then no useful information can be obtained from a study of temperature effects on the enzyme-catalysed reaction.

6 Multi-substrate enzyme systems

6.1. Introduction

In chapter 2, we developed kinetic equations that were applicable to a single substrate–single enzyme system. These equations were then further developed in chapter 3 to include the effects of inhibitors. Although many enzyme systems can be described by these simple two- and three-step mechanisms, there are many more for which these kinetic equations are not applicable. The development of interest in enzymes that catalyse reaction between two or more substrates required a more extensive treatment. Early work by Alberty (1953), Dalziel (1957) and Wong & Hanes (1962) was elegantly developed into a shorthand method of expressing multi-substrate reactions by Cleland (1963).

The most common enzymatic reactions are those with two or three substrates and as many products. In order to elucidate the mechanism of catalysis of these enzymes, appropriate experiments must be designed to determine the *order* (if any) of addition of the substrates and the *order* (if any) of release of the products. Although the mechanisms for these multi-substrate reactions are complex, their kinetics can often be described by an equation of the form $v = V[A]/(K + [A])$ if the concentrations of all the other substrates except A are held constant. The 'constants' V and K will be functions of the concentrations of all substrates other than A. Thus, the velocity of the reaction, v, is commonly determined for a variety of concentrations of A but using the same initial concentration of the other substrates.

6.2. Types of reaction sequences

Cleland (1963) has classified enzyme-catalysed reactions according to the number of substrates and products and the reaction mechanism. The *reactancy* is equal to the number of kinetically significant substrates or products; it is designated by the syllables Uni, Bi, Ter, Quad. For example:

$$\begin{array}{rl} A \rightleftharpoons P & \text{Uni Uni} \\ A \rightleftharpoons P+Q & \text{Uni Bi} \\ A+B \rightleftharpoons P & \text{Bi Uni} \\ A+B \rightleftharpoons P+Q & \text{Bi Bi} \\ A+B+C \rightleftharpoons P+Q & \text{Ter Bi} \end{array}$$

Mechanisms where all the substrates must add to the enzyme before any products are released are designated *sequential.* If substrates add in an obligatory order and products leave similarly, the mechanism is said to be *ordered.* If there is no obligatory order of addition of substrates and departure of products, the mechanism is said to be *random.* When one or more products are released before all the substrates have added, the enzyme will exist in two or more stable forms between which it oscillates during the reaction. This type of mechanism is called *Ping Pong.* If isomerisation of stable, as distinct from transitory, enzyme forms occurs, the term *Iso* is added to the designation of the mechanism. The Ordered Bi Bi mechanism can either be described by the following equations:

$$E + A \underset{k_{-1}}{\overset{k_1}{\rightleftharpoons}} EA \quad EA + B \underset{k_{-2}}{\overset{k_2}{\rightleftharpoons}} EAB \quad \left(EAB \underset{k_{-3}}{\overset{k_3}{\rightleftharpoons}} EPQ\right)$$

$$EPQ \underset{k_{-4}}{\overset{k_4}{\rightleftharpoons}} EQ + P \quad EQ \underset{k_{-5}}{\overset{k_5}{\rightleftharpoons}} E + Q \tag{6.1}$$

or by the simpler diagrammatical method proposed by Cleland

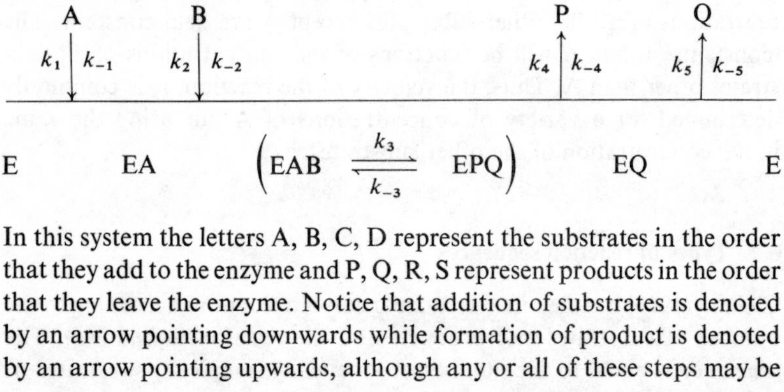

$$\begin{array}{cccccc} A & B & & & P & Q \\ k_1 \downarrow k_{-1} & k_2 \downarrow k_{-2} & & & k_4 \uparrow k_{-4} & k_5 \uparrow k_{-5} \\ \hline E & EA & \left(EAB \underset{k_{-3}}{\overset{k_3}{\rightleftharpoons}} EPQ\right) & & EQ & E \end{array}$$

In this system the letters A, B, C, D represent the substrates in the order that they add to the enzyme and P, Q, R, S represent products in the order that they leave the enzyme. Notice that addition of substrates is denoted by an arrow pointing downwards while formation of product is denoted by an arrow pointing upwards, although any or all of these steps may be reversible. Enzyme forms, including stable isomerisation forms, are represented by the letters E, F, G, H. Transitory complexes, which only undergo unimolecular reactions are enclosed in brackets and the number of these transitory forms cannot be deduced from steady-state kinetics. The base line represents the enzyme in all its forms as it undergoes reaction.

6.2.1. Ordered Uni Bi mechanism. Whilst this mechanism is not a multi-substrate reaction, the derivation of the rate equation does allow the concept of product inhibition studies to be introduced with relatively simple equations. The mechanism can be described by

A (k_1 ↓ k_{-1}) P (k_3 ↑ k_{-3}) Q (k_4 ↑ k_{-4})

$$\mathrm{E} \qquad \left(\mathrm{EA} \underset{k_{-2}}{\overset{k_2}{\rightleftharpoons}} \mathrm{EPQ}\right) \qquad \mathrm{EQ} \qquad \mathrm{E}$$

Using the King–Altman procedure,

$$\begin{array}{ccc} \mathrm{E} & \underset{k_{-1}}{\overset{k_1[\mathrm{A}]}{\rightleftharpoons}} & \mathrm{EA} \\ k_4 \uparrow\downarrow k_{-4}[\mathrm{Q}] & & k_{-2} \uparrow\downarrow k_2 \\ \mathrm{EQ} & \underset{k_3}{\overset{k_{-3}[\mathrm{P}]}{\rightleftharpoons}} & \mathrm{EPQ} \end{array}$$

we can express

$$\frac{[\mathrm{E}]}{[\mathrm{E}_0]} = \frac{k_{-1}k_{-2}k_{-3}[\mathrm{P}] + k_{-1}k_{-2}k_4 + k_{-1}k_3k_4 + k_2k_3k_4}{\Sigma}$$

$$\frac{[\mathrm{EA}]}{[\mathrm{E}_0]} = \{k_1k_3k_4[\mathrm{A}] + k_1k_{-2}k_4[\mathrm{A}] + k_1k_{-2}k_{-3}[\mathrm{A}][\mathrm{P}] + k_{-2}k_{-3}k_{-4}[\mathrm{P}][\mathrm{Q}]\}/\Sigma$$

$$\frac{[\mathrm{EPQ}]}{[\mathrm{E}_0]} = \{k_1k_2k_4[\mathrm{A}] + k_1k_2k_{-3}[\mathrm{A}][\mathrm{P}] + k_2k_{-3}k_{-4}[\mathrm{P}][\mathrm{Q}] + k_{-3}k_{-4}k_{-1}[\mathrm{P}][\mathrm{Q}]\}/\Sigma$$

$$\frac{[EQ]}{[E_0]} = \{k_1 k_2 k_3[A] + k_2 k_3 k_{-4}[Q] + k_3 k_{-4} k_{-1}[Q] + k_{-4} k_{-1} k_{-2}[Q]\}/\Sigma$$

where

$$\Sigma = k_4(k_{-1}k_{-2} + k_{-1}k_3 + k_2 k_3) + k_1(k_3 k_4 + k_2 k_4 + k_{-2}k_4 + k_2 k_3)[A] + k_{-1}k_{-2}k_{-3}[P] + k_{-4}(k_2 k_3 + k_{-1}k_3 + k_{-1}k_{-2})[Q] + k_1 k_{-3}(k_{-2} + k_2)[A][P] + k_{-3}k_{-4}(k_{-1} + k_{-2} + k_2)[P][Q];$$

the summation term, Σ, can be written in coefficient form as

$$\Sigma = \textit{constant} + (\textit{coef}\,\mathrm{A})[A] + (\textit{coef}\,\mathrm{P})[P] + (\textit{coef}\,\mathrm{Q})[Q] + (\textit{coef}\,\mathrm{AP})[A][P] + (\textit{coef}\,\mathrm{PQ})[P][Q]$$

where $\textit{constant} = k_4(k_{-1}k_{-2} + k_{-1}k_3 + k_2 k_3)$

$$\textit{coef}\,\mathrm{A} = k_1(k_3 k_4 + k_2 k_4 + k_{-2}k_4 + k_2 k_3)$$

$$\textit{coef}\,\mathrm{P} = k_{-1}k_{-2}k_{-3}$$

$$\textit{coef}\,\mathrm{Q} = k_{-4}(k_2 k_3 + k_{-1}k_3 + k_{-1}k_{-2})$$

$$\textit{coef}\,\mathrm{AP} = k_1 k_{-3}(k_{-2} + k_2)$$

$$\textit{coef}\,\mathrm{PQ} = k_{-3}k_{-4}(k_{-1} + k_{-2} + k_2).$$

The velocity of the reaction is given by

$$v = k_4[EQ] - k_{-4}[E][Q].$$

Hence,

$$v = \frac{(k_1 k_2 k_3 k_4[A] - k_{-1}k_{-2}k_{-3}k_{-4}[P][Q])[E_0]}{\Sigma} \tag{6.2}$$

which can be written as

$$v = \{(\textit{numerator}\ 1)[A] - (\textit{numerator}\ 2)[P][Q]\}/\Sigma, \tag{6.3}$$

where

$$\textit{numerator}\ 1 = k_1 k_2 k_3 k_4[E_0]$$

$$\textit{numerator}\ 2 = k_{-1}k_{-2}k_{-3}k_{-4}[E_0].$$

The numerator of the rate equation (6.2) contains two terms, one positive term that consists of the product of the concentrations of all the reactants and associated rate constants for the forward reaction and one negative term that contains the product of the concentrations of all the reactants and rate constants for the reverse reaction. This applies to all *non-random* mechanisms. The denominator consists of a number of terms; some are only products of rate constants and summed in the *constant* term, others are products of rate constants and a concentration or a product of concentrations. The collections of rate constants are termed coefficients and are named after their associated concentration term(s). We can define the maximum velocity for the reaction in a particular direction as the numerator term for that direction divided by the coefficient for the term containing the product of all the reactants in that direction. In this case we can define, V_1, the maximum velocity in the forward direction when the substrate, A, is at saturating concentration as

$$V_1 = \frac{\textit{numerator } 1}{\textit{coef } \mathrm{A}}$$

and similarly the maximum velocity in the reverse direction, V_2, at saturating levels of P and Q is

$$V_2 = \frac{\textit{numerator } 2}{\textit{coef } \mathrm{PQ}}.$$

We can also define a number of Michaelis and inhibition constants as ratios of various coefficients in the denominator. These kinetic constants are named after the remaining letter in the denominator after cancelling letters common to the coefficients in the numerator and denominator. In this particular scheme, we can define the following constants relating to the reactants A, P and Q.

$$K_A = \frac{\textit{constant}}{\textit{coef } \mathrm{A}} \quad \text{or} \quad K_A = \frac{\textit{coef } \mathrm{P}}{\textit{coef } \mathrm{AP}}. \tag{6.4}$$

$$K_P = \frac{\textit{constant}}{\textit{coef } \mathrm{P}} \quad \text{or} \quad K_P = \frac{\textit{coef } \mathrm{A}}{\textit{coef } \mathrm{AP}} \quad \text{or } K_P = \frac{\textit{coef } \mathrm{Q}}{\textit{coef } \mathrm{PQ}}. \tag{6.5}$$

$$K_Q = \frac{\textit{constant}}{\textit{coef } \mathrm{Q}} \quad \text{or} \quad K_Q = \frac{\textit{coef } \mathrm{P}}{\textit{coef } \mathrm{PQ}}. \tag{6.6}$$

The definition of the Michaelis constant for any reactant must have the same coefficient in the denominator as the definition for the maximum velocity in the direction for which it is a substrate. There is only one substrate in the forward direction, A, and the definition for V_1 contains (*coef* A) in the denominator so that of the two possibilities listed in equation (6.4), K_A must be defined as

$$K_A = \frac{constant}{coef\,\mathrm{A}} = \frac{k_4(k_{-1}k_{-2}+k_{-1}k_3+k_2k_3)}{k_1(k_3k_4+k_{-2}k_4+k_2k_4+k_2k_3)}.$$

The remaining kinetic constant that can be defined for A will then be classed as an inhibition constant, K_{IA}.

$$K_{IA} = \frac{coef\,\mathrm{P}}{coef\,\mathrm{AP}} = \frac{k_{-1}k_{-2}}{k_1(k_2+k_{-2})}.$$

Similarly the definitions for the Michaelis constants that refer to P and Q are

$$K_P = \frac{coef\,\mathrm{Q}}{coef\,\mathrm{PQ}} = \frac{(k_{-1}k_{-2}+k_{-1}k_3+k_2k_3)}{k_{-3}(k_2+k_{-1}+k_{-2})}$$

and

$$K_Q = \frac{coef\,\mathrm{P}}{coef\,\mathrm{PQ}} = \frac{k_{-1}k_{-2}}{k_{-4}(k_2+k_{-1}+k_{-2})},$$

since the denominator must contain the same coefficient term as in the definition for V_2. The remaining kinetic constants in equations (6.5) and (6.6) that refer to P and Q are classed as inhibition constants:

$$K'_{IP} = \frac{constant}{coef\,\mathrm{P}} = \frac{k_4(k_{-1}k_{-2}+k_{-1}k_3+k_2k_3)}{k_{-1}k_{-2}k_{-3}}$$

$$K_{IP} = \frac{coef\,\mathrm{A}}{coef\,\mathrm{AP}} = \frac{(k_3k_4+k_{-2}k_4+k_2k_4+k_2k_3)}{(k_{-3}k_2+k_{-2}k_{-3})}$$

$$K_{IQ} = \frac{constant}{coef\,\mathrm{Q}} = \frac{k_4(k_{-1}k_{-2}+k_{-1}k_3+k_2k_3)}{k_{-4}(k_{-1}k_{-2}+k_{-1}k_3+k_2k_3)} = \frac{k_4}{k_{-4}}.$$

There may be more than one way of defining an inhibition constant for a particular reactant, as in the case of P, where we can define two constants K_{IP} and K'_{IP}. In some cases the different possible ways of defining a particular inhibition constant in terms of coefficients are the same when written out in rate constant form. If this is not the case then, as we shall see later, one particular definition may be more suitable for simplifying a particular equation.

Since all the steps in the reaction mechanism of equation (6.1) are reversible, the reaction will eventually reach an equilibrium where the overall velocity for the reaction is zero. For the velocity to be zero the numerator in equation (6.2) must be zero and hence

$$k_1 k_2 k_3 k_4 [A_e] = k_{-1} k_{-2} k_{-3} k_{-4} [P_e][Q_e]$$

where $[A_e]$, $[P_e]$ and $[Q_e]$ are the concentrations of the reactants at equilibrium. We can therefore define the equilibrium constant K_e for the reaction as

$$K_e = \frac{[P_e][Q_e]}{[A_e]} = \frac{k_1 k_2 k_3 k_4}{k_{-1} k_{-2} k_{-3} k_{-4}} = \frac{\textit{numerator } 1}{\textit{numerator } 2} = \frac{V_1 K_{IP} K_Q}{V_2 K_{IA}}$$

$$= \frac{V_1 K_P K_{IQ}}{V_2 K_A}. \tag{6.7}$$

The relationships between K_e, V_1, V_2 and the various Michaelis and inhibition constants in equation (6.7) are called Haldane equations. There are other and often more complicated Haldane equations for different reaction mechanisms.

If we multiply equation (6.3) by the factor *numerator* 2/(*coef* A) (*coef* PQ) and use the relationship in equation (6.7) that *numerator* 2 = *numerator* $1/K_e$, we obtain

$$v = \left\{ \frac{(\textit{numerator } 1)(\textit{numerator } 2)[A]}{(\textit{coef } A)(\textit{coef } PQ)} - \frac{(\textit{numerator } 1)(\textit{numerator } 2)[P][Q]}{(\textit{coef } A)(\textit{coef } PQ) K_e} \right\} \Big/ \sum$$

$$\text{where } \sum = \left\{ \frac{(\textit{constant})(\textit{numerator } 2)}{(\textit{coef } A)(\textit{coef } PQ)} + \frac{(\textit{numerator } 2)[A]}{(\textit{coef } PQ)} + \frac{(\textit{coef } AP)(\textit{numerator } 2)[A][P]}{(\textit{coef } A)(\textit{coef } PQ)} + \frac{(\textit{numerator } 1)[P][Q]}{(\textit{coef } A) K_e} + \frac{(\textit{coef } P)(\textit{numerator } 1)[P]}{(\textit{coef } PQ)(\textit{coef } A) K_e} + \frac{(\textit{coef } Q)(\textit{numerator } 1)[Q]}{(\textit{coef } PQ)(\textit{coef } A) K_e} \right\}$$

and hence

$$v = \frac{V_1 V_2[\text{A}] - V_1 V_2[\text{P}][\text{Q}]/K_e}{\{V_2 K_A + V_2[\text{A}] + (V_2/K_{IP})[\text{A}][\text{P}] + (V_1/K_e)[\text{P}][\text{Q}] + (V_1 K_Q/K_e)[\text{P}] + (V_1 K_P/K_e)[\text{Q}]\}} \tag{6.8}$$

Using the two Haldane relationships, equation (6.8) can be further simplified to give

$$v = \frac{V_1[\text{A}] - (V_1/K_e)[\text{P}][\text{Q}]}{\{K_A + [\text{A}] + ([\text{A}][\text{P}]/K_{IP}) + (K_{IA}/K_{IP}K_Q)[\text{P}][\text{Q}] + (K_{IA}/K_{IP})[\text{P}] + (K_A/K_{IQ})[\text{Q}]\}} \tag{6.9}$$

If we carry out only initial-velocity studies, we may eliminate all the terms involving the concentrations of the products and hence equation (6.9) simplifies to $v = V_1[\text{A}]/(K_A + [\text{A}])$ which is of course the Michaelis–Menten equation for the simple single substrate–single enzyme mechanism that was derived in chapter 2.

The form of equation (6.9) is useful since it allows us to determine the type of inhibition that would occur on addition of either product P or Q to the reaction. Suppose we add Q as an inhibitor when there is no product, P, we can write equation (6.9) as follows:

$$v = \frac{V_1[\text{A}]}{K_A + [\text{A}] + (K_A/K_{IQ})[\text{Q}]}$$

or

$$v = \frac{V_1[\text{A}]}{K_A\{1 + ([\text{Q}]/K_{IQ})\} + [\text{A}]}. \tag{6.10}$$

Equation (6.10) is analogous to equation (3.1) for competitive inhibition so that Q behaves as a competitive inhibitor. In the reciprocal form

$$\frac{1}{v} = \frac{K_A\{1 + ([\text{Q}]/K_{IQ})\}}{V_1}\frac{1}{[\text{A}]} + \frac{1}{V_1}$$

which shows more clearly that the ordinal intercept $(1/V_1)$ is unaffected by the addition of product Q.

If we add P as an inhibitor to the reaction with Q absent, we obtain

$$v = \frac{V_1[\text{A}]}{K_A\{1 + (K_{IA}[\text{P}]/K_A K_{IP})\} + [\text{A}]\{1 + ([\text{P}]/K_{IP})\}}$$

which upon inversion gives

$$\frac{1}{v} = \frac{K_A\{1 + (K_{IA}[\text{P}]/K_A K_{IP})\}}{V_1}\frac{1}{[\text{A}]} + \frac{\{1 + ([\text{P}]/K_{IP})\}}{V_1}$$

which shows that inhibition by P is mixed, i.e. both K_A and V_1 are changed but by different amounts. If by chance $K_{IA} = K_A$ then the inhibition by P reduces to non-competitive inhibition (equation (3.11)). The difference between the inhibition patterns obtained with different products is used to determine the mechanism and order of addition of substrates and release of products.

6.2.2. Bi Uni mechanism

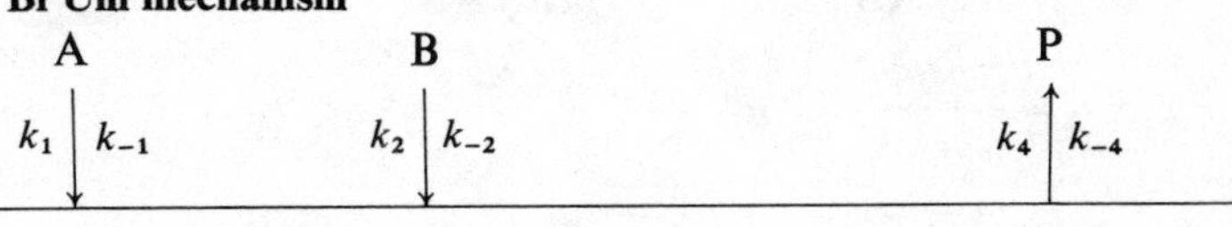

$$\text{E} \qquad \text{EA} \qquad \text{EAB} \underset{k_{-3}}{\overset{k_3}{\rightleftharpoons}} \text{EP} \qquad \text{E}$$

Using the King–Altman method, we obtain

$$v = \frac{(k_1 k_2 k_3 k_4 [\text{A}][\text{B}] - k_{-1} k_{-2} k_{-3} k_{-4} [\text{P}])[\text{E}_0]}{\sum}$$

where

$$\sum = \textit{constant} + (\textit{coef}\ \text{A})[\text{A}] + (\textit{coef}\ \text{B})[\text{B}] + (\textit{coef}\ \text{BP})[\text{B}][\text{P}] + (\textit{coef}\ \text{AB})[\text{A}][\text{B}] + (\textit{coef}\ \text{P})[\text{P}]$$

and

$$\textit{constant} = k_{-1}(k_{-2} k_{-3} + k_{-2} k_4 + k_3 k_4)$$

$$\textit{coef}\ \text{A} = k_1(k_3 k_4 + k_{-2} k_4 + k_{-2} k_{-3})$$

$$\textit{coef}\ \text{B} = k_2 k_3 k_4$$

$$\textit{coef}\ \text{BP} = k_2 k_{-4}(k_3 + k_{-3})$$

$$\textit{coef}\ \text{AB} = k_1 k_2 (k_4 + k_{-3} + k_3)$$

$$\textit{coef}\ \text{P} = k_{-4}(k_{-1} k_3 + k_{-1} k_{-2} + k_{-1} k_{-3} + k_{-2} k_{-3})$$

$$v = \{(\textit{numerator}\ 1)[\text{A}][\text{B}] - (\textit{numerator}\ 2)[\text{P}]\}/\sum.$$

Again if we multiply by the factor (*numerator* 2)/(*coef* AB) (*coef* P) and utilise the relation $K_e = (\textit{numerator}\ 1)/(\textit{numerator}\ 2)$, we obtain

$$v = \frac{V_1 V_2 [\text{A}][\text{B}] - (V_1 V_2 [\text{P}]/K_e)}{\{(K_P V_1/K_e) + V_2 K_B [\text{A}] + V_2 K_A [\text{B}] + (V_1 [\text{B}][\text{P}]/K_{IB} K_e) + V_2 [\text{A}][\text{B}] + (V_1 [\text{P}]/K_e)\}}$$

where $V_1 = \dfrac{\textit{numerator } 1}{\textit{coef}\,\mathrm{AB}}$, $\quad V_2 = \dfrac{\textit{numerator } 2}{\textit{coef}\,\mathrm{P}}$,

$$K_P = \frac{\textit{constant}}{\textit{coef}\,\mathrm{P}}$$

$$K_B = \frac{\textit{coef}\,\mathrm{A}}{\textit{coef}\,\mathrm{AB}}, \qquad K_A = \frac{\textit{coef}\,\mathrm{B}}{\textit{coef}\,\mathrm{AB}},$$

$$K_{IA} = \frac{\textit{constant}}{\textit{coef}\,\mathrm{A}}$$

$$K_{IB} = \frac{\textit{coef}\,\mathrm{P}}{\textit{coef}\,\mathrm{BP}}, \; K'_{IB} = \frac{\textit{constant}}{\textit{coef}\,\mathrm{B}}, \; K_{IP} = \frac{\textit{coef}\,\mathrm{B}}{\textit{coef}\,\mathrm{BP}}.$$

There are two Haldane relationships:

$$K_e = \frac{V_1 K_P}{V_2 K_B K_{IA}} = \frac{V_1 K_{IP}}{V_2 K_{IB} K_A},$$

which give

$$v = \frac{V_1[\mathrm{A}][\mathrm{B}] - (V_1[\mathrm{P}]/K_e)}{\{K_B K_{IA} + K_B[\mathrm{A}] + K_A[\mathrm{B}] + (K_A/K_{IP})[\mathrm{B}][\mathrm{P}] + [\mathrm{A}][\mathrm{B}] + (K_{IB} K_A/K_{IP})[\mathrm{P}]\}} \tag{6.11}$$

Using initial-velocity studies where $[\mathrm{P}] = 0$, equation (6.11) simplifies to

$$v = \frac{V_1[\mathrm{A}][\mathrm{B}]}{K_B K_{IA} + K_B[\mathrm{A}] + K_A[\mathrm{B}] + [\mathrm{A}][\mathrm{B}]}. \tag{6.12}$$

This equation is true for all ordered sequential bireactant mechanisms, e.g. Bi Uni, Bi Bi, Bi Ter. There are two ways of carrying out kinetic experiments with two substrates; one can keep [B] at a fixed initial value and determine the initial velocity of the reaction for a number of different concentrations of A, or one can vary [B] with [A] fixed. By 'fixed', one means the same initial concentration of B is used for different initial values of [A]; the concentration of B will of course decrease during the reaction, but for initial-velocity studies it is the initial concentrations of A and B that are important. Equation (6.12) can be inverted for plotting $1/v$ against $1/[\mathrm{A}]$ with [B] fixed:

$$\frac{1}{v} = \frac{K_A}{V_1}\left(1 + \frac{K_B K_{IA}}{K_A[\mathrm{B}]}\right)\frac{1}{[\mathrm{A}]} + \frac{1}{V_1}\left(1 + \frac{K_B}{[\mathrm{B}]}\right). \tag{6.13}$$

This will give a series of straight lines, for different initial values of [B], with a slope of $K_A(1 + K_B K_{IA}/K_A[\mathrm{B}])/V_1$ and an ordinal intercept of $(1 + K_B/[\mathrm{B}])/V_1$ (fig. 6.1). The point of intersection of two lines with initial concentrations $[\mathrm{B}_1]$ and $[\mathrm{B}_2]$ can be obtained by equating

$$\frac{K_A}{V_1}\left(1+\frac{K_B K_{IA}}{K_A[\mathrm{B}_1]}\right)\frac{1}{[\mathrm{A}]}+\frac{1}{V_1}\left(1+\frac{K_B}{[\mathrm{B}_1]}\right)$$
$$=\frac{K_A}{V_1}\left(1+\frac{K_B K_{IA}}{K_A[\mathrm{B}_2]}\right)\frac{1}{[\mathrm{A}]}+\frac{1}{V_1}\left(1+\frac{K_B}{[\mathrm{B}_2]}\right)$$

so that

$$\frac{1}{[\mathrm{B}_1]}\left(1+\frac{K_{IA}}{[\mathrm{A}]}\right)=\frac{1}{[\mathrm{B}_2]}\left(1+\frac{K_{IA}}{[\mathrm{A}]}\right)$$

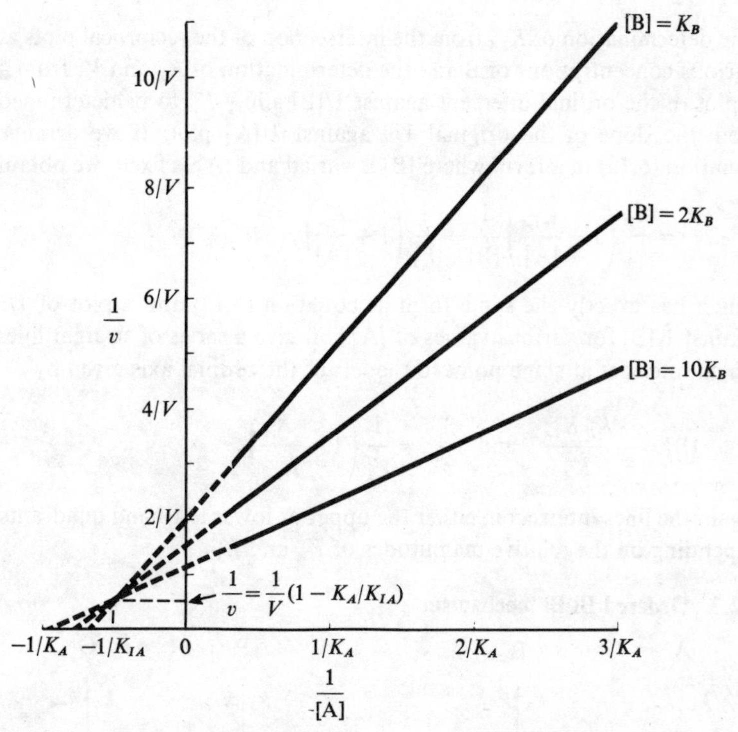

Fig. 6.1. Reciprocal plots of $1/v$ against $1/[\mathrm{A}]$ for varying initial concentrations of [B] for a sequential mechanism described by equations (6.12) and (6.13) ($K_A < K_{IA}$).

and since $[B_1] \neq [B_2]$, then $[A] = -K_{IA}$. Replacing this value of [A] in equation (6.13), we obtain

$$\frac{1}{v} = \frac{1}{V_1}\left(1 - \frac{K_A}{K_{IA}}\right).$$

Consequently, the linear plots of $1/v$ against $1/[A]$ at various constant values of [B] will intersect in the upper left-hand quandrant if $K_A < K_{IA}$ (fig. 6.1) and in the lower left-hand quadrant if $K_A > K_{IA}$. Notice that the maximum velocity observed, $V_1' = V_1/(1 + K_B/[B])$, will only be the true maximum velocity when $[B] \gg K_B$, i.e. at a saturating concentration of B. The true maximum velocity, V_1, could be obtained by plotting $1/V_1'$ against $1/[B]$:

$$\frac{1}{V_1'} = \frac{1}{V_1} + \frac{K_B}{V_1[B]}.$$

The determination of K_{IA} from the intersection of the reciprocal plots at various concentrations of B and the determination of K_B and V_1 from a replot of the ordinal intercept against $1/[B]$ allow K_A to be determined from the slope of the original $1/v$ against $1/[A]$ plot. If we arrange equation (6.12) in a form where [B] is varied and [A] is fixed, we obtain

$$\frac{1}{v} = \frac{K_B}{V_1}\left(1 + \frac{K_{IA}}{[A]}\right)\frac{1}{[B]} + \frac{1}{V_1}\left(1 + \frac{K_A}{[A]}\right),$$

which has exactly the same form as equation (6.13) and a plot of $1/v$ against $1/[B]$ for various values of [A] will give a series of straight lines which intersect at some point to the left of the ordinal axis given by

$$[B] = -\frac{K_B K_{IA}}{K_A} \quad \text{and} \quad \frac{1}{v} = \frac{1}{V_1}\left(1 - \frac{K_A}{K_{IA}}\right).$$

Again the lines intersect in either the upper or lower left-hand quadrants depending on the relative magnitudes of K_A and K_{IA}.

6.2.3. Ordered Bi Bi mechanism

$$\begin{array}{ccccccccc} & \text{A} & & \text{B} & & \text{P} & & \text{Q} & \\ & k_1 \downarrow k_{-1} & & k_2 \downarrow k_{-2} & & k_3 \uparrow k_{-3} & & k_4 \uparrow k_{-4} & \\ \hline \text{E} & & \text{EA} & & (\text{EAB} \rightleftharpoons \text{EPQ}) & & \text{EQ} & & \text{E} \end{array}$$

Since the number of transitory isomerisation steps between EAB and EPQ cannot be determined, the rate constants for this step are omitted to simplify the determination of the rate equation. Using the King–Altman method, we find that the velocity of the reaction is

$$v = \{(numerator\ 1)[\mathrm{A}][\mathrm{B}] - (numerator\ 2)[\mathrm{P}][\mathrm{Q}]\}/\Sigma$$

where $\Sigma = constant + (coef\ \mathrm{A})[\mathrm{A}] + (coef\ \mathrm{B})[\mathrm{B}] + (coef\ \mathrm{AP})[\mathrm{A}][\mathrm{P}] + (coef\ \mathrm{PQ})[\mathrm{P}][\mathrm{Q}] + (coef\ \mathrm{BPQ})[\mathrm{B}][\mathrm{P}][\mathrm{Q}] + (coef\ \mathrm{ABP})\ [\mathrm{A}][\mathrm{B}][\mathrm{P}] + (coef\,\mathrm{AB})[\mathrm{A}][\mathrm{B}] + (coef\,\mathrm{BQ})[\mathrm{B}][\mathrm{Q}] + (coef\,\mathrm{P})[\mathrm{P}] + (coef\,\mathrm{Q})\ [\mathrm{Q}]$

and $constant = k_{-1}k_4(k_{-2}+k_3)$ $\quad coef\,\mathrm{A} = k_1k_4(k_{-2}+k_3)$

$coef\,\mathrm{B} = k_2k_3k_4$ $\quad coef\ \mathrm{AP} = k_1k_{-2}k_{-3}$

$coef\,\mathrm{PQ} = k_{-3}k_{-4}(k_{-1}+k_{-2})$ $\quad coef\,\mathrm{BPQ} = k_2k_{-3}k_{-4}$

$coef\,\mathrm{ABP} = k_1k_2k_{-3}$ $\quad coef\ \mathrm{AB} = k_1k_2(k_3+k_4)$

$coef\,\mathrm{BQ} = k_2k_3k_{-4}$ $\quad coef\,\mathrm{P} = k_{-1}k_{-2}k_{-3}$

$coef\,\mathrm{Q} = k_{-1}k_{-4}(k_3+k_{-2})$

$numerator\ 1 = k_1k_2k_3k_4[\mathrm{E}_0]$ $\quad numerator\ 2 = k_{-1}k_{-2}k_{-3}k_{-4}[\mathrm{E}_0]$.

If we use the factors $(numerator\ 2)/(coef\,\mathrm{AB})\,(coef\,\mathrm{PQ})$ and $(numerator\ 2) = (numerator\ 1)/K_e$ as in the previous examples, we obtain

$$v = \left\{\frac{(numerator\ 1)(numerator\ 2)[\mathrm{A}][\mathrm{B}]}{(coef\,\mathrm{AB})(coef\,\mathrm{PQ})} - \frac{(numerator\ 1)(numerator\ 2)[\mathrm{P}][\mathrm{Q}]}{(coef\,\mathrm{AB})(coef\,\mathrm{PQ})K_e}\right\}\bigg/\sum \tag{6.14}$$

where

$$\sum = \left\{\frac{(constant)(numerator\ 2)}{(coef\,\mathrm{AB})(coef\,\mathrm{PQ})} + \frac{(coef\,\mathrm{A})(numerator\ 2)[\mathrm{A}]}{(coef\,\mathrm{AB})(coef\,\mathrm{PQ})} + \frac{(coef\,\mathrm{B})(numerator\ 2)[\mathrm{B}]}{(coef\,\mathrm{AB})(coef\,\mathrm{PQ})} + \frac{(coef\,\mathrm{AP})(numerator\ 1)[\mathrm{A}][\mathrm{P}]}{(coef\,\mathrm{AB})(coef\,\mathrm{PQ})K_e} + \frac{(numerator\ 1)[\mathrm{P}][\mathrm{Q}]}{(coef\,\mathrm{AB})K_e} + \frac{(coef\,\mathrm{BPQ})(numerator\ 1)[\mathrm{B}][\mathrm{P}][\mathrm{Q}]}{(coef\,\mathrm{AB})(coef\,\mathrm{PQ})K_e}\right.$$

$$+\frac{(\textit{coef}\ \mathrm{ABP})(\textit{numerator}\ 2)[\mathrm{A}][\mathrm{B}][\mathrm{P}]}{(\textit{coef}\ \mathrm{AB})(\textit{coef}\ \mathrm{PQ})}$$
$$+\frac{(\textit{numerator}\ 2)[\mathrm{A}][\mathrm{B}]}{(\textit{coef}\ \mathrm{PQ})}$$
$$+\frac{(\textit{coef}\ \mathrm{BQ})(\textit{numerator}\ 2)[\mathrm{B}][\mathrm{Q}]}{(\textit{coef}\ \mathrm{AB})(\textit{coef}\ \mathrm{PQ})}$$
$$+\frac{(\textit{coef}\ \mathrm{P})(\textit{numerator}\ 1)[\mathrm{P}]}{(\textit{coef}\ \mathrm{AB})(\textit{coef}\ \mathrm{PQ})K_e}$$
$$\left.+\frac{(\textit{coef}\ \mathrm{Q})(\textit{numerator}\ 1)[\mathrm{Q}]}{(\textit{coef}\ \mathrm{AB})(\textit{coef}\ \mathrm{PQ})K_e}\right\}.$$

Equation (6.14) now simplifies to:

$$v=\frac{V_1V_2[\mathrm{A}][\mathrm{B}]-(V_1V_2[\mathrm{P}][\mathrm{Q}]/K_e)}{\begin{array}{r}\{K_{IA}K_BV_2+K_BV_2[\mathrm{A}]+K_AV_2[\mathrm{B}]+(K_QV_1/K_{IA}K_e)[\mathrm{A}][\mathrm{P}]\\ +(V_1/K_e)[\mathrm{P}][\mathrm{Q}]+(V_1/K_{IB}K_e)[\mathrm{B}][\mathrm{P}][\mathrm{Q}][\mathrm{A}][\mathrm{B}][\mathrm{P}]\\ +(V_2/K_{IP})+V_2[\mathrm{A}][\mathrm{B}]+(V_2K_A/K_{IQ})[\mathrm{B}][\mathrm{Q}]+(K_QV_1/K_e)[\mathrm{P}]\\ +(K_PV_1/K_e)[\mathrm{Q}]\}\end{array}} \quad (6.15)$$

where

$$V_1=\frac{\textit{numerator}\ 1}{\textit{coef}\ \mathrm{AB}},\ V_2=\frac{\textit{numerator}\ 2}{\textit{coef}\ \mathrm{PQ}},\ K_A=\frac{\textit{coef}\ \mathrm{B}}{\textit{coef}\ \mathrm{AB}},\ K_B=\frac{\textit{coef}\ \mathrm{A}}{\textit{coef}\ \mathrm{AB}}$$

$$K_P=\frac{\textit{coef}\ \mathrm{Q}}{\textit{coef}\ \mathrm{PQ}},\qquad K_Q=\frac{\textit{coef}\ \mathrm{P}}{\textit{coef}\ \mathrm{PQ}},$$

$$K_{IA}=\frac{\textit{constant}}{\textit{coef}\ \mathrm{A}}=\frac{\textit{coef}\ \mathrm{P}}{\textit{coef}\ \mathrm{AP}}=\frac{k_{-1}}{k_1}$$

$$K_{IQ}=\frac{\textit{constant}}{\textit{coef}\ \mathrm{Q}}=\frac{\textit{coef}\ \mathrm{B}}{\textit{coef}\ \mathrm{BQ}}=\frac{k_4}{k_{-4}},\qquad K_{IB}=\frac{\textit{coef}\ \mathrm{PQ}}{\textit{coef}\ \mathrm{BPQ}},$$

$$K_{IP}=\frac{\textit{coef}\ \mathrm{AB}}{\textit{coef}\ \mathrm{ABP}}.$$

Other inhibition constants that could be defined are

$$K'_{IB}=\frac{\textit{constant}}{\textit{coef}\ \mathrm{B}},\qquad K''_{IB}=\frac{\textit{coef}\ \mathrm{AP}}{\textit{coef}\ \mathrm{ABP}},\qquad K'''_{IB}=\frac{\textit{coef}\ \mathrm{Q}}{\textit{coef}\ \mathrm{BQ}},$$

$$K'_{IP}=\frac{\textit{coef}\ \mathrm{A}}{\textit{coef}\ \mathrm{AP}},\qquad K''_{IP}=\frac{\textit{coef}\ \mathrm{BQ}}{\textit{coef}\ \mathrm{BPQ}},\qquad K'''_{IP}=\frac{\textit{constant}}{\textit{coef}\ \mathrm{P}}.$$

For initial-velocity studies, [P] = [Q] = 0, and equation (6.15) simplifies to

$$v = \frac{V_1[\mathrm{A}][\mathrm{B}]}{K_{IA}K_B + K_B[\mathrm{A}] + K_A[\mathrm{B}] + [\mathrm{A}][\mathrm{B}]}, \tag{6.16}$$

which is exactly the same equation as for the Bi Uni reaction and so reciprocal plots are exactly the same as in fig. 6.1. For the Bi Bi mechanism there are two Haldane equations:

$$K_e = \frac{V_1 K_P K_{IQ}}{V_2 K_B K_{IA}} = \left(\frac{V_1}{V_2}\right)^2 \frac{K_{IP} K_Q}{K_{IB} K_A}.$$

The second Haldane equation can be deduced by noting that

$$\frac{(\textit{numerator } 1)}{(\textit{numerator } 2)} = \frac{k_1 k_2 k_3 k_4}{k_{-1} k_{-2} k_{-3} k_{-4}} = \frac{(\textit{coef}\ \mathrm{B})(\textit{coef}\ \mathrm{ABP})}{(\textit{coef}\ \mathrm{P})(\textit{coef}\ \mathrm{BPQ})}.$$

Using the first Haldane relationship we can eliminate V_2 from equation (6.15) and obtain

$$v = \frac{V_1\{[\mathrm{A}][\mathrm{B}] - ([\mathrm{P}][\mathrm{Q}]/K_e)\}}{\begin{aligned}\{K_{IA}K_B + K_B[\mathrm{A}] + K_A[\mathrm{B}] + (K_B K_Q/K_P K_{IQ})[\mathrm{A}][\mathrm{P}] \\ + (K_B K_{IA}/K_P K_{IQ})[\mathrm{P}][\mathrm{Q}] + (K_B K_{IA}/K_P K_{IQ} K_{IB})[\mathrm{B}][\mathrm{P}][\mathrm{Q}] \\ + ([\mathrm{A}][\mathrm{B}][\mathrm{P}]/K_{IP}) + [\mathrm{A}][\mathrm{B}] + (K_A/K_{IQ})[\mathrm{B}][\mathrm{Q}] \\ + (K_Q K_B K_{IA}/K_P K_{IQ})[\mathrm{P}] + (K_B K_{IA}/K_{IQ})[\mathrm{Q}]\}\end{aligned}}. \tag{6.17}$$

With the equation in this form it allows us to determine the effect of adding either product, P or Q, as inhibitors.

(*a*) If P is added as the inhibitor and [Q] = 0, equation (6.17) becomes

$$v = \frac{V_1[\mathrm{A}][\mathrm{B}]}{\begin{aligned}\{K_{IA}K_B + K_B[\mathrm{A}] + K_A[\mathrm{B}] + (K_B K_Q/K_P K_{IQ})[\mathrm{A}][\mathrm{P}] \\ + ([\mathrm{A}][\mathrm{B}][\mathrm{P}]/K_{IP}) + [\mathrm{A}][\mathrm{B}] + (K_Q K_B K_{IA}/K_P K_{IQ})[\mathrm{P}]\}\end{aligned}}. \tag{6.18}$$

If we arrange equation (6.18) in the reciprocal form when [A] is varied and [B] is fixed,

$$\frac{1}{v} = \frac{K_A}{V_1}\left[1 + \frac{K_{IA}K_B}{K_A[\mathrm{B}]}\left(1 + \frac{K_Q[\mathrm{P}]}{K_P K_{IQ}}\right)\right]\frac{1}{[\mathrm{A}]} + \frac{1}{V_1}\left[1 + \frac{[\mathrm{P}]}{K_{IP}} + \frac{K_B}{[\mathrm{B}]}\left(1 + \frac{K_Q[\mathrm{P}]}{K_P K_{IQ}}\right)\right].$$

At non-saturating concentrations of B, the inhibition by P is mixed, whilst at saturating concentrations of B ([B] $\to \infty, 1/[B] \to 0$), P behaves in an uncompetitive manner:

$$\frac{1}{v} = \frac{K_A}{V_1}\frac{1}{[A]} + \frac{1}{V_1}\left(1 + \frac{[P]}{K_{IP}}\right).$$

(*b*) Conversely, rearranging equation (6.18) in the reciprocal form when [B] is varied and [A] is fixed gives

$$\frac{1}{v} = \frac{K_B}{V_1}\left(1 + \frac{K_{IA}}{[A]}\right)\left(1 + \frac{K_Q[P]}{K_P K_{IQ}}\right)\frac{1}{[B]} + \frac{1}{V_1}\left(1 + \frac{K_A}{[A]} + \frac{[P]}{K_{IP}}\right).$$

Thus, the inhibition by P is mixed both at low and saturating concentrations of A.

(*c*) If we add Q as an inhibitor and [P] = 0, equation (6.17) becomes

$$v = \frac{V_1[A][B]}{\{K_{IA}K_B + K_B[A] + K_A[B] + [A][B] + (K_A/K_{IQ})[B][Q] + (K_B K_{IA}/K_{IQ})[Q]\}}.$$

When [A] is varied and [B] is fixed,

$$\frac{1}{v} = \frac{K_A}{V_1}\left(1 + \frac{K_{IA}K_B}{K_A[B]}\right)\left(1 + \frac{[Q]}{K_{IQ}}\right)\frac{1}{[A]} + \frac{1}{V_1}\left(1 + \frac{K_B}{[B]}\right)$$

which shows that Q behaves as a competitive inhibitor at both low and saturating concentrations of B.

(*d*) When [B] is varied and [A] is fixed,

$$\frac{1}{v} = \frac{K_B}{V_1}\left[1 + \frac{K_{IA}}{[A]}\left(1 + \frac{[Q]}{K_{IQ}}\right)\right]\frac{1}{[B]} + \frac{1}{V_1}\left[1 + \frac{K_A}{[A]}\left(1 + \frac{[Q]}{K_{IQ}}\right)\right]$$

which shows that the inhibition by Q is mixed at non-saturating concentrations of A. The inhibition will be non-competitive if $K_{IA} = K_A$. There is no inhibition by Q, however, at saturating concentrations of A.

6.2.4. Theorell–Chance mechanism

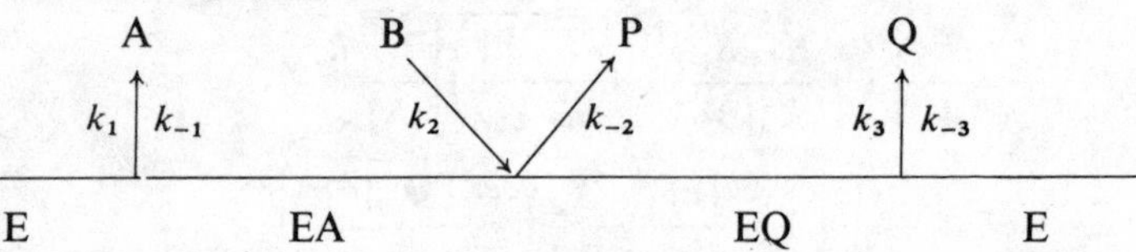

The above scheme, which was proposed by Theorell & Chance (1951) to explain the results obtained for alcohol dehydrogenase, is really a limiting case of the Ordered Bi Bi mechanism in which the steady-state concentration of the central complex is very low. Using the King–Altman diagram

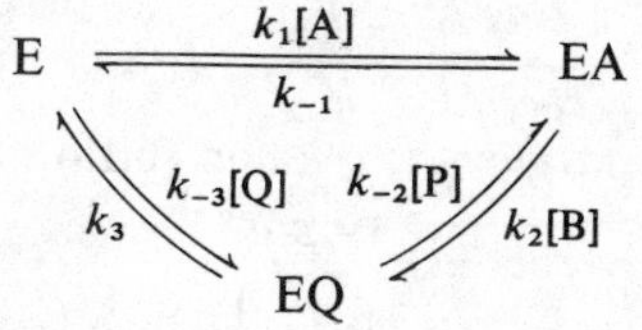

$$\mathrm{E} \underset{k_{-1}}{\overset{k_1[\mathrm{A}]}{\rightleftharpoons}} \mathrm{EA}$$

k_3 $k_{-3}[\mathrm{Q}]$ (E ⇌ EQ) $k_{-2}[\mathrm{P}]$ $k_2[\mathrm{B}]$ (EA ⇌ EQ)

EQ

we obtain the rate equation

$$v = \{(numerator\ 1)\,[\mathrm{A}]\,[\mathrm{B}] - (numerator\ 2)\,[\mathrm{P}]\,[\mathrm{Q}]\}/\Sigma$$

where

$$\begin{aligned}\Sigma = \{&constant + (coef\,\mathrm{A})\,[\mathrm{A}] + (coef\,\mathrm{B})\,[\mathrm{B}] + (coef\,\mathrm{P})\,[\mathrm{P}]\\ &+ (coef\,\mathrm{Q})\,[\mathrm{Q}] + (coef\,\mathrm{PQ})\,[\mathrm{P}]\,[\mathrm{Q}]\\ &+ (coef\,\mathrm{BQ})\,[\mathrm{B}]\,[\mathrm{Q}] + (coef\,\mathrm{AB})\,[\mathrm{A}]\,[\mathrm{B}]\\ &+ (coef\,\mathrm{AP})\,[\mathrm{A}]\,[\mathrm{P}]\}\end{aligned}$$

and $numerator\ 1 = k_1 k_2 k_3[\mathrm{E_0}]$, $numerator\ 2 = k_{-1} k_{-2} k_{-3}[\mathrm{E_0}]$,

$$constant = k_{-1} k_3, \quad coef\,\mathrm{A} = k_1 k_3, \quad coef\,\mathrm{B} = k_2 k_3,$$

$$coef\,\mathrm{P} = k_{-1} k_{-2}, \; coef\,\mathrm{Q} = k_{-1} k_{-3}, \; coef\,\mathrm{PQ} = k_{-2} k_{-3},$$

$$coef\,\mathrm{BQ} = k_2 k_{-3}, \; coef\,\mathrm{AB} = k_1 k_2, \quad coef\,\mathrm{AP} = k_1 k_{-2}.$$

Using the same factorising procedure as for the previous examples, we obtain

$$v = \frac{V_1 V_2[\mathrm{A}]\,[\mathrm{B}] - (V_1 V_2[\mathrm{P}]\,[\mathrm{Q}]/K_e)}{\begin{aligned}\{&K_{IB} K_A V_2 + K_B V_2[\mathrm{A}] + K_A V_2[\mathrm{B}] + (V_1 K_Q[\mathrm{P}]/K_e)\\ &+ (V_1 K_P[\mathrm{Q}]/K_e) + (V_1[\mathrm{P}]\,[\mathrm{Q}]/K_e) + (K_A V_2[\mathrm{B}]\,[\mathrm{Q}]/K_{IQ})\\ &+ V_2[\mathrm{A}]\,[\mathrm{B}] + (V_1 K_Q[\mathrm{A}]\,[\mathrm{P}]/K_e K_{IA})\}\end{aligned}} \tag{6.19}$$

where $K_A = \dfrac{coef\,\mathrm{B}}{coef\,\mathrm{AB}} = \dfrac{k_3}{k_1}$, $\quad K_B = \dfrac{coef\,\mathrm{A}}{coef\,\mathrm{AB}} = \dfrac{k_3}{k_2}$,

$$K_P = \frac{coef\,\mathrm{Q}}{coef\,\mathrm{PQ}} = \frac{k_{-1}}{k_{-2}}, \quad K_Q = \frac{coef\,\mathrm{P}}{coef\,\mathrm{PQ}} = \frac{k_{-1}}{k_{-3}},$$

$$K_{IA} = \frac{constant}{coef\,\mathrm{A}} = \frac{coef\,\mathrm{P}}{coef\,\mathrm{AP}} = \frac{k_{-1}}{k_1},$$

$$K_{IB} = \frac{constant}{coef\,\mathrm{B}} = \frac{coef\,\mathrm{Q}}{coef\,\mathrm{BQ}} = \frac{k_{-1}}{k_2},$$

$$K_{IP} = \frac{constant}{coef\,\mathrm{P}} = \frac{coef\,\mathrm{A}}{coef\,\mathrm{AP}} = \frac{k_3}{k_{-2}},$$

$$K_{IQ} = \frac{constant}{coef\,\mathrm{Q}} = \frac{coef\,\mathrm{B}}{coef\,\mathrm{BQ}} = \frac{k_3}{k_{-3}}.$$

Note that equation (6.19) is the same as equation (6.15) ($K_A K_{IB} = K_{IA} K_B$) except that terms in [A] [B] [P] and [B] [P] [Q] are absent. For initial-velocity studies, equation (6.19) reduces to equation (6.16) and so the Theorell–Chance mechanism cannot be distinguished from Ordered Bi Bi mechanism by initial-velocity studies. There are four Haldane equations for this mechanism,

$$K_e = \frac{V_1 K_P K_{IQ}}{V_2 K_{IA} K_B} = \frac{V_1 K_P K_{IQ}}{V_2 K_A K_{IB}} = \frac{V_1 K_{IP} K_Q}{V_2 K_A K_{IB}} = \frac{V_1 K_{IP} K_Q}{V_2 K_{IA} K_B}.$$

Using these relationships, equation (6.19) gives

$$v = \frac{V_1[\mathrm{A}][\mathrm{B}] - (V_1[\mathrm{P}][\mathrm{Q}]/K_e)}{\begin{aligned}\{K_{IB} K_A + K_B[\mathrm{A}] + K_A[\mathrm{B}] + (K_A K_{IB}[\mathrm{P}]/K_{IP}) + (K_A K_{IB}[\mathrm{Q}]/K_{IQ}) \\ + (K_A K_{IB}[\mathrm{P}][\mathrm{Q}]/K_P K_{IQ}) + (K_A [\mathrm{B}][\mathrm{Q}]/K_{IQ}) \\ + [\mathrm{A}][\mathrm{B}] + (K_B[\mathrm{A}][\mathrm{P}]/K_{IP})\}\end{aligned}}$$

which can be used for the analysis of product inhibition studies.

(*a*) When P is the inhibitor, [A] is varied and [B] is fixed,

$$\frac{1}{v} = \frac{K_A}{V_1}\left[1 + \frac{K_{IB}}{[\mathrm{B}]}\left(1 + \frac{[\mathrm{P}]}{K_{IP}}\right)\right]\frac{1}{[\mathrm{A}]} + \frac{1}{V_1}\left[1 + \frac{K_B}{[\mathrm{B}]}\left(1 + \frac{[\mathrm{P}]}{K_{IP}}\right)\right].$$

Thus, the inhibition by P is mixed at low concentrations of B and there is no effect at saturating concentrations of B. If $K_{IB} = K_B$ the inhibition will be non-competitive at non-saturating concentrations of B.

(*b*) When P is the inhibitor, [B] is varied and [A] is fixed,

$$\frac{1}{v} = \frac{K_B}{V_1}\left(1 + \frac{[\mathrm{P}]}{K_{IP}}\right)\left(1 + \frac{K_{IA}}{[\mathrm{A}]}\right)\frac{1}{[\mathrm{B}]} + \frac{1}{V_1}\left(1 + \frac{K_A}{[\mathrm{A}]}\right)$$

$$(K_{IB} K_A / K_B = K_{IA})$$

which shows that P inhibits competitively with A in both low and saturating concentrations.

(*c*) When Q is the inhibitor, [A] is varied and [B] is fixed,

$$\frac{1}{v}=\frac{K_A}{V_1}\left(1+\frac{[Q]}{K_{IQ}}\right)\left(1+\frac{K_{IB}}{[B]}\right)\frac{1}{[A]}+\frac{1}{V_1}\left(1+\frac{K_B}{[B]}\right)$$

and hence there is competitive inhibition by Q at both low and high concentrations of B.

(*d*) Finally, when [B] is varied and [A] is fixed,

$$\frac{1}{v}=\frac{K_B}{V_1}\left[1+\frac{K_{IA}}{[A]}\left(1+\frac{[Q]}{K_{IQ}}\right)\right]\frac{1}{[B]}+\frac{1}{V_1}\left[1+\frac{K_A}{[A]}\left(1+\frac{[Q]}{K_{IQ}}\right)\right]$$

$(K_{IB}K_A/K_B=K_{IA})$.

In this case when [A] is low, the inhibition by Q is mixed and there is no inhibition at saturating concentrations of A. If $K_{IA}=K_A$ the inhibition by Q will be non-competitive at non-saturating concentrations of A.

6.2.5. Random mechanisms. There are three possible completely or partly Random Bi Bi mechanisms.

(*a*) Ordered addition of substrates and random release of products.

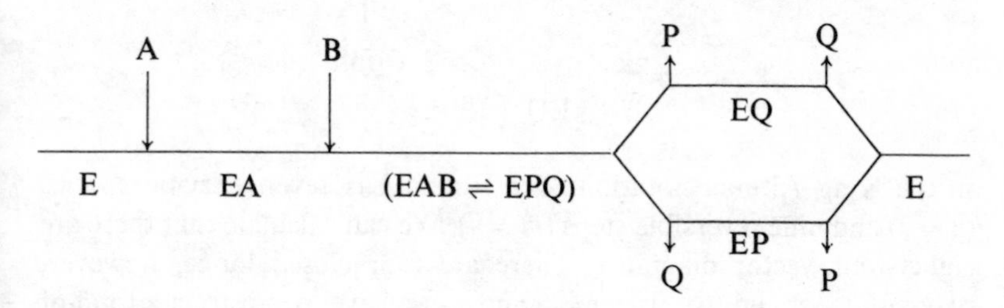

(*b*) Random addition of substrates and ordered release of products.

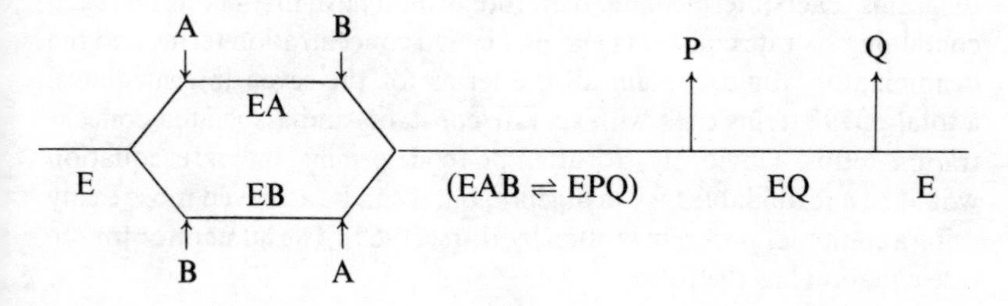

(*c*) Random addition of substrates and release of products.

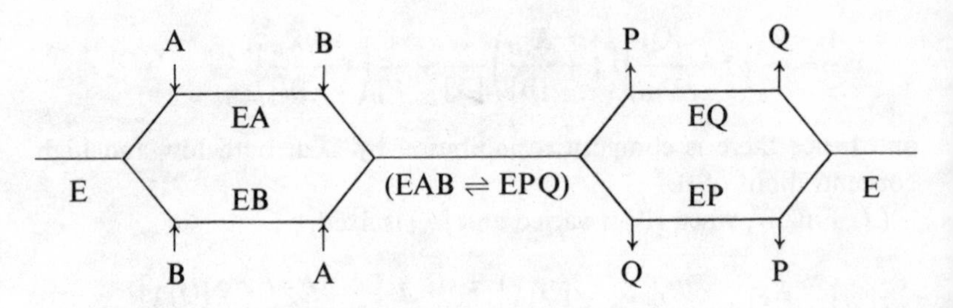

Kinetic equations for random mechanisms are complex to derive using the King–Altman method. We can draw the completely random mechanism (*c*) as

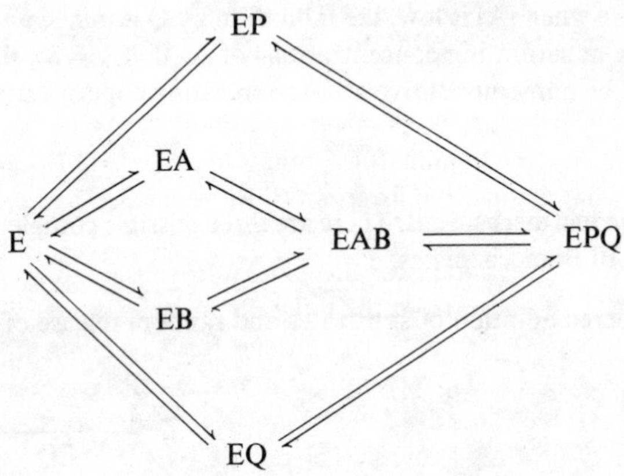

in the King–Altman notation. The system has seven enzyme species ($n = 7$) and nine reversible steps ($m = 9$). We can calculate that there are eighty-four vector diagrams. There are four closed loops, however, two with $r = 4$ and two with $r = 5$ and so we have to substract a total of twenty-eight vector diagrams, to give a total of fifty-six unique vector diagrams. Each intermediate, therefore, would have fifty-six terms, each containing six rate constants and associated concentration terms and the denominator would contain all the terms for the seven intermediates, a total of 392 terms each with six rate constants and associated concentration terms. Obviously, to attempt to determine the rate equation would be a formidable task in algebra, but it can be achieved more easily using a computer program written by Hurst (1969). The numerator for the rate equation has the form

$$\left([A][B] - \frac{[P][Q]}{K_e}\right)(a + b[A] + c[B] + d[P] + e[Q] + f[A][P]$$
$$+ g[A][Q] + h[B][P] + i[B][Q])[E_0]$$

where $a, b, c, \cdots i$ are sums of products of rate constants. The denominator has many other terms in addition to those that are present in the Ordered Bi Bi rate equation. The initial-velocity equation has the form

$$v = \frac{V\{K_a[A][B] + K_b[A]^2[B] + [A][B]^2\}}{\{K_c + K_d[A] + K_e[B] + K_f[A]^2 + K_g[B]^2 + K_h[A][B] + K_i[A]^2[B] + [A][B]^2\}}.$$

The definition of the various kinetic constants, K_a–K_i, differs from those in the other mechanisms. A plot of velocity against concentration of either substrate does not follow the normal Michaelis–Menten hyperbolic equation, nor are the reciprocal plots of $1/v$ against $1/[A]$ or $1/[B]$ linear unless the other substrate is at a saturating concentration. Product inhibition is always mixed or non-competitive but of a hyperbolic nature as the denominator contains terms in [P], $[P]^2$, [Q] and $[Q]^2$.

A simpler version of the Random Bi Bi mechanism is one in which all steps other than the central conversion of EAB $\rightleftharpoons$ EPQ are in rapid equilibrium. The equation for this system (Rapid Equilibrium Random Bi Bi) can be obtained from the more complex equation by eliminating all the terms that contain either of the rate constants for the central isomerisation step. The rate equation is then identical in form to the Ordered Bi Bi but there are no terms in [A] [P], [B] [Q], [A] [B] [P] and [B] [P] [Q] in the denominator. The definitions for the kinetic constants are the same except that $K_{IB} = constant/coef\,\mathrm{B}$ and $K_{IP} = constant/coef\,\mathrm{P}$. The Haldane equations are identical to those of the Theorell–Chance mechanism. Product inhibition by either product against either substrate is competitive and is overcome by saturation with the other substrate.

6.2.6. Non-sequential Ping Pong Bi Bi mechanism. In the Ping Pong mechanism the enzyme oscillates between two stable forms. It is described by the following mechanism:

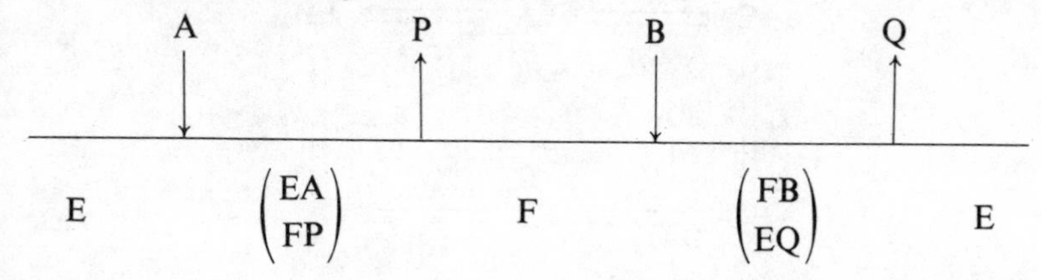

The mechanism can equally well be written as

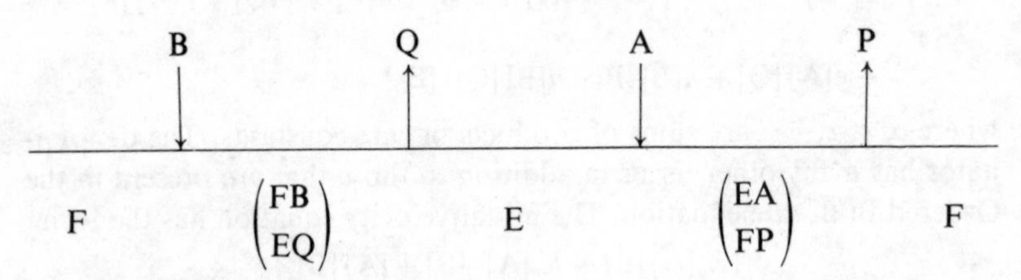

If isomerisation of the enzyme occurs (as distinct from transitory complexes) the mechanism is called Iso Ping Pong Bi Bi.

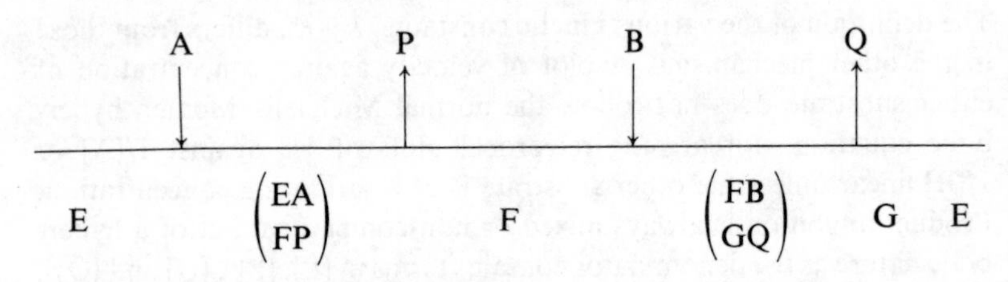

In chapter 2, mention was made of the hydrolytic enzymes, trypsin and chymotrypsin. The mechanism of these enzymes is really Ping Pong Bi Bi with water as the second substrate. Due to the high concentration of water (~55 M), the mechanism may be described as a Crypto Ping Pong Uni Bi mechanism.

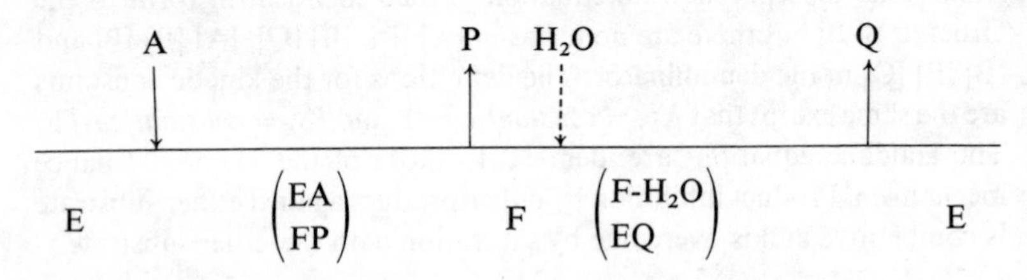

In King–Altman form, the Ping Pong Bi Bi mechanism can be described by

$$
\begin{array}{ccccc}
E & \underset{k_{-1}}{\overset{k_1[A]}{\rightleftharpoons}} & EA & \underset{k_{-2}}{\overset{k_2}{\rightleftharpoons}} & FP \\
k_6 \Big\uparrow\Big\downarrow k_{-6}[Q] & & & & k_{-3}[P] \Big\uparrow\Big\downarrow k_3 \\
EQ & \underset{k_5}{\overset{k_{-5}}{\rightleftharpoons}} & FB & \underset{k_4[B]}{\overset{k_{-4}}{\rightleftharpoons}} & F
\end{array}
$$

The velocity equation has the form

$$v = \{(numerator\ 1)[A][B] - (numerator\ 2)[P][Q]\}/\Sigma$$

where $\Sigma = \{(coef\ A)[A] + (coef\ B)[B] + (coef\ P)[P] + (coef\ Q)[Q] + (coef\ AB)[A][B] + (coef\ PQ)[P][Q] + (coef\ AP)[A][P] + (coef\ BQ)[B][Q]\}$.

Note that this is similar to the equation for Ordered Bi Bi but without the *constant* term and terms in [B] [P] [Q] and [A] [B] [P]. For this equation

$$numerator\ 1 = k_1 k_2 k_3 k_4 k_5 k_6 [E_0],$$
$$numerator\ 2 = k_{-1} k_{-2} k_{-3} k_{-4} k_{-5} k_{-6} [E_0],$$
$$coef\ A = k_1 k_2 k_3 (k_5 k_6 + k_{-4} k_6 + k_{-4} k_{-5}),$$
$$coef\ B = k_4 k_5 k_6 (k_{-1} k_{-2} + k_{-1} k_3 + k_2 k_3),$$
$$coef\ P = k_{-1} k_{-2} k_{-3} (k_{-4} k_{-5} + k_{-4} k_6 + k_5 k_6),$$
$$coef\ Q = k_{-4} k_{-5} k_{-6} (k_2 k_3 + k_{-1} k_3 + k_{-1} k_{-2}),$$
$$coef\ AB = k_1 k_4 (k_3 k_5 k_6 + k_{-2} k_5 k_6 + k_2 k_5 k_6 + k_2 k_3 k_6 + k_2 k_3 k_{-5} + k_2 k_3 k_5),$$
$$coef\ AP = k_1 k_{-3} (k_{-2} k_5 k_6 + k_{-2} k_{-4} k_6 + k_{-2} k_{-4} k_{-5} + k_2 k_{-4} k_6 + k_2 k_5 k_6 + k_2 k_{-4} k_{-5}),$$
$$coef\ PQ = k_{-3} k_{-6} (k_{-2} k_{-4} k_{-5} + k_2 k_{-4} k_{-5} + k_{-1} k_{-4} k_{-5} + k_{-1} k_{-2} k_{-5} + k_{-1} k_{-2} k_{-4} + k_{-1} k_{-2} k_5),$$
$$coef\ BQ = k_4 k_{-6} (k_{-1} k_3 k_{-5} + k_{-1} k_{-2} k_{-5} + k_{-1} k_{-2} k_5 + k_2 k_3 k_5 + k_{-1} k_3 k_5 + k_2 k_3 k_{-5}).$$

Using the same factorising procedure described for the earlier mechanisms, we obtain

$$v = \frac{V_1 V_2 [A][B] - (V_1 V_2 [P][Q]/K_e)}{\{V_2 K_B [A] + V_2 K_A [B] + (V_1 K_Q [P]/K_e) + (V_1 K_P [Q]/K_e) + V_2 [A][B] + (V_1 K_P [B][Q]/K_e K_{IB}) + (V_1 K_Q [A][P]/K_e K_{IA})\}}. \quad (6.20)$$

For initial-velocity studies ([P] = [Q] = 0), this equation becomes

$$v = \frac{V_1 [A][B]}{K_B [A] + K_A [B] + [A][B]}.$$

Inversion to plot $1/v$ against $1/[A]$ gives

$$\frac{1}{v} = \frac{K_A}{V_1} \frac{1}{[A]} + \frac{1}{V_1}\left(1 + \frac{K_B}{[B]}\right), \quad (6.21)$$

whereas if we plot $1/v$ against $1/[\mathrm{B}]$, we obtain

$$\frac{1}{v}=\frac{K_B}{V_1}\frac{1}{[\mathrm{B}]}+\frac{1}{V_1}\left(1+\frac{K_A}{[\mathrm{A}]}\right). \tag{6.22}$$

Equations (6.21) and (6.22) show that plots of $1/v$ against the reciprocal of either substrate form a series of parallel lines (fig. 6.2). For this scheme we can define the following Haldane equations:

$$K_e=\frac{K_{IP}K_{IQ}}{K_{IA}K_{IB}}=\frac{V_1K_{IP}K_Q}{V_2K_{IA}K_B}=\frac{V_1K_PK_{IQ}}{V_2K_AK_{IB}}=\left(\frac{V_1}{V_2}\right)^2\frac{K_PK_Q}{K_AK_B}$$

where

$$K_A = coef\,\mathrm{B}/coef\,\mathrm{AB}, \quad K_B = coef\,\mathrm{A}/coef\,\mathrm{AB},$$

$$K_P = coef\,\mathrm{Q}/coef\,\mathrm{PQ}, \quad K_Q = coef\,\mathrm{P}/coef\,\mathrm{PQ},$$

$$K_{IA} = coef\,\mathrm{P}/coef\,\mathrm{AP}, \quad K_{IB} = coef\,\mathrm{Q}/coef\,\mathrm{BQ},$$

$$K_{IQ} = coef\,\mathrm{B}/coef\,\mathrm{BQ}, \quad K_{IP} = coef\,\mathrm{A}/coef\,\mathrm{AP}.$$

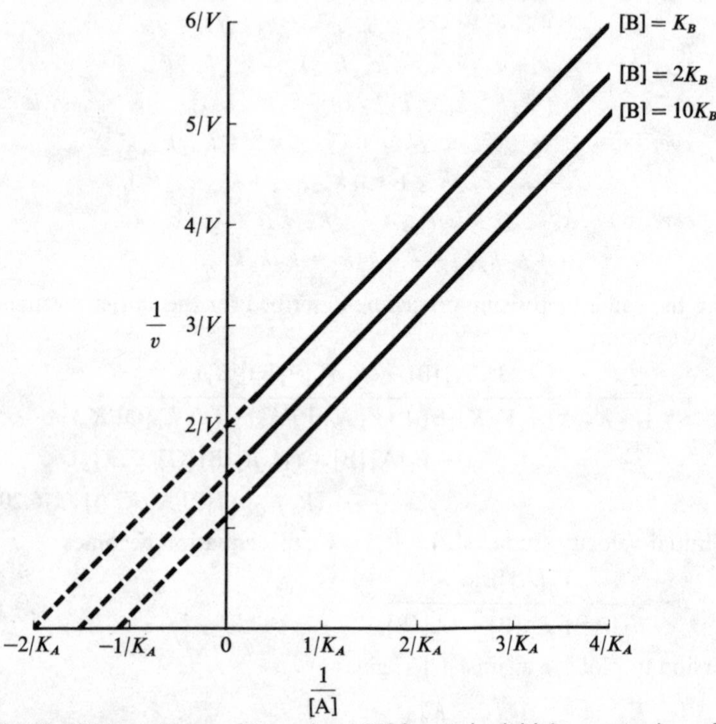

Fig. 6.2. Reciprocal plots of $1/v$ against $1/[\mathrm{A}]$ for varying initial concentrations of [B] for a Ping Pong Bi Bi mechanism described by equations (6.20) and (6.21).

For product inhibition studies where only one product at a time is added, we can simplify equation (6.20) by eliminating terms in [P] [Q], and by using the most suitable Haldane equation we obtain

$$v = \frac{V_1[\mathrm{A}][\mathrm{B}]}{\{K_B[\mathrm{A}] + K_A[\mathrm{B}] + (K_{IA}K_B[\mathrm{P}]/K_{IP}) + (K_A K_{IB}[\mathrm{Q}]/K_{IQ}) + [\mathrm{A}][\mathrm{B}] + (K_A[\mathrm{B}][\mathrm{Q}]/K_{IQ}) + (K_B[\mathrm{A}][\mathrm{P}]/K_{IP})\}}.$$

(*a*) When [A] is varied, [B] is fixed and P is the inhibitor, ([Q] = 0), we get

$$\frac{1}{v} = \frac{K_A}{V_1}\left(1 + \frac{K_{IA}K_B[\mathrm{P}]}{K_A[\mathrm{B}]K_{IP}}\right)\frac{1}{[\mathrm{A}]} + \frac{1}{V_1}\left[1 + \frac{K_B}{[\mathrm{B}]}\left(1 + \frac{[\mathrm{P}]}{K_{IP}}\right)\right].$$

Thus, the inhibition by P is of a mixed nature at non-saturating concentrations of B, and there is no inhibition at high concentrations of B.

(*b*) When [B] is varied, [A] is fixed and P is the inhibitor, we get

$$\frac{1}{v} = \frac{K_B}{V_1}\left[1 + \frac{[\mathrm{P}]}{K_{IP}}\left(1 + \frac{K_{IA}}{[\mathrm{A}]}\right)\right]\frac{1}{[\mathrm{B}]} + \frac{1}{V_1}\left(1 + \frac{K_A}{[\mathrm{A}]}\right)$$

and hence inhibition is competitive both at high and low concentrations of A.

(*c*) When [A] is varied, [B] is fixed, and Q is the inhibitor ([P] = 0), we obtain

$$\frac{1}{v} = \frac{K_A}{V_1}\left[1 + \frac{[\mathrm{Q}]}{K_{IQ}}\left(1 + \frac{K_{IB}}{[\mathrm{B}]}\right)\right]\frac{1}{[\mathrm{A}]} + \frac{1}{V_1}\left(1 + \frac{K_B}{[\mathrm{B}]}\right).$$

In this case Q is a competitive inhibitor at all concentrations of B.

(*d*) Finally, when [B] is varied, [A] is fixed and Q is the inhibitor, we get

$$\frac{1}{v} = \frac{K_B}{V_1}\left(1 + \frac{K_A K_{IB}}{K_B[\mathrm{A}]}\frac{[\mathrm{Q}]}{K_{IQ}}\right)\frac{1}{[\mathrm{B}]} + \frac{1}{V_1}\left[1 + \frac{K_A}{[\mathrm{A}]}\left(1 + \frac{[\mathrm{Q}]}{K_{IQ}}\right)\right].$$

The inhibition by Q is mixed at low concentrations of A and there is no inhibition at saturating concentrations of A.

6.3. Summary

Using the King–Altman notation, it is possible to derive velocity equations for any reaction scheme. As the number of enzyme forms increases, these equations soon become very cumbersome and unwieldy when written in rate constant form. Using the Cleland nomenclature, these reaction schemes can be converted from the tedious rate constant form into a more convenient kinetic constant form. It has been the aim of this chapter to present the Cleland approach for the simpler reaction schemes involving two substrates in a reasonable detailed form so that the reader should not at any time become overwhelmed by the algebraic manipulations involved. Whilst this chapter has only dealt with two-substrate reaction schemes, it should be possible for the reader to develop and understand the rate equations for the four possible Ordered Ter Ter mechanisms given below.

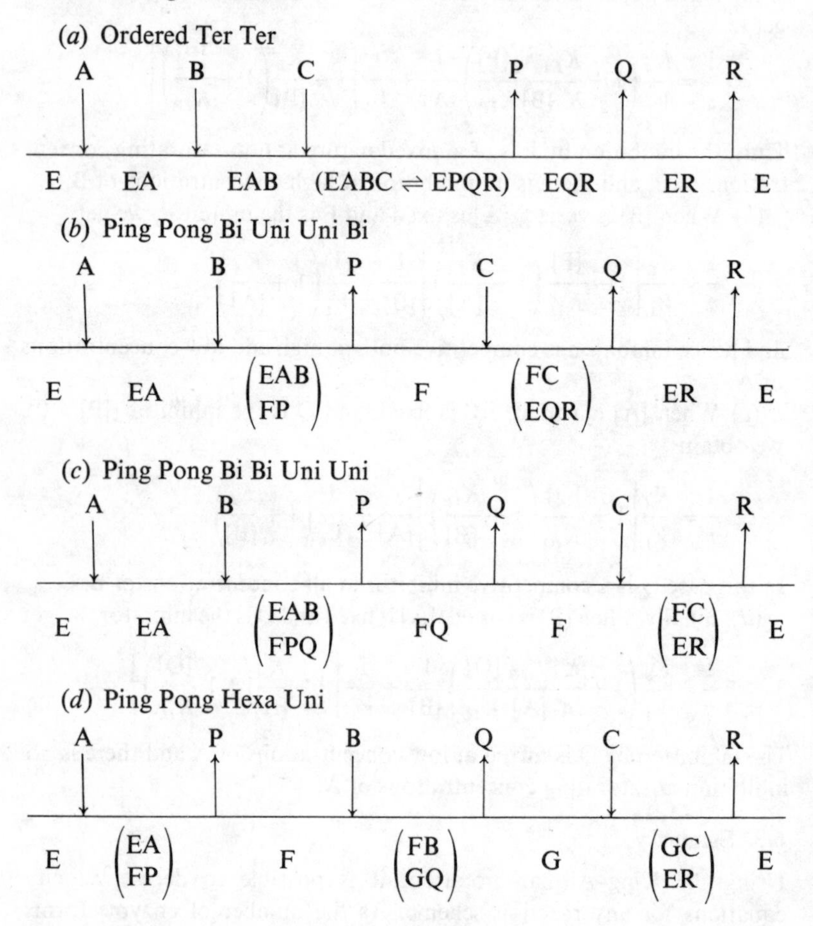

Of the four basic mechanisms mentioned in this chapter, it is only possible to distinguish the Random and Ping Pong mechanisms on the basis of initial-velocity studies; the Ordered, Rapid Random and

TABLE 6.1. *Summary of the product inhibition behaviour for two-substrate reactions*

Mechanism	Inhibitory product	Substrate A varies: Non-saturating [B]	Substrate A varies: Saturating [B]	Substrate B varies: Non-saturating [A]	Substrate B varies: Saturating [A]
Ordered Bi Bi	P	Mixed	Uncompetitive	Mixed	Mixed
	Q	Competitive	Competitive	Mixed (non-competitive if $K_{IA} = K_A$)	None
Theorell–Chance	P	Mixed (non-competitive if $K_{IB} = K_B$)	None	Competitive	Competitive
	Q	Competitive	Competitive	Mixed (non-competitive if $K_{IA} = K_A$)	None
Random	P	Mixed	Mixed	Mixed	Mixed
	Q	Mixed	Mixed	Mixed	Mixed
Rapid Random	P	Competitive	None	Competitive	None
	Q	Competitive	None	Competitive	None
Ping Pong Bi Bi	P	Mixed	None	Competitive	Competitive
	Q	Competitive	Competitive	Mixed	None

Competitive: ordinal intercept unchanged, slope and abscissal intercept vary.
Uncompetitive: ordinal intercept and abscissal intercept vary by the same amount, giving parallel lines.
Non-competitive: Abscissal intercept unchanged, slope and ordinal intercept vary.
Mixed: ordinal and abscissal intercepts vary by different amounts. The intersection of any two lines of a double reciprocal plot may lie above or below the abscissal axis.

Theorell–Chance mechanisms cannot be distinguished from each other because they have the same form of initial-velocity equation but as a group they can be distinguished from the other two mechanisms. Using the additional information provided by product inhibition studies, it is possible to distinguish each mechanism. A summary of the inhibition patterns is given in Table 6.1.

The kinetic techniques required to study multi-substrate reactions are little different from those used in the single substrate–single enzyme system outlined in chapter 2. If the equilibrium constant for the overall reaction is large and the reaction lies mainly to the right, the products have such a low affinity for any of the enzyme species that product inhibition studies are not feasible. One possible way to overcome this problem is to utilise inhibitors which are substrate analogues and to determine their effect on the velocity of the reaction. Finally, in some circumstances, it may be possible to determine individual rate constants for some of the reaction steps in the mechanism from the ratio of some of the kinetic constants, but if the number of transitory central complexes is not known and cannot be determined by some other means, the observed rate constants may in fact be more complex.

7 Fast reactions

7.1. Introduction

The conditions of most kinetic experiments are arranged so that measurements can be obtained within a period of 30 seconds to 1 hour. Reactions that take place in less than 30 seconds present problems for the experimenter of mixing and accurately defining zero time for the reaction. Reactions that take place over periods of 1 hour or more cause fatigue in the experimenter and can introduce questions regarding the long-term stability of enzyme preparations and instruments. Up to now we have dealt only with equations that pertain to the steady-state region of an enzyme-catalysed reaction. In steady-state experiments, the concentration of enzyme used is very low (e.g. a range of 10^{-7} M to 10^{-10} M is common), so that the overall rate of the reaction is slow enough to be conveniently followed. Under these conditions, where $[S] \gg [E]$, all enzyme species rapidly reach a steady state and the only parameters that can be determined are V and K_m. Both these parameters, as indicated earlier, are usually complex functions of a number of rate constants. Information about individual steps in the mechanism can only be obtained under certain circumstances, e.g. conditions where there is a rate-limiting step or by techniques like adding other nucleophiles to the reaction (see section 2.5). It may also be possible to equate individual rate constants with the ratio of various coefficients using the methods of

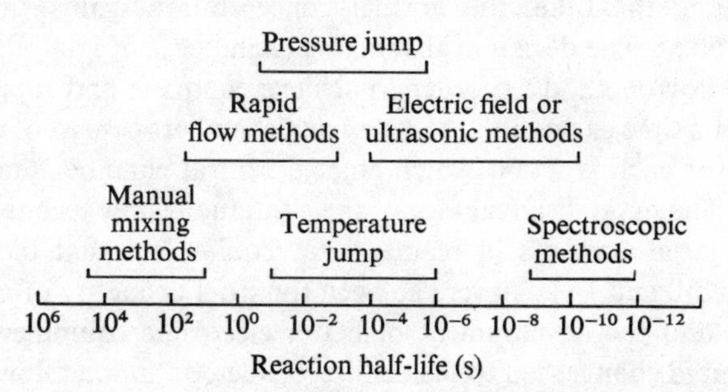

Fig. 7.1. Time scale of enzymic reactions in relation to experimental methods.

Cleland, but since the number of transitory complexes cannot be determined by steady-state techniques, these 'individual' rate constants may in fact be more complex. If we hope to be able to describe accurately the processes that take place during an enzyme reaction from the time the enzyme is added to the substrate to the end of the reaction, we must be able to measure transient species whose lifetimes may approach the period of molecular vibrations (10^{-13} s). Obviously specialised techniques must be used to study such fast reactions. These can be arbitrarily divided into rapid flow techniques for reactions that take place in 1–2 milliseconds and longer and relaxation techniques for reactions that take place in shorter times. There are other specialised techniques, e.g. flash induced reactions, which cannot be categorised by reaction times. Fig. 7.1 lists some of the methods that are available and the reaction times over which they are applicable.

7.2. Experimental methods

7.2.1. Rapid flow techniques. Until 1923, there was no way of following reactions whose half lifetimes were less than about 15 s. In that year, Hartridge & Roughton designed and constructed an apparatus that allowed the measurement of reactions with half lifetimes of about 1–2 ms. The apparatus was designed to study the fast reaction between haemoglobin and oxygen. The two reactant solutions were passed under constant pressure through a suitable mixer (M) into an observation cell (C) at a constant velocity u cm s^{-1} (fig. 7.2). If d is the distance between the mixer (M) and the observation cell (C) then the elapsed time before any observations can take place is d/u s, i.e. if d is 3 cm and $u =$ 1000 cm s^{-1}, the elapsed time is 3 ms. The value of u is obtained from the volume flow velocity (v ml s^{-1}) divided by the cross-sectional area (A) of the observation tube, $u = v/A$. If observations are made at various points along the tube, the normal concentration-against-time curve can be plotted. The design of the mixing chamber is of vital importance in all the flow methods in order to achieve complete and rapid mixing in as short a time as possible. Most mixing chambers consist of a number of inlets for each reactant which enter a central chamber tangentially (fig. 7.3). The main disadvantage of the continuous flow technique is the need for large amounts of reactants, a requirement that biochemists often find difficult to achieve. The need for strict economy of biological materials and the development of better electronic equipment for recording rapid changes led to the demise of the continuous flow method in favour of the accelerated and stopped flow methods devised by Chance and co-workers in the late 1950s (Chance, 1963).

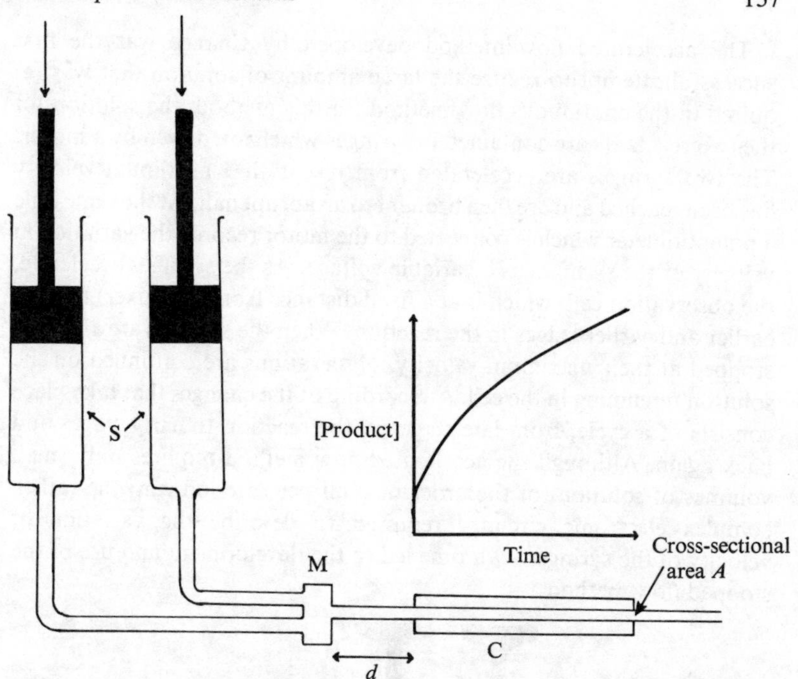

Fig. 7.2. Diagrammatic representation of a continuous flow apparatus. The reactants are driven at a constant velocity from the two syringes (S) into the mixing chamber (M) and the reaction is monitored along the observation cell (C). A product/time plot is thus obtained.

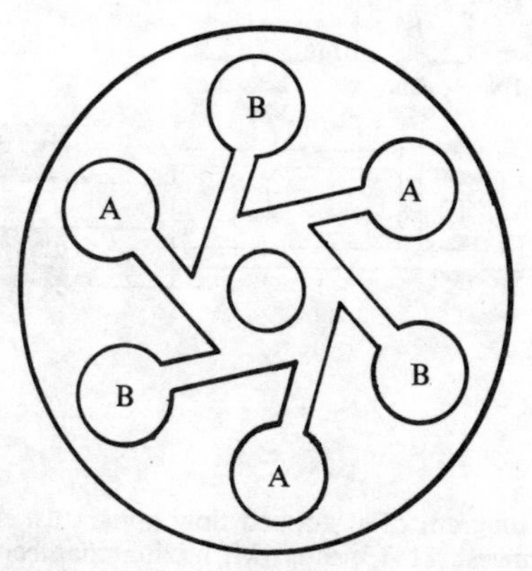

Fig. 7.3. Diagrammatic representation of a typical mixing chamber used in a flow apparatus. The reactants (A, B) enter tangentially from a number of ports, thus ensuring rapid and complete mixing. The mixer shown has six ports, three for each reactant.

The accelerated flow method developed by Chance was the first successful attempt to reduce the large amount of solution that was required in the continuous flow method. In this method, the solutions of the two reactants are contained in syringes which are driven by a motor. The two syringes are accelerated from rest until a maximum velocity has been reached and are then brought to an abrupt halt. At the same time a potentiometer which is connected to the motor records the variation in velocity of the syringes as a variable voltage. As the syringes accelerate, the observation cell (which is at a fixed distance from the mixer) records earlier and earlier stages in the reaction. When the syringes are abruptly stopped at their maximum velocity, observations are continued on the solution remaining in the cell. A recording of the changes that take place consists of a cycle, from late stages in the reaction to early stages and back again. Although the accelerated flow method requires only small volumes of solution (of the order of 1 ml per reaction run) the rather complex electronic circuitry required to describe the variation of velocity of the syringes with time led to the development and use of the stopped flow method.

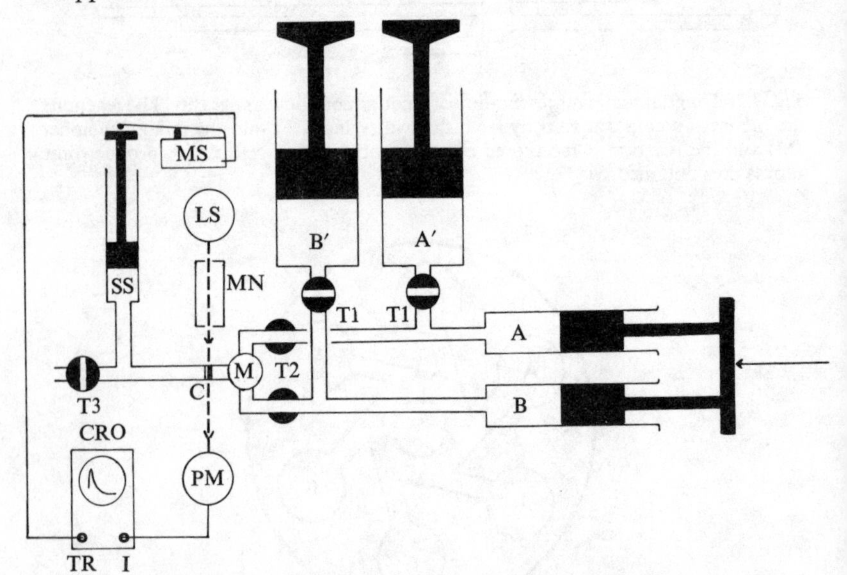

Fig. 7.4. Schematic diagram of a stopped flow apparatus. A, B, driven syringes; A′, B′, reservoir syringes; T1–T3, taps; M, mixing chamber; C, observation cell; LS, light source; MN, monochromator; PM, photomultiplier; SS, stopping syringe; MS, microswitch for triggering the recording device; CRO, storage oscilloscope; TR, trigger circuit; I, input to the oscilloscope from the photomultiplier.

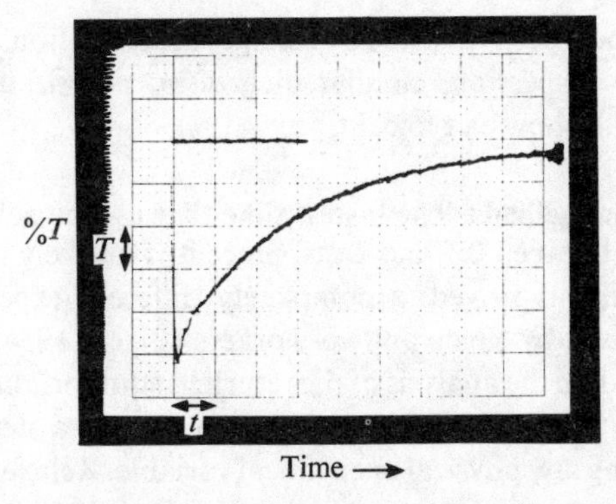

Fig. 7.5. A typical trace obtained from the rapid reaction of *p*-nitrophenyl-N^2-acetyl-N^1-*m*-methoxybenzylcarbazate (2.45 μM) and α-chymotrypsin (12.75 μM) at pH 8.0 and 25 °C (0.05 M potassium phosphate containing 0.5 M sodium chloride and 5% tetrahydrofuran). The *x*-axis corresponds to 500 ms cm^{-1}, the *y*-axis corresponds to 4% *T* cm^{-1}. (Adapted with permission of the author from Orr, 1973.)

The most widely used stopped flow instruments are based on the design of Gibson & Milnes (1964). Although these instruments are now available commercially, many have been built in the laboratory (fig. 7.4). The two solutions are kept in two syringes, A and B, of equal volumes which can be filled easily from the reservoir syringes A′ and B′. The syringes are pushed by a hydraulically driven block running in a parallel track. The solutions are driven at high speed under the force of the driving ram and are mixed and passed through the observation cell. The ejected solution passes into a stopping syringe and forces the plunger back against a microswitch. The microswitch triggers the recording device, normally a storage oscilloscope, and also arrests the driving ram. The change in concentration of reactants in the small volume of solution that remains in the observation cell is then followed with time. The lower limit of this method is generally between 0.5 and 2 ms before reaction changes can be monitored. This is to ensure that by the time the solutions enter the observation cell they are thoroughly mixed. This lower limit cannot be improved by increasing the hydraulic driving force as this rapidly leads to mechanical failure of syringes, cell windows and also creates hydrodynamic problems such as laminar flow which causes poor mixing. Approximately 1–2 ml of solution is used for each run; of this only 0.06–0.1 ml is used for observation purposes, the remainder being required to ensure that the reactants from the previous run are thoroughly flushed out before new observations are made. The various taps (T1–T3) allow the rapid refilling of the driven syringes in preparation for the next run. Reactions have been followed by changes in absorption, fluorescence, optical rotatory dispersion, circular dichroism, polarisation and conductance. Fig. 7.5 shows a typical trace.

7.2.2. Relaxation methods. The fastest time that can be achieved by the flow methods is between 0.5 and 2 ms. Since it is unlikely that this limit will be materially improved, a completely different experimental approach was devised by Eigen and co-workers (Eigen, 1954). Relaxation kinetics are based on the analysis of the return to equilibrium of a reaction after it has been perturbed. The perturbation may be a step or periodic function involving any physical or chemical variable. Relaxation methods permit the observation of reaction rates with half lifetimes as short as 10^{-10} s. The equilibrium position of a chemical reaction generally depends upon one or more thermodynamic variables, e.g. temperature, pressure, electric field intensity (equations (7.1), (7.2) and (7.3)).
For a temperature change at constant pressure

$$\left(\frac{\partial \ln K}{\partial T}\right)_P = \frac{\Delta H^\circ}{RT^2}. \tag{7.1}$$

For a pressure change at constant temperature

$$\left(\frac{\partial \ln K}{\partial P}\right)_T = \frac{\Delta V^\circ}{RT}, \tag{7.2}$$

where ΔV° is the standard volume change.
For an electric field change at constant temperature and pressure

$$\left(\frac{\partial \ln K}{\partial \mathrm{H}}\right)_{T,P} = \frac{\Delta M}{RT}, \tag{7.3}$$

where ΔM is the difference in dipole moment between a mole of reactant and a mole of product.

Temperature jump is the most frequently used perturbation in biochemical studies. The temperature of a solution can be increased by a few degrees in a few microseconds. The rise in temperature can be achieved by the discharge of a capacitor (C), charged to a high voltage (e), through the solution, by microwave heating and by laser flashes. Fig. 7.6 shows the arrangement of a typical temperature jump apparatus. The temperature rise depends on the energy stored in the capacitor, $E = \frac{1}{2}Ce^2$, and the time taken to discharge this through the cell. The rise time for the temperature pulse depends on the time constant for the

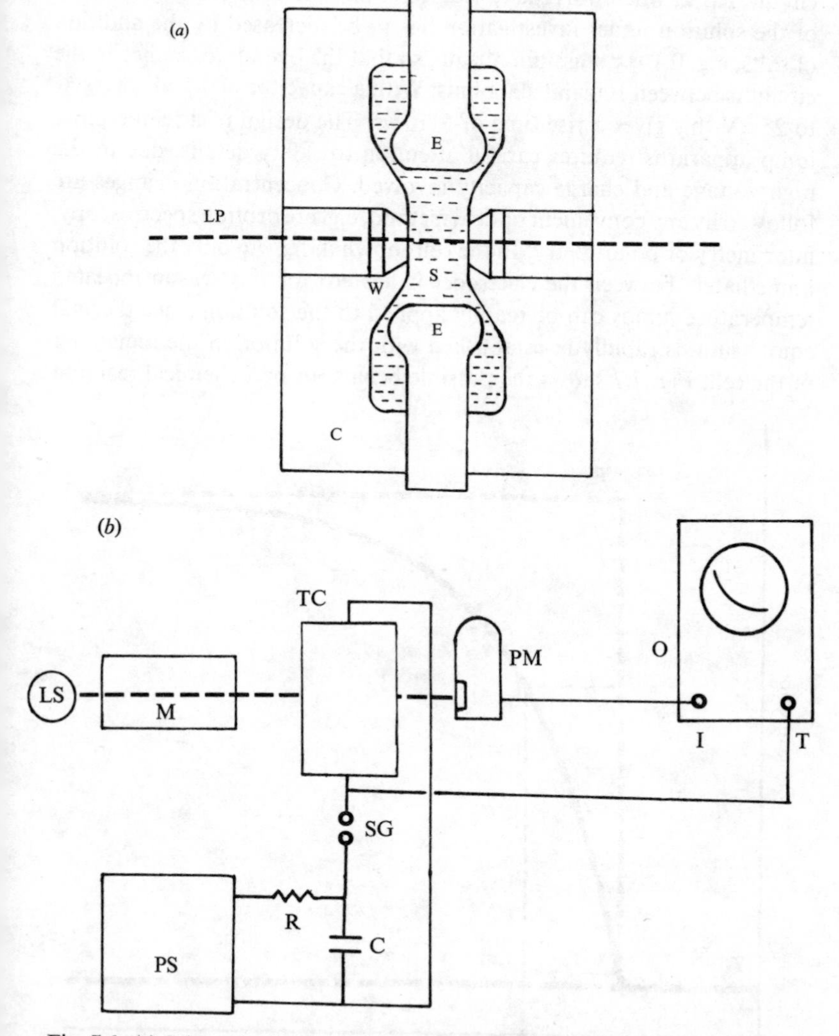

Fig. 7.6. (*a*) Schematic diagram of a typical capacitor discharge temperature jump cell. E, electrodes; S, sample; W, optical windows; LP, light path; C, main body of the cell. (*b*) Schematic diagram of a temperature jump apparatus which uses joule heating caused by the discharge of the energy stored in a capacitor through the solution in the cell. TC, temperature jump cell (see (*a*)); LS, light source; M, monochromator; PM, photomultiplier; O, oscilloscope; PS, high voltage power supply; R, resistor to control the recharge rate of the capacitor; C, high voltage storage capacitor; SG, spark gap for triggering the discharge of the capacitor; T, trigger circuit for the oscilloscope; I, input from the photomultiplier.

circuit, $RC/2$, where R is the resistance of the solution. The conductivity of the solution under investigation has to be increased by the addition of salts, e.g. 0.1 M potassium nitrate, so that the overall resistance in the circuit is between 100 and 200 ohms. With a capacitor of 0.1 μF charged to 25 kV this gives a rise time of 5–10 μs. The design of a temperature jump apparatus requires careful attention to safety details, due to the high voltage and charge capacity involved. Concentration changes are followed by any convenient optical method, e.g. absorption spectroscopy, fluorimetry or polarimetry. In the cell shown in fig. 7.6 only the solution immediately between the electrodes is heated; for this reason repeated temperature jumps can be readily applied to the solution since thermal equilibrium is rapidly re-established with the solution in the remainder of the cell. Fig. 7.7 shows the possible behaviour of a chemical reaction

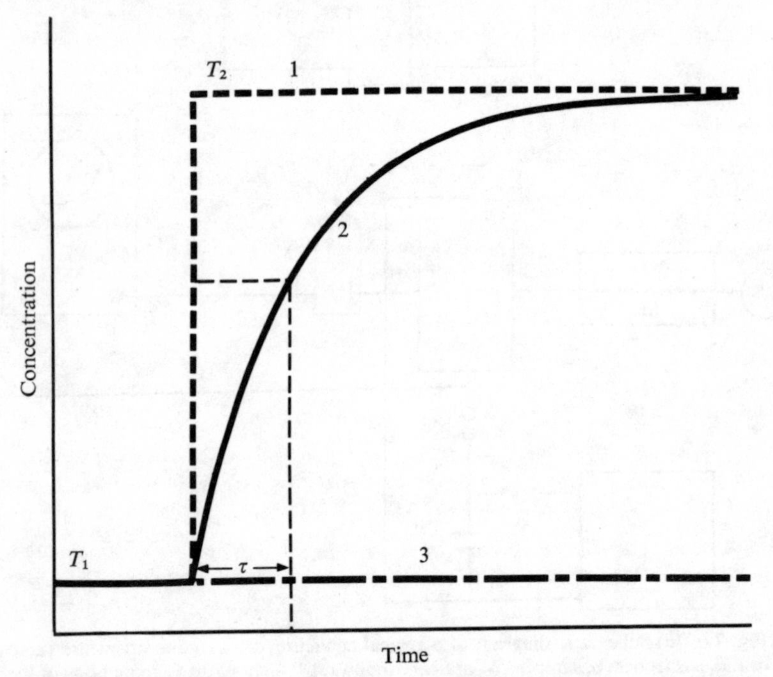

Fig. 7.7. Possible behaviour of a chemical system subject to a temperature jump in the form of a step function. Curve 1 represents the change in temperature from T_1 to T_2. Curve 1 also represents a chemical relaxation that is very fast, i.e. it follows the perturbation. Curve 3 represents a chemical reaction whose relaxation time is long compared with the time scale of the experiment and hence there is essentially no change in the equilibrium. Curve 2 represents a chemical relaxation intermediate between examples 1 and 3, and the chemical readjustments can therefore be followed.

after subjection to a temperature jump. In 1, the new equilibrium is established in a time faster than the rise time of the temperature jump. In 2, the new equilibrium is achieved more slowly and can thus be observed. In 3, the re-equilibration is very slow and essentially no change is observed on the time scale of the experiment.

In the pressure jump method, the forcing function is obtained by a sudden increase or decrease in the pressure acting on the system. The rise time is limited by the velocity of propagation of the pressure wave in the system. An increase in pressure is usually accompanied by a rise in temperature as a result of adiabatic compression. The change in pressure is achieved by puncturing a thin membrane separating the high and low pressure regions of the apparatus. Pressure differences of 50–60 atmospheres are commonly used. The temperature jump method is more widely used than the pressure jump because of its higher sensitivity. For a 10-degree shift in temperature (at 300 K) a value of ΔH° as low as 4 kJmol^{-1} will produce a change in equilibrium position of about 6 %. The pressure jump method relies on there being a change in volume during the reaction. For a pressure change of 65 atmospheres to bring about a change equivalent to a 10-degree shift in temperature, ΔV° would have to be 22 cm^3 mol^{-1}. This is a rather large value and is only found in reactions in which there is a charge neutralisation, e.g. $H^+ + OH^- \rightarrow H_2O$. Most reactions have a ΔV° of about 1 cm^3 mol^{-1} and so are considerably less sensitive to pressure changes.

Instead of using a single-step forcing function, it is possible to apply a periodic or sinusoidal forcing function to the system, usually as an electric field or ultrasonic wave. Fig. 7.8 shows three possible results obtained with a periodic forcing function. In 1, the equilibrium is re-established at a rate which is faster than the period of the forcing function and consequently the system follows the forcing function. In 2 and 3, the rate of re-establishment of equilibrium is comparable to the rate of change of the forcing function and there is a lag phase in the system. In 4, the re-establishment of the equilibrium is slow and the system is unable to follow the perturbation. The energy absorbed by the system is at a maximum when the frequency, ω, of the forcing function is equal to the reciprocal of the relaxation time. Relaxation methods are only applicable to systems which are in equilibrium. Many enzyme reactions are irreversible, i.e. the equilibrium is displaced very much to one side, and these reactions are very difficult to study by relaxation methods. This problem can be overcome in two ways. The first method uses substrate analogues which bind but do not take part in the catalytic steps. The

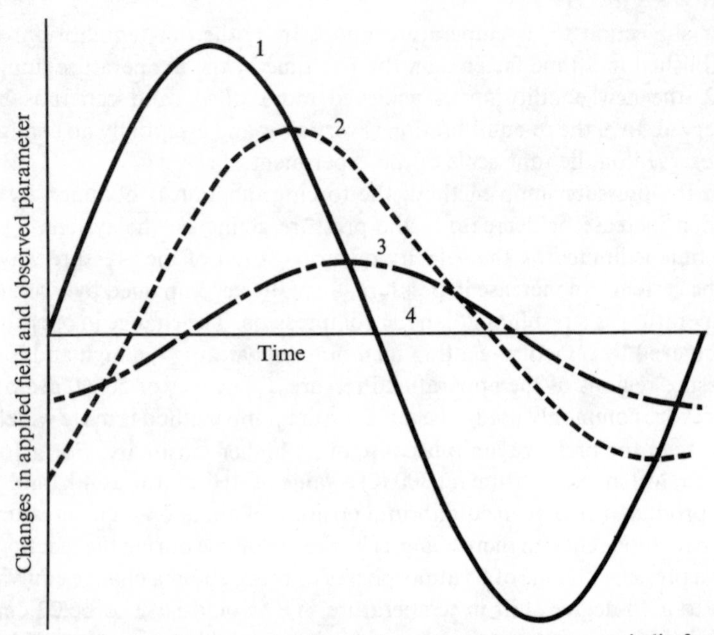

Fig. 7.8. Possible behaviour of a chemical system subjected to a periodic forcing function. Curve 1 represents the periodic forcing function, and also a system whose relaxation time is faster than the period of the forcing function and which therefore follows closely the variation in field strength. Curves 2 and 3 represent systems whose relaxation times are comparable with the period of the forcing function. The curves lag behind the forcing function and also have a smaller amplitude. Curve 4 represents a system whose relaxation time is much longer than the period of the forcing function and which thus cannot follow the changes in field strength.

second method is to combine the stopped flow and temperature jump methods. In this method, a temperature jump is applied to a system shortly after the reaction has been initiated by mixing. The perturbation is applied at the beginning of the steady-state region of the reaction and the approach to the new steady state is observed.

7.2.3. Flash-induced reactions. Some of the other methods that have been used to follow fast reactions are (1) flash photolysis, (2) polarography and (3) nuclear magnetic resonance. In flash photolysis, the reactants are subjected to a light flash of very high energy. A typical apparatus is shown in fig. 7.9. The reactants are contained in a long quartz cell (V). The reaction is monitored using a low powered light source (LS) and a detector (D) (spectrograph or photomultiplier). The cell is subjected to a high energy flash at right angles from the flash tube (FT). This

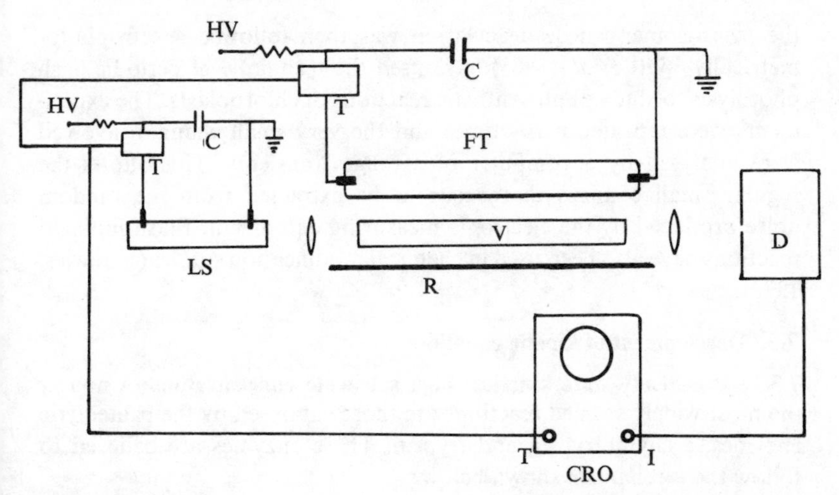

Fig. 7.9. Schematic diagram of a flash photolysis apparatus. V, reaction vessel; R, magnesium oxide coated reflector; FT, photolysis flash tube; T, trigger circuit; C, storage capacitor for the high voltage discharge tube; HV, high voltage source; LS, light source, either a continuous source or a low power flash tube; D, detector (photomultiplier or spectrograph); CRO, storage oscilloscope; I, input to the oscilloscope from the detector.

is usually a capacitor discharge lamp but laser flashes have also been used. A reflector is usually placed behind the cell to increase the amount of light available. The reaction to be observed is initiated by the high energy flash and changes in the concentration of the reactants are then followed by normal spectroscopic means. The light output from an electronic flash tube rises very rapidly and then decays exponentially. The duration of this decay determines the fastest reaction that can be studied. Normally this decay time is of the order of 1–2 μs which enables reactions with half lifetimes of >5 μs to be studied.

Berezin, Varfolomeyev & Martinek (1970) used the light-induced stereoisomerisation of *cis-* → *trans-p*-nitrocinnamoyl-α-chymotrypsin to study the deacylation of *trans-p*-nitrocinnamoyl-α-chymotrypsin. They firstly produced the *p*-nitrophenol ester of *trans*-cinnamic acid. This was subjected to ultraviolet irradiation to produce the photostationary mixture of *cis-* and *trans*-stereoisomers. The mixture was then allowed to react with α-chymotrypsin. The *trans*-substrate was completely hydrolysed whereas the *cis*-stereoisomer reacted more slowly and formed a stable acyl-enzyme. This *cis*-acyl-enzyme was then subjected to a powerful light flash which caused the conversion of the *cis*-isomer into the *trans*-isomer whose deacylation was then followed spectrophotometrically. Witt *et al.* (1965) have used the technique of periodic flash photolysis to study photosynthetic reactions of chloroplasts. The experiments were repeated many times and the very small changes involved were analysed by a computer of average transients. This allows the regular small changes that occur to be extracted from the random noise produced by the electronic measuring equipment. Flash-initiated reactions have also been used in chemical modification studies (Knowles, 1972).

7.3. Development of kinetic equations

7.3.1. Pre-steady-state kinetics: high substrate concentration. Amongst the most widely studied reactions are those catalysed by the proteolytic enzymes, α-chymotrypsin and trypsin. These enzymes are believed to follow the mechanism shown below:

$$E + S \underset{k_{-1}}{\overset{k_1}{\rightleftharpoons}} ES \xrightarrow[\downarrow P_1]{k_2} EP \xrightarrow{k_3} E + P_2. \tag{7.4}$$

If the substrate S is an ester, then the first product P_1 is an alcohol or phenol and the second product P_2 is an acid. In steady-state kinetics, it is possible to follow either the release of P_2 by means of a pH-stat or, in the case of an aryl ester, the release of P_1 by spectrophotometric methods. In the pre-steady-state region of the reaction, the concentrations of P_1 and P_2 are of the same order of magnitude as the enzyme concentration. In order to achieve the high sensitivity required to follow this early stage of the reaction, it is necessary to use either spectrophotometric or spectrofluorimetric methods. Most reactions that have been studied have involved substrates that are *p*-nitrophenyl esters and have been followed by the liberation of *p*-nitrophenol with time (Hartley & Kilby, 1952; Gutfreund, 1955; Elmore & Smyth, 1968*a*,*b*). For some two-substrate reactions, the conversion of NADH to NAD^+ (Hess & Wurster, 1970) and FAD to $FADH_2$ (Palmer & Massey, 1968) have also been used to follow the fast region of these reactions. The mechanism proposed in equation (7.4) can be described by the following kinetic equations:

$$[E_0] = [E] + [ES] + [EP] \tag{7.5}$$

$$\frac{d[ES]}{dt} = k_1[E][S] - (k_{-1} + k_2)[ES] \tag{7.6}$$

$$\frac{d[EP]}{dt} = k_2[ES] - k_3[EP] \tag{7.7}$$

$$\frac{d[P_1]}{dt} = k_2[ES] \tag{7.8}$$

$$\frac{d[P_2]}{dt} = k_3[EP]. \tag{7.9}$$

Differentiating equation (7.5) (remembering $[E_0]$ is a constant) and substituting for $d[EP]/dt$ in equation (7.7) gives

$$\frac{d[E]}{dt} = -\frac{d[ES]}{dt} - k_2[ES] + k_3[EP]. \tag{7.10}$$

Substitution for [EP] from equation (7.5) in equation (7.10) gives

$$\frac{d[E]}{dt} = -\frac{d[ES]}{dt} - k_2[ES] + k_3[E_0] - k_3[E] - k_3[ES]. \tag{7.11}$$

By differentiation of equation (7.6),

$$\frac{d^2[ES]}{dt^2} = k_1[S]\frac{d[E]}{dt} - (k_{-1} + k_2)\frac{d[ES]}{dt}. \tag{7.12}$$

In deriving equation (7.12), we must impose firstly the restriction that $[S] \gg [E]$ and secondly that t is sufficiently small so that the amount of product P_2 that is formed is small in relation to [S], i.e. $[P_2] \ll [S]$. Under these conditions the conservation equation that could be written for S, namely $[S_0] = [S] + [ES] + [EP] + [P_2]$ becomes simply $[S_0] = [S]$ and hence we can regard [S] as a constant. Substitution for $d[E]/dt$ and [E] from equations (7.11) and (7.6) in equation (7.12) gives

$$\frac{d^2[ES]}{dt^2} + (k_1[S_0] + k_{-1} + k_2 + k_3)\frac{d[ES]}{dt} + k_1(k_2 + k_3)([S_0] + K_m) = k_1 k_3[S_0][E_0], \tag{7.13}$$

where

$$K_m = \frac{k_3(k_{-1} + k_2)}{k_1(k_2 + k_3)}.$$

This is a second order differential equation of the type

$$\frac{d^2 y}{dt^2} + a\frac{dy}{dt} + by = \text{constant}. \tag{7.14}$$

The solution to an equation of this type consists of two parts $y = u_1 + u_2$

where $u_1 = u_1(t)$ is the complementary function obtained as the solution to the equation

$$\frac{d^2 y}{dt^2} + a\frac{dy}{dt} + by = 0$$

and $u_2 = u_2(t)$ is a particular integral of the original equation that contains no constants of integration. We can write equation (7.13) in differential operator form where $D = d/dt$ and $D^2 = d^2/dt^2$ as

$$(D^2 + aD + b)\,[ES] = k_1 k_3[E_0]\,[S_0] \tag{7.15}$$

where $a = k_1[S_0] + k_{-1} + k_2 + k_3$ and $b = k_1(k_2 + k_3)\,([S_0] + K_m)$. The complementary function is obtained as the solution to

$$(D^2 + aD + b)\,[ES] = 0.$$

This can be written as

$$(D + \lambda_1)(D + \lambda_2)\,[ES] = 0 \tag{7.16}$$

where

$$\lambda_1 = \{a + \sqrt{(a^2 - 4b)}\}/2$$

and $\lambda_2 = \{a - \sqrt{(a^2 - 4b)}\}/2$

are the roots of the quadratic in D. The solution to equation (7.16) is of the form

$$[ES] = Ae^{-\lambda_1 t} + Be^{-\lambda_2 t}$$

where A and B are the constants of integration. One way of obtaining the particular integral is to assume that the solution for [ES] for a second order differential equation is of the form

$$[ES] = \alpha t^2 + \beta t + \gamma,$$

hence $D[ES] = 2\alpha t + \beta$ and $D^2[ES] = 2\alpha$. Substitution in equation (7.15) and collecting coefficients in t and t^2 gives

$$\alpha a t^2 + t(b\beta + 2\alpha a) + (2\alpha + a\beta + b\gamma) = k_1 k_3[S_0][E_0].$$

Obviously α and β are both zero and hence

$$[ES] = \gamma = \frac{k_3[E_0]\,[S_0]}{(k_2 + k_3)\,([S_0] + K_m)}.$$

The solution for [ES] at this stage is now

$$[ES] = Ae^{-\lambda_1 t} + Be^{-\lambda_2 t} + \frac{k_3[E_0]\,[S_0]}{(k_2 + k_3)\,([S_0] + K_m)}. \tag{7.17}$$

The two integration constants, A and B, must be determined from the boundary conditions. When $t = 0$, we know that $[ES] = 0$ and $d[ES]/dt = k_1[E_0][S_0]$ (from equation (7.6)). From equation (7.17) when $t = 0$

$$A + B = -\frac{k_3[E_0][S_0]}{(k_2 + k_3)([S_0] + K_m)}. \tag{7.18}$$

Differentiating equation (7.17) and putting $t = 0$, we have

$$\frac{d[ES]}{dt} = -\lambda_1 A - \lambda_2 B = k_1[E_0][S_0]. \tag{7.19}$$

From equations (7.18) and (7.19) we can therefore solve for A and B and obtain

$$A = -\frac{k_3[S_0][E_0]}{(k_2 + k_3)([S_0] + K_m)} - \frac{k_1[S_0][E_0]}{(\lambda_1 - \lambda_2)} - \frac{k_3\lambda_1[S_0][E_0]}{(k_2 + k_3)([S_0] + K_m)(\lambda_2 - \lambda_1)}$$

$$B = \frac{k_1[S_0][E_0]}{(\lambda_1 - \lambda_2)} + \frac{k_3\lambda_1[S_0][E_0]}{(k_2 + k_3)([S_0] + K_m)(\lambda_2 - \lambda_1)}.$$

From equations (7.8) and (7.17), we know that

$$\frac{d[P_1]}{dt} = \frac{k_2 k_3[E_0][S_0]}{([S_0] + K_m)(k_2 + k_3)} + k_2 A e^{-\lambda_1 t} + k_2 B e^{-\lambda_2 t},$$

which is a first order differential equation that can be integrated directly to give

$$[P_1] = \frac{k_2 k_3 [E_0][S_0] t}{(k_2 + k_3)([S_0] + K_m)} - \frac{k_2 A e^{-\lambda_1 t}}{\lambda_1} - \frac{k_2 B e^{-\lambda_2 t}}{\lambda_2} + C.$$

when $t = 0$, $[P_1] = 0$, hence

$$C = \frac{k_2 A}{\lambda_1} + \frac{k_2 B}{\lambda_2}$$

and the complete solution is

$$[P_1] = \frac{k_{cat}[E_0][S_0] t}{([S_0] + K_m)} + \frac{k_2 A}{\lambda_1}(1 - e^{-\lambda_1 t}) + \frac{k_2 B}{\lambda_2}(1 - e^{-\lambda_2 t}), \tag{7.20}$$

where $k_{cat} = k_2 k_3/(k_2 + k_3)$. Both λ_1 and λ_2 are positive so that as t becomes large the exponential terms become vanishingly small, and

$$[P_1] \approx \frac{k_{cat}[S_0][E_0] t}{([S_0] + K_m)} + \frac{k_2 A}{\lambda_1} + \frac{k_2 B}{\lambda_2}. \tag{7.21}$$

This is the equation of a straight line with an ordinal intercept π,

$$[P_1]_{t=0} = \pi = \frac{k_2 A}{\lambda_1} + \frac{k_2 B}{\lambda_2} \quad \text{and slope} \quad \frac{k_{cat}[E_0][S_0]}{([S_0]+K_m)}.$$

The intercept, τ, on the t-axis is obtained by putting $[P_1] = 0$ in equation (7.21) and hence

$$\tau = -\left(\frac{k_2 A}{\lambda_1} + \frac{k_2 B}{\lambda_2}\right)\left(\frac{[S_0]+K_m}{k_{cat}[E_0][S_0]}\right).$$

If we substitute for A, B, λ_1 and λ_2,

$$\tau = \frac{k_3^2 - k_1 k_2[S_0]}{k_1 k_3(k_2+k_3)([S_0]+K_m)}$$

which will be negative, i.e. π will be positive, when $k_1k_2[S_0] > k_3^2$. The variation in the rate of production of P_1 with time for a particular set of

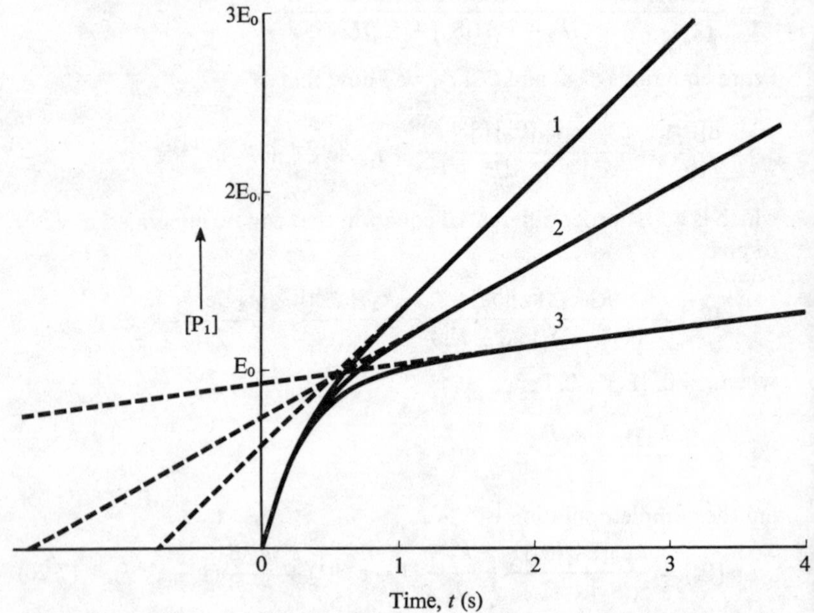

Fig. 7.10. The variation with time of the concentration of the first product, P_1, released in a double-intermediate three-step mechanism with $[S_0] \gg [E_0]$. The curves show the effect of changes in the value of k_3, the rate constant for the breakdown of the EP complex. Note as $k_3 \to 0$, the intercept on the $[P_1]$ axis tends to $[E_0]$. The parameters used are $[E_0] = 10^{-6}$ M, $[S_0] = 10^{-3}$ M, $k_1 = 10^4$ M^{-1} s^{-1}, $k_{-1} = 10^2$ s^{-1}, $k_2 = 50$ s^{-1}. (1), $k_3 = 1$ s^{-1}; (2), $k_3 = 0.5$ s^{-1}; (3), $k_3 = 0.1$ s^{-1}.

rate constants and concentrations is shown in fig. 7.10. The actual shape of the curve will depend considerably on the ratio of the various rate constants.

At this point, having obtained the solution for [ES], (equation (7.17)), we could substitute for [ES] in equation (7.7) and solve the first order equation in [EP]. The solution for [EP] would then allow us to solve for $[P_2]$ by substituting for [EP] in equation (7.9) and integrating. However, unless a lot of extra algebraic manipulation is carried out, the solutions obtained appear to be more complex than they really are. Thus, the solution for [EP] obtained by this systematic approach is

$$[\mathrm{EP}] = \frac{k_{\mathrm{cat}}[\mathrm{S_0}][\mathrm{E_0}]}{([\mathrm{S_0}] + K_m)k_3}(1 - \mathrm{e}^{-k_3 t}) + \frac{k_2 A}{(k_3 - \lambda_1)}(\mathrm{e}^{-\lambda_1 t} - \mathrm{e}^{-k_3 t}) + \frac{k_2 B}{(k_3 - \lambda_2)}(\mathrm{e}^{-\lambda_2 t} - \mathrm{e}^{-k_3 t}).$$

It is instructive to solve the same set of equations using the Laplace–Carson integral transform method (Rodiguin & Rodiguina, 1964). In this method we aim to produce an integral operator transform, the solutions of which are tabulated in many mathematics books. The essential features of this operator method are set out in appendix I. The differential equations that describe the three-step mechanism are given below in which the differential $(\mathrm{d}/\mathrm{d}t)$ has been replaced by the Laplace–Carson operator ψ and the transformed time-dependent species by lower case letters.

$$\psi[\mathrm{es}] = k_1[\mathrm{e}][\mathrm{S_0}] - (k_{-1} + k_2)[\mathrm{es}] \quad (7.22)$$

$$\psi[\mathrm{ep}] = k_2[\mathrm{es}] - k_3[\mathrm{ep}] \quad (7.23)$$

$$\psi[\mathrm{p_1}] = k_2[\mathrm{es}] \quad (7.24)$$

$$\psi[\mathrm{p_2}] = k_3[\mathrm{ep}] \quad (7.25)$$

and for [E] which has a value $[\mathrm{E_0}]$ at $t = 0$

$$\psi[\mathrm{e}] - \psi[\mathrm{E_0}] = -k_1[\mathrm{e}][\mathrm{S_0}] + k_{-1}[\mathrm{es}] + k_3[\mathrm{ep}]. \quad (7.26)$$

Since the operator, ψ, is independent of the variable, (t), it can be manipulated algebraically just like a constant so that equation (7.23) can be written as

$$(\psi + k_3)[\mathrm{ep}] = k_2[\mathrm{es}]$$

or $$[\mathrm{ep}] = \frac{k_2[\mathrm{es}]}{(\psi + k_3)}. \quad (7.27)$$

Substituting for [ep] in equation (7.26) we can express [e] as

$$[\mathrm{e}] = \frac{\psi[\mathrm{E_0}]}{(\psi + k_1[\mathrm{S_0}])} + \frac{k_{-1}[\mathrm{es}]}{(\psi + k_1[\mathrm{S_0}])} + \frac{k_2 k_3[\mathrm{es}]}{(\psi + k_1[\mathrm{S_0}])(\psi + k_3)},$$

which can be substituted for [e] in equation (7.22) to give

$$\{\psi^2 + (k_1[\mathrm{S_0}] + k_{-1} + k_2 + k_3)\psi + k_1(k_2 + k_3)([\mathrm{S_0}] + K_m)\}[\mathrm{es}]$$
$$= k_1[\mathrm{E_0}][\mathrm{S_0}](\psi + k_3)$$

or

$$[\mathrm{es}] = \frac{k_1[\mathrm{E_0}][\mathrm{S_0}](\psi + k_3)}{(\psi + \lambda_1)(\psi + \lambda_2)} \tag{7.28}$$

where

$$\lambda_1 = \{A + \sqrt{(A^2 - 4B)}\}/2$$
$$\lambda_2 = \{A - \sqrt{(A^2 - 4B)}\}/2$$
$$A = (k_1[\mathrm{S_0}] + k_{-1} + k_2 + k_3)$$
$$B = k_1(k_2 + k_3)([\mathrm{S_0}] + k_m).$$

Equation (7.28) is a Laplace–Carson transform, the solution (or original) of which is number (8) in table I.1 of appendix I.

$$[\mathrm{ES}] = k_1[\mathrm{E_0}][\mathrm{S_0}]\left[\frac{k_3}{\lambda_1\lambda_2} - \frac{(k_3 - \lambda_1)\mathrm{e}^{-\lambda_1 t}}{\lambda_1(\lambda_2 - \lambda_1)} - \frac{(k_3 - \lambda_2)\mathrm{e}^{-\lambda_2 t}}{\lambda_2(\lambda_1 - \lambda_2)}\right]$$

Using the transform for [es], (equation (7.28)) and equation (7.27) we can readily obtain the transform for [ep]:

$$[\mathrm{ep}] = \frac{k_1 k_2[\mathrm{E_0}][\mathrm{S_0}]}{(\psi + \lambda_1)(\psi + \lambda_2)},$$

the solution for which is

$$[\mathrm{EP}] = k_1 k_2[\mathrm{E_0}][\mathrm{S_0}]\left[\frac{1}{\lambda_1\lambda_2} - \frac{\mathrm{e}^{-\lambda_1 t}}{\lambda_1(\lambda_2 - \lambda_1)} - \frac{\mathrm{e}^{-\lambda_2 t}}{\lambda_2(\lambda_1 - \lambda_2)}\right].$$

Similarly we can obtain transforms for $[\mathrm{p_1}]$ and $[\mathrm{p_2}]$ as

$$[\mathrm{p_1}] = \frac{k_1 k_2[\mathrm{E_0}][\mathrm{S_0}](\psi + k_3)}{\psi(\psi + \lambda_1)(\psi + \lambda_2)} \tag{7.29}$$

and

$$[\mathrm{p_2}] = \frac{k_1 k_2 k_3[\mathrm{E_0}][\mathrm{S_0}]}{\psi(\psi + \lambda_1)(\psi + \lambda_2)}.$$

To obtain the solutions to these last two transforms in which the operator ψ appears in the denominator, they must first be simplified using partial fractions;
equating

$$\frac{(\psi+k_3)}{\psi(\psi+\lambda_1)(\psi+\lambda_2)}=\frac{A}{\psi}+\frac{B}{(\psi+\lambda_1)}+\frac{C}{(\psi+\lambda_2)},$$

equation (7.29) becomes:

$$[\mathrm{p}_1]=k_1k_2[\mathrm{E}_0][\mathrm{S}_0]\left[\frac{k_3}{\lambda_1\lambda_2\psi}-\frac{(k_3-\lambda_1)}{\lambda_1(\lambda_2-\lambda_1)(\psi+\lambda_1)}-\frac{(k_3-\lambda_2)}{\lambda_2(\lambda_1-\lambda_2)(\psi+\lambda_2)}\right].$$

The complete solution for $[\mathrm{P}_1]$ is therefore obtained from the originals for $1/\psi$, $1/(\psi+\lambda_1)$, $1/(\psi+\lambda_2)$ and is

$$[\mathrm{P}_1]=k_1k_2[\mathrm{E}_0][\mathrm{S}_0]\left[\frac{k_3t}{\lambda_1\lambda_2}-\frac{(k_3-\lambda_1)}{\lambda_1^2(\lambda_2-\lambda_1)}(1-\mathrm{e}^{-\lambda_1t})-\frac{(k_3-\lambda_2)}{\lambda_2^2(\lambda_1-\lambda_2)}(1-\mathrm{e}^{-\lambda_2t})\right]. \tag{7.30}$$

Similarly,

$$[\mathrm{P}_2]=k_1k_2k_3[\mathrm{E}_0][\mathrm{S}_0]\left[\frac{1}{\lambda_1\lambda_2\psi}-\frac{1}{\lambda_1(\lambda_2-\lambda_1)(\psi+\lambda_1)}-\frac{1}{\lambda_2(\lambda_1-\lambda_2)(\psi+\lambda_2)}\right].$$

Hence

$$[\mathrm{P}_2]=k_1k_2k_3[\mathrm{E}_0][\mathrm{S}_0]\left[\frac{t}{\lambda_1\lambda_2}-\frac{(1-\mathrm{e}^{-\lambda_1t})}{\lambda_1^2(\lambda_2-\lambda_1)}-\frac{(1-\mathrm{e}^{-\lambda_2t})}{\lambda_2^2(\lambda_1-\lambda_2)}\right]. \tag{7.31}$$

If we differentiate equation (7.31) with respect to time, we obtain

$$\frac{\mathrm{d}[\mathrm{P}_2]}{\mathrm{d}t}=k_1k_2k_3[\mathrm{E}_0][\mathrm{S}_0]\left[\frac{1}{\lambda_1\lambda_2}-\frac{\mathrm{e}^{-\lambda_1t}}{\lambda_1(\lambda_2-\lambda_1)}-\frac{\mathrm{e}^{-\lambda_2t}}{\lambda_2(\lambda_1-\lambda_2)}\right].$$

At large values of t when the steady-state condition exists, this reduces to

$$\frac{\mathrm{d}[\mathrm{P}_2]}{\mathrm{d}t}=\frac{k_1k_2k_3[\mathrm{E}_0][\mathrm{S}_0]}{\lambda_1\lambda_2}.$$

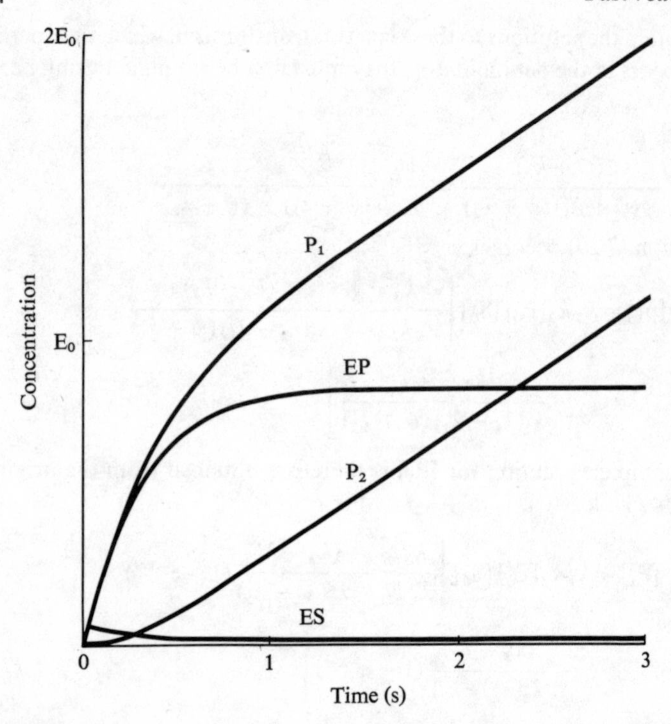

Fig. 7.11. The variation with time of the concentration of products P_1 and P_2 and enzyme complexes ES and EP in the pre-steady-state and early steady-state region of the reaction under the conditions $[S_0] \gg [E_0]$. The parameters are those used in fig. 7.10 with $k_3 = 0.5\ s^{-1}$.

Substitution for $\lambda_1\lambda_2$ gives

$$\frac{d[P_2]}{dt} = \frac{k_2 k_3 [E_0][S_0]}{(k_2+k_3)([S_0]+K_m)}$$

$$\text{or} \quad v = \frac{V[S_0]}{[S_0]+K_m} \qquad \left(V = \frac{k_2 k_3 [E_0]}{(k_2+k_3)}\right)$$

which is of course the normal Michaelis–Menten equation that describes the steady-state region of the reaction. These equations for [ES], [EP], $[P_1]$ and $[P_2]$ will only be true while the original assumption that $[S] \approx [S_0]$ remains true. Fig. 7.11 shows the variation of [ES], [EP], $[P_1]$ and $[P_2]$ during the pre-steady-state region of the reaction for a particular set of rate constants.

7.3.2. Pre-steady-state kinetics: high enzyme concentration. Usually the experimental conditions for studying the pre-steady-state region of an enzyme reaction are those that applied in the derivation of the equations in section 7.3.1, namely $[S_0] \gg [E_0]$. It is preferable to study reactions when this restriction applies as it is usually easier to obtain substantial amounts of substrate rather than enzyme. Sometimes, however, it may not be possible to satisfy this condition. This may be due to the low solubility of the substrate which means that for $[S_0] \gg [E_0]$, the enzyme concentration must be quite low. Since the concentrations of products P_1 and P_2 liberated during the pre-steady-state region of the reaction are comparable to the enzyme concentration, the use of a low enzyme concentration results in a considerable loss in sensitivity. Frequently, the substrates that are used are aryl esters and these may have quite high rates of non-enzymic hydrolysis. With $[S_0] \gg [E_0]$, the liberation of the products P_1 and P_2 due to the enzymic reaction could be masked by the high background due to the non-enzymic reaction. In contrast, if $[E_0] \gg [S_0]$, the formation of the [ES] complex is very fast and there is then virtually no unbound substrate left which could undergo non-enzymic reaction. The same reaction scheme outlined in section 7.3.1 applies and we can treat the equations in an analogous manner using the Laplace–Carson transform procedure. In this case, the free enzyme concentration, [E], is regarded as a constant $[E_0]$ and we require an equation for [S] which has a value, $[S_0]$, when $t = 0$, and hence in terms of the transformed time-dependent species [s] and the operator ψ, the equation is

$$\psi[s] - \psi[S_0] = -k_1[s][E_0] + k_{-1}[es].$$

Hence

$$[s] = \frac{\psi[S_0]}{(\psi + k_1[E_0])} + \frac{k_{-1}[es]}{(\psi + k_1[E_0])}. \tag{7.32}$$

Equation (7.22), with $[e] = [E_0]$, and [S] variable, can now be written as

$$(\psi + k_{-1} + k_2)[es] = k_1[E_0][s]$$

so that substitution for [s] from equation (7.32) leads to

$$(\psi^2 + \psi(k_1[E_0] + k_{-1} + k_2) + k_1 k_2[E_0])[es] = k_1[E_0][S_0]\psi$$

or $$[es] = \frac{k_1[E_0][S_0]\psi}{(\psi + \lambda_1)(\psi + \lambda_2)} \tag{7.33}$$

where

$$\lambda_1 = \frac{(k_1[\mathrm{E_0}] + k_{-1} + k_2) + \sqrt{\{(k_1[\mathrm{E_0}] + k_{-1} + k_2)^2 - 4k_1 k_2 [\mathrm{E_0}]\}}}{2}$$

and $$\lambda_2 = \frac{(k_1[\mathrm{E_0}] + k_{-1} + k_2) - \sqrt{\{(k_1[\mathrm{E_0}] + k_{-1} + k_2)^2 - 4k_1 k_2 [\mathrm{E_0}]\}}}{2}.$$

The solution for equation (7.33) is

$$[\mathrm{ES}] = k_1[\mathrm{E_0}][\mathrm{S_0}]\left[\frac{\mathrm{e}^{-\lambda_1 t}}{(\lambda_2 - \lambda_1)} + \frac{\mathrm{e}^{-\lambda_2 t}}{(\lambda_1 - \lambda_2)}\right].$$

Using the transform for [es] (equation (7.33)) and the equation for [ep] (equation (7.27)), we can readily obtain the transform for [ep] as

$$[\mathrm{ep}] = \frac{k_1 k_2 [\mathrm{E_0}][\mathrm{S_0}]}{(\psi + k_3)(\psi + \lambda_1)(\psi + \lambda_2)} \tag{7.34}$$

for which the solution is

$$[\mathrm{EP}] = k_1 k_2 [\mathrm{E_0}][\mathrm{S_0}]\left[\frac{\mathrm{e}^{-\lambda_1 t}}{(\lambda_2 - \lambda_1)(k_3 - \lambda_1)} + \frac{\mathrm{e}^{-\lambda_2 t}}{(\lambda_1 - \lambda_2)(k_3 - \lambda_2)} + \frac{\mathrm{e}^{-k_3 t}}{(\lambda_1 - k_3)(\lambda_2 - k_3)}\right].$$

The transforms for products P_1 and P_2 under these conditions ($[\mathrm{E_0}] \gg [\mathrm{S_0}]$) are

$$[\mathrm{p_1}] = \frac{k_1 k_2 [\mathrm{E_0}][\mathrm{S_0}]}{(\psi + \lambda_1)(\psi + \lambda_2)}$$ (from equations (7.24) and (7.33))

and $$[\mathrm{p_2}] = \frac{k_1 k_2 k_3 [\mathrm{E_0}][\mathrm{S_0}]}{(\psi + \lambda_1)(\psi + \lambda_2)(\psi + k_3)}$$ (from equations (7.25) and (7.34)).

The solution for $[\mathrm{P_1}]$ is therefore

$$[\mathrm{P_1}] = k_1 k_2 [\mathrm{E_0}][\mathrm{S_0}]\left[\frac{1}{\lambda_1 \lambda_2} - \frac{\mathrm{e}^{-\lambda_1 t}}{\lambda_1(\lambda_2 - \lambda_1)} - \frac{\mathrm{e}^{-\lambda_2 t}}{\lambda_2(\lambda_1 - \lambda_2)}\right]$$

and for $[\mathrm{p_2}]$

$$[\mathrm{P_2}] = k_1 k_2 k_3 [\mathrm{E_0}][\mathrm{S_0}]\left[\frac{1}{\lambda_1 \lambda_2 k_3} - \frac{\mathrm{e}^{-\lambda_1 t}}{\lambda_1(\lambda_2 - \lambda_1)(k_3 - \lambda_1)} - \frac{\mathrm{e}^{-\lambda_2 t}}{\lambda_2(\lambda_1 - \lambda_2)(k_3 - \lambda_2)} - \frac{\mathrm{e}^{-k_3 t}}{k_3(\lambda_1 - k_3)(\lambda_2 - k_3)}\right].$$

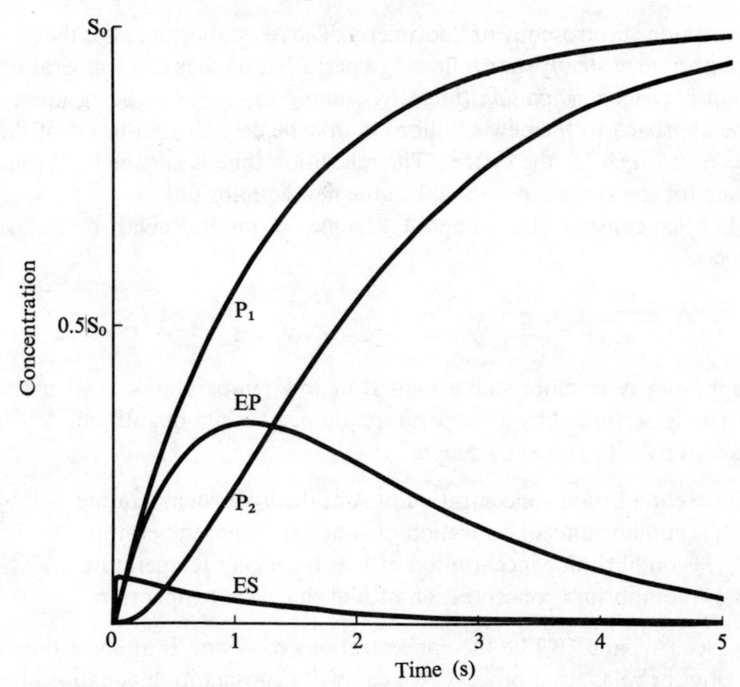

Fig. 7.12. The variation with time of the concentration of products P_1 and P_2 and enzyme complexes ES and EP under the conditions $[E_0] \gg [S_0]$. The parameters are the same as those in fig. 7.10 with the exception that $[E_0] = 10^{-5}$ M, $[S_0] = 10^{-6}$ M and $k_3 = 1$ s^{-1}.

An examination of the equations for [ES], [EP], $[P_1]$ and $[P_2]$ shows that there is no steady-state region under the conditions of $[E_0] \gg [S_0]$. As t becomes large the products P_1 and P_2 rise in a complicated exponential fashion to a final concentration $[S_0]$. Fig. 7.12 shows the variation of [ES], [EP], $[P_1]$ and $[P_2]$ with time for a particular set of conditions.

Using this systematic transform approach to the solution of the pre-steady-state equations, it is quite possible to derive equations for more complicated systems such as two-substrate reactions and reactions in the presence of inhibitors or other modifiers.

7.3.3. The kinetics of relaxation processes. If a system which is initially in equilibrium or a steady state is perturbed, a new equilibrium or steady state will be established. The progress towards this new equilibrium position can be followed by some suitable physical technique such as absorption spectroscopy or fluorimetry. The re-establishment of the new equilibrium or steady state following a perturbation such as a temperature jump is termed *relaxation* (literally, coming to rest after being upset). The approach to the new equilibrium may be described in terms of the *relaxation time* for the system. The relaxation time is *not* the total time taken for the system to re-establish the new equilibrium.

Let us consider the simplest scheme, a unimolecular reversible process:

$$\mathrm{A} \underset{k_{-1}}{\overset{k_1}{\rightleftharpoons}} \mathrm{B}$$

Supposing A is more stable than B at low temperatures, then if the system is perturbed by a temperature jump, a certain quantity of A will be converted to B. Let us denote

$[A_e]$ = equilibrium concentration of A at the *higher* temperature
$[A'_e]$ = equilibrium concentration of A at the *lower* temperature
$[B_e]$ = equilibrium concentration of B at the *higher* temperature
$[B'_e]$ = equilibrium concentration of B at the *lower* temperature

and let $[A_t]$ and $[B_t]$ be the concentrations of A and B at some time t during the relaxation process. We can write conservation equations for the system

$$[S_0] = [A_e] + [B_e] = [A_t] + [B_t] = [A'_e] + [B'_e].$$

After the temperature jump, more B will be formed according to the ordinary rate equation

$$\frac{d[B_t]}{dt} = k_1[A_t] - k_{-1}[B_t]. \tag{7.35}$$

Similarly,

$$\frac{d[A_t]}{dt} = -k_1[A_t] + k_{-1}[B_t]$$

when $t = 0$, $[B] = [B'_e]$ and $[A] = [A'_e]$ and we also know that $[A_t] = [S_0] - [B_t]$ so we can write equation (7.35) in the operator form as

$$\psi[B_t] - \psi[B'_e] = k_1[S_0] - (k_1 + k_{-1})[B_t]$$

or $$[B_t] = \frac{k_1[S_0]}{(\psi + k_1 + k_{-1})} + \frac{\psi[B'_e]}{(\psi + k_1 + k_{-1})},$$

and hence the solution is

$$[B_t] = k_1[S_0]\left[\frac{1}{(k_1+k_{-1})} - \frac{e^{-(k_1+k_{-1})t}}{(k_1+k_{-1})}\right] + [B'_e]\,e^{-(k_1+k_{-1})t}.$$

But

$$\frac{k_1[S_0]}{(k_1+k_{-1})} = [B_e]$$

therefore

$$\frac{[B_t]-[B_e]}{[B'_e]-[B_e]} = e^{-(k_1+k_{-1})t}. \qquad (7.36)$$

The relaxation time, τ, is defined as the reciprocal of the apparent first order rate constant governing the relaxation process. For this particular scheme

$$\tau = 1/(k_1+k_{-1}).$$

The dimensions of k_1 and k_{-1} are reciprocal time, so τ has dimensions of time.

There is a close analogy between the relaxation time and the time constant of an electrical circuit. For example, if a capacitor C, initially at e_0 volts, is discharged through a resistor R, the voltage e_t at any time t is given by

$$\frac{e_t}{e_0} = e^{-t/RC},$$

where RC is defined as the time constant of the circuit and has dimensions of time. If we put t equal to τ, then equation (7.36) becomes

$$\frac{[B_t]-[B_e]}{[B'_e]-[B_e]} = e^{-1}.$$

Hence the difference between the actual concentration $[B_t]$ and the new equilibrium concentration $[B_e]$ has been reduced to $1/e$ of the original difference $([B'_e]-[B_e])$ after a time τ. The reaction is therefore 63.2% complete ($1/e = 0.368$) (see fig. 7.7). If we take logarithms of equation (7.36), then

$$\ln([B_t]-[B_e]) = \ln([B'_e]-[B_e]) - t/\tau,$$

so that a plot of $\ln([B_t]-[B_e])$ against time will be linear with a slope of $-1/\tau$ or $-(k_1+k_{-1})$. If we determine the new equilibrium constant at the higher temperature, $K = k_1/k_{-1} = [B_e]/[A_e]$, we can then determine the individual rate constants, k_1 and k_{-1}, from the two equations.

Let us now consider an equilibrium, such as complex formation or ion association, which we can represent as

$$A + B \underset{k_{-1}}{\overset{k_1}{\rightleftharpoons}} C.$$

We can use the same nomenclature as for the previous example but with the additional terms $\Delta[A_t]$, $\Delta[B_t]$, $\Delta[C_t]$ which represent the perturbations of concentration from the equilibrium concentrations at time t:

$$\Delta[A_t] = [A_t] - [A_e], \tag{7.37}$$

$$\Delta[B_t] = [B_t] - [B_e], \tag{7.38}$$

$$\Delta[C_t] = [C_t] - [C_e]. \tag{7.39}$$

From the stoichiometry of the process

$$\Delta[A_t] = \Delta[B_t] = -\Delta[C_t].$$

After the perturbation, the relaxation process can be described by the equation

$$-\frac{d[A_t]}{dt} = -\frac{d[B_t]}{dt} = \frac{d[C_t]}{dt} = k_1[A_t][B_t] - k_{-1}[C_t]. \tag{7.40}$$

Using the relationships in equations (7.37)–(7.39), we can rewrite equation (7.40):

$$\frac{d[A_t]}{dt} = \frac{d\Delta[A_t]}{dt} = -k_1[A_e][B_e] - k_1\,\Delta[A_t]([A_e] + [B_e])$$

$$+ k_{-1}[C_e] - k_{-1}\,\Delta[A_t] - k_1\,\Delta[A_t]\,\Delta[A_t]. \tag{7.41}$$

If the perturbation is kept small, then $k_1\ \Delta[A_t]\Delta[A_t]$ is negligible. Also $-k_1[A_e]\,[B_e] + k_{-1}[C_e]$ is zero since these are the equilibrium concentrations at the final temperature. Finally, when $t = 0$, $\Delta[A_t] = [A_e'] - [A_e]$. Hence we can write equation (7.41) as

$$\psi\Delta[A_t] - \psi([A_e'] - [A_e]) = -k_1\,\Delta[A_t]([A_e] + [B_e]) - k_{-1}\,\Delta[A_t]$$

and the transform for $\Delta[A_t]$ is

$$\Delta[A_t] = \frac{\psi([A_e'] - [A_e])}{\{\psi + k_1([A_e] + [B_e] + k_{-1})\}}.$$

The solution or original of this transform is

$$\Delta[A_t] = ([A'_e] - [A_e])\,e^{-(k_1([A_e]+[B_e])+k_{-1})t}$$

or
$$\frac{[A_t] - [A_e]}{[A'_e] - [A_e]} = e^{-(k_1([A_e]+[B_e])+k_{-1})t}.$$

In this case we can put $\tau = 1/(k_1([A_e] + [B_e]) + k_{-1})$. Notice here that τ now depends on the final equilibrium concentrations of the reactants. If one of the reactants is in large excess, i.e. $[A_e] \gg [B_e]$, then $\tau = 1/(k_1[A_e] + k_{-1})$, so if we determine τ for a series of concentrations of [A], all of which satisfy the condition $[A_e] \gg [B_e]$, then a plot of $1/\tau$ against $[A_e]$ will give k_1 and k_{-1}. For a four-component system,

$$A + B \underset{k_{-1}}{\overset{k_1}{\rightleftharpoons}} C + D,$$

a similar treatment gives the relaxation time as

$$\tau = 1/\{k_1([A_e] + [B_e]) + k_{-1}([C_e] + [D_e])\}.$$

For a consecutive reaction such as

$$A \underset{k_{-1}}{\overset{k_1}{\rightleftharpoons}} B \underset{k_{-2}}{\overset{k_2}{\rightleftharpoons}} C$$

there will be two relaxation processes. For these to be observed experimentally, the first must be considerably faster than the second, i.e. $\tau_1 \ll \tau_2$. After the perturbation,

$$-\frac{d[A_t]}{dt} = -\frac{d\Delta[A_t]}{dt} = k_1[A_t] - k_{-1}[B_t]$$

$$\frac{d[B_t]}{dt} = \frac{d\Delta[B_t]}{dt} = k_1[A_t] - k_{-1}[B_t] - k_2[B_t] + k_{-2}[C_t]$$

$$\frac{d[C_t]}{dt} = \frac{d\Delta[C_t]}{dt} = k_2[B_t] - k_{-2}[C_t].$$

Only two of these equations are independent so we will take the first and last for simplicity:

$$-\frac{d[A_t]}{dt} = -\frac{d\Delta[A_t]}{dt} = k_1([A_e] + \Delta[A_t]) - k_{-1}([B_e] + \Delta[B_t]).$$

The stoichiometry of the process requires that $\Delta[B_t] = -\Delta[A_t] - \Delta[C_t]$.

Using this and the identities $k_1[A_e]=k_{-1}[B_e]$ and $k_2[B_e]=k_{-2}[C_e]$ we arrive at the equations

$$\frac{d\Delta[A_t]}{dt}=-(k_1+k_{-1})\,\Delta[A_t]-k_{-1}\,\Delta[C_t],$$

$$\frac{d\Delta[C_t]}{dt}=-k_2\,\Delta[A_t]-(k_2+k_{-2})\,\Delta[C_t].$$

At $t=0$, $\Delta[A_t]=[A_e']-[A_e]$ and $\Delta[C_t]=[C_e']-[C_e]$ so we can write these equations as

$$\psi\Delta[A_t]-\psi([A_e']-[A_e])=-(k_1+k_{-1})\,\Delta[A_t]-k_{-1}\,\Delta[C_t], \quad (7.42)$$

$$\psi\Delta[C_t]-\psi([C_e']-[C_e])=-k_2\,\Delta[A_t]-(k_2+k_{-2})\,\Delta[C_t]. \quad (7.43)$$

From equation (7.43)

$$\Delta[C_t]=\frac{\psi([C_e']-[C_e])}{(\psi+k_2+k_{-2})}-\frac{k_2\,\Delta[A_t]}{(\psi+k_2+k_{-2})}$$

which upon substituting for $\Delta[C_t]$ in equation (7.42) gives

$$\{\psi^2+\psi(k_1+k_{-1}+k_2+k_{-2})+k_1k_2+k_1k_{-2}+k_{-1}k_{-2}\}\,\Delta[A_t]$$
$$=\psi(\psi+k_2+k_{-2})([A_e']-[A_e])-k_{-1}\psi([C_e']-[C_e])$$

or
$$\Delta[A_t]=\frac{\psi(\psi+k_2+k_{-2})([A_e']-[A_e])}{(\psi+\lambda_1)(\psi+\lambda_2)}-\frac{k_{-1}\,\psi([C_e']-[C_e])}{(\psi+\lambda_1)(\psi+\lambda_2)} \quad (7.44)$$

where

$$\lambda_1=\{A+\surd(A-4B)\}/2$$
$$\lambda_2=\{A-\surd(A-4B)\}/2$$
$$A=(k_1+k_{-1}+k_2+k_{-2})$$
$$B=(k_1k_2+k_1k_{-2}+k_{-1}k_{-2}).$$

The original for equation (7.44) is

$$\Delta[A_t]=([A_e']-[A_e])\left[\frac{(k_2+k_{-2}-\lambda_1)\,e^{-\lambda_1t}}{(\lambda_2-\lambda_1)}+\frac{(k_2+k_{-2}-\lambda_2)\,e^{-\lambda_2t}}{(\lambda_1-\lambda_2)}\right]$$
$$-\;k_{-1}([C_e']-[C_e])\left[\frac{e^{-\lambda_1t}}{(\lambda_2-\lambda_1)}+\frac{e^{-\lambda_2t}}{(\lambda_1-\lambda_2)}\right]$$

which can be rearranged into

$$\frac{[A_t]-[A_e]}{[A_e']-[A_e]}=\alpha e^{-\lambda_1t}+\beta e^{-\lambda_2t}$$

where

$$\alpha = (k_2 + k_{-2} - \lambda_1 - k_1 k_2/k_{-2})/(\lambda_2 - \lambda_1)$$

and $\beta = (k_2 + k_{-2} - \lambda_2 - k_1 k_2/k_{-2})/(\lambda_1 - \lambda_2)$.

The relaxation process therefore has two relaxation times, $\tau_1 = 1/\lambda_1$ and $\tau_2 = 1/\lambda_2$. These are related to the kinetic constants by

$$\frac{1}{\tau_1} + \frac{1}{\tau_2} = k_1 + k_{-1} + k_2 + k_{-2}$$

and $$\frac{1}{\tau_1 \tau_2} = k_1 k_2 + k_1 k_{-2} + k_{-1} k_{-2}.$$

Having measured τ_1 and τ_2 experimentally, these two equations along with the equilibrium relationships $k_1[A_e] = k_{-1}[B_e]$ and $k_2[B_e] = k_{-2}[C_e]$ allow all four rate constants to be determined.

7.3.4. Analysis of relaxation data. For a simple system involving only a single relaxation the change in the observed parameter, Δ_t is given by

$$\Delta_t = \Delta_0 e^{-t/\tau},$$

where Δ_t is the difference between the observed parameter at time t and $t = \infty$, and Δ_0 is the difference between the parameter at $t = 0$ and $t = \infty$. In this case a plot of $\ln \Delta_t$ versus t is linear with a slope $-1/\tau$. For more complicated systems the change in the observed parameter Δ_t can be described by

$$\Delta_t = \Delta_0 \sum_{i=1}^{n} \alpha_i e^{-t/\tau_i}. \tag{7.45}$$

In this case a plot of $\ln \Delta_t$ versus t is curved for $i > 1$. Where t is large, however, all the terms in the series except that having the longest relaxation time will go to zero. Hence the curve will become linear and the relaxation time τ_n and $\Delta_0 \alpha_n$ can be obtained. These terms can then be subtracted from equation (7.45) and the process repeated for $n = n - 1$. The method outlined above is the well known 'peeling' technique and becomes very unreliable for $n > 2$. Numerous methods for numerically analysing multi-exponential processes have been proposed, including a Fourier transform method, but none of the procedures is particularly accurate unless the differences between τ_1, τ_2, τ_3, etc. are large and n is small. An alternative procedure is to utilise computer simulation techniques. It is necessary first of all to estimate the number of relaxation times from the experimental data. This number is an indication of the number of steps in a particular process, for example, in the scheme below

$$E + S \underset{k_{-1}}{\overset{k_1}{\rightleftharpoons}} ES \underset{k_{-2}}{\overset{k_2}{\rightleftharpoons}} EP \underset{k_{-3}}{\overset{k_3}{\rightleftharpoons}} E + P$$

there would be three relaxation times corresponding to the three states of the enzyme, E, ES and EP. Initial estimates of the rate constants are made and the mechanism simulated using either an analogue or digital computer. The initial estimates are refined until one obtains the best fit to the experimental data. The estimates for the rate constants would have to be consistent with other data available from steady-state and pre-steady-state kinetics.

7.4. Determination of the operational molarity of enzyme solutions

Anyone interested in enzyme kinetics is eventually presented with the problem of determining the enzyme concentration of a solution that he is using. The research worker may resort to using a method for determining the *quantity of protein* present in the solution; if the molecular weight of the enzyme is known, the molarity of the enzyme solution can be calculated. This method depends on the *total* concentration of protein present in the solution, some of which is likely to be inert protein. An alternative approach is to use a rate assay based on the rate of reaction of a standard substrate under standard conditions with a known quantity of enzyme solution. This method will not, however, differentiate between an enzyme preparation that contains 100% of 50% active molecules and one that contains 50% of 100% active molecules. The fact that some molecules are only 50% active may have resulted from some process in the separation technique. If possible, a method is needed that will assay only those molecules that are active towards the system being studied. What is required is a specific enzyme titrant. This will be a compound that when added to an enzyme solution will liberate a group or produce a change in the physical parameters of the enzyme, in proportion to the quantity of enzyme present. The liberation of the group can be measured by any suitable physical method, e.g. pH change, spectrophotometry, spectrofluorimetry or radioactivity.

Several spectrophotometric and spectrofluorimetric titrants have been described for determining the operational molarity of solutions of chymotrypsin, trypsin, thrombin, elastase and factor Xa (Bender *et al.*, 1966; Chase & Shaw, 1967; Baird & Elmore, 1968; Elmore & Smyth, 1968*a*,*b*; Knowles & Preston, 1968; Tanizawa, Ishii & Kanaoka, 1968,

1970; Vaughan & Westheimer, 1969; Roberts *et al.*, 1971; Jameson *et al.*, 1973). For all these enzymes, there is overwhelming evidence for the formation of a covalent enzyme intermediate as one step in the reaction pathway. Bell & Koshland (1971) list over sixty examples of other enzymes for which there is good evidence for covalent enzyme intermediates and it is possible that for some of these enzymes suitable titrants could be synthesised.

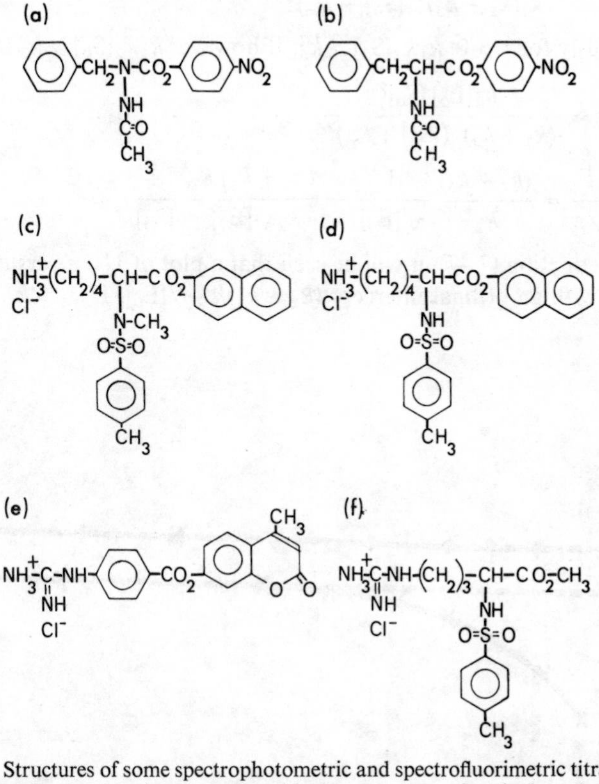

Fig. 7.13. Structures of some spectrophotometric and spectrofluorimetric titrants for α-chymotrypsin and trypsin compared with the structures of typical synthetic substrates. (*a*) Spectrophotometric titrant for α-chymotrypsin: *p*-nitrophenyl-N^2-acetyl-N^1-benzyl carbazate (Elmore & Smyth, 1968*b*). (*b*) Substrate for α-chymotrypsin: *N*-acetyl-L-phenylalanine-*p*-nitrophenyl ester. (*c*) Spectrophotometric or spectrofluorimetric titrant for trypsin: *N*-methyl-*N*-toluene-*p*-sulphonyl-L-lysine β-naphthyl ester (Elmore & Smyth, 1968*a*). (*d*) Substrate for trypsin: *N*-toluene-*p*-sulphonyl-L-lysine β-naphthyl ester (Elmore, Roberts & Smyth, 1967). (*e*) Spectrofluorimetric titrant for trypsin: 4-methylumbelliferyl-*p*-guanidinobenzoate (Roberts *et al.*, 1971; Jameson *et al.*, 1973). (*f*) Substrate for trypsin: toluene-*p*-sulphonyl-L-arginine methyl ester (Schwert, Neurath, Kaufman & Snoke, 1948).

In the earlier section on pre-steady-state kinetics, we derived the solutions for the differential equations that described the variation of the enzyme intermediates and products with time. Equation (7.20) gives the variation of product P_1 with time and a typical plot is shown in fig. 7.10. We also obtained expressions for the intercept τ on the t axis and for π on the P_1 axis. After substitution for A, B, λ_1 and λ_2,

$$\pi = \frac{k_2[E_0]\,[S_0](k_1 k_2[S_0] - k_3^2)}{k_1(k_2+k_3)^2([S_0]+K_m)^2}.$$

Obviously for $\pi > 0$, $k_1 k_2[S_0] > k_3^2$. If however $k_1 k_2[S_0] \gg k_3^2$ then

$$\pi \approx \frac{k_2^2[E_0]\,[S_0]^2}{(k_2+k_3)^2([S_0]+K_m)^2} \tag{7.46}$$

$$\text{or} \quad \frac{1}{\sqrt{\pi}} = \frac{(k_2+k_3)}{k_2}\frac{1}{\sqrt{[E_0]}} + \frac{(k_2+k_3)\,K_m}{k_2\sqrt{[E_0]}}\frac{1}{[S_0]}. \tag{7.47}$$

From equation (7.47), it will be seen that a plot of $1/\sqrt{\pi}$ versus $1/[S_0]$ is linear with an ordinal intercept $(k_2+k_3)/k_2\sqrt{[E_0]}$. If $k_2 \gg k_3$, then the

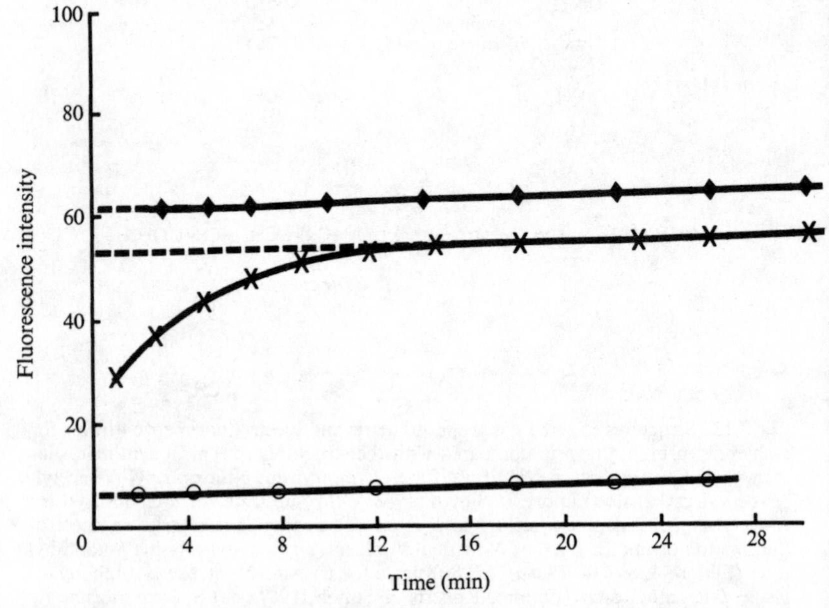

Fig. 7.14. Time course of the reaction between 4-methylumbelliferyl-*p*-guanidino-benzoate (fig. 7.13*e*) with α-trypsin (×) and β-trypsin (◆). The non-enzymic hydrolysis of the reagent is shown by the lower trace (○). (Adapted from Jameson *et al.* (1973) *Biochem. J.* **131**, 107–17.)

ordinal intercept approximates to $1/\sqrt{[E_0]}$. If also $[S_0] \gg K_m$ then equation (7.46) reduces to $\pi = [E_0]$. Since $K_m = k_3(k_{-1} + k_2)/k_1(k_2 + k_3)$ the condition $k_2 \gg k_3$ will almost certainly ensure that K_m is small and so the further condition $[S_0] \gg K_m$ will be easily satisfied. A titrant for a particular enzyme must be designed so that it is sufficiently similar to a normal substrate to be bound by the enzyme, but should be altered in such a way that there is a considerable reduction in k_3 but not k_2 and hence the variation of $[P_1]$ with time can be determined with reasonable precision without the need to use rapid mixing techniques. The compounds shown in fig. 7.13 are examples of titrants for trypsin and chymotrypsin compared with their normal substrates. Fig. 7.14 shows the reaction of α- and β-trypsin with the spectrofluorimetric titrant 4-methylumbelliferyl-*p*-guanidinobenzoate. For those enzymes that form covalent intermediates it is, in theory, possible to produce an enzyme titrant. For enzymes in which a covalent enzyme intermediate is not formed, it may still be possible to obtain a titrant using a compound that forms a very tightly bound complex with the enzyme. An example of such a titrant is the use of 5-fluoro-2′-deoxyuridylate as an active site titrant for thymidylate synthetase (Santi, McHenry & Perriard, 1974). They showed that 5-fluoro-2′-deoxyuridylate (FdUMP) formed an extremely stable complex with the enzyme in the presence of the co-factor 5,10-methylene-tetrahydrofolic acid. Using tritiated FdUMP of high specific activity, it was possible to trap the enzyme complex on nitrocellulose membranes and elute any unbound FdUMP. The remaining radioactivity was then a measure of the active enzyme. The affinity constant is sufficiently high, $>10^9$, that there is negligible loss of radioactivity upon washing the bound complex. No complex is formed in the absence of the co-factor 5,10-methylene-tetrahydrofolic acid.

Whether the amount of work involved in producing a usable titrant is justified depends on the work that is being undertaken. Obviously for clinical work, say in the routine analysis of phosphatases, blood enzymes etc., the search for a specific titrant that might result in increased sensitivity or easier analysis, may well be rewarding.

8 Regulatory enzymes and their kinetic behaviour

8.1. Regulatory processes

So far, in the earlier chapters, we have derived the kinetic equations that describe the behaviour of a single enzyme system catalysing the reaction of one or more substrates. Very little has been said about the mechanism, the number of active sites or the physical structure of any particular enzyme, nor have we looked at the kinetics of an enzyme-catalysed reaction in the overall context of a metabolic pathway. The experimental conditions that are frequently used for studying a particular enzyme system *in vitro* are rather artificial. They are not the conditions that would be likely to occur *in vivo* but are chosen purely for experimental reasons so that the reaction is slow enough to be conveniently followed by, for example, pH-stat or spectrophotometric techniques. The enzyme concentration is usually very low and hence any interactions between enzyme molecules are unlikely to be significant purely on the basis of mass action considerations. In contrast, the enzyme concentration in the cell may be several orders of magnitude higher and interactions between enzyme molecules or enzyme molecules and other proteins may be very important.

The kinetics that we have described so far can be called Michaelis–Menten kinetics, since a plot of reciprocal velocity against reciprocal substrate concentration is linear. Many reactions involving two or more substrates also give such linear plots with respect to one substrate at a fixed initial concentration of the other substrate(s). Whilst there are many enzymes that obey Michaelis–Menten kinetics, there is also a significant number that do not. These enzymes typically give velocity versus substrate curves which are sigmoidal rather than hyperbolic. These enzymes are called *regulatory* or *control* enzymes and an examination of metabolic maps will show that they are frequently found at the beginning or at a branch point of a metabolic process. The need for some sort of regulatory process becomes more apparent if one considers some of the operations undertaken by a human being during the day. Food is ingested at irregular times during the day so that the level of the various breakdown products will rise and fall between meals. We therefore need both a positive control mechanism to speed up the process of digestion at high levels of food and also a means of slowing down or stopping the process at low levels of food. Similarly the energy requirements of the body vary considerably from a period of high activity, such as running, to a period of low activity during rest periods. If one is suddenly subjected to physiological stress such as fear, then epinephrine (adrenalin) is rapidly secreted into the blood stream by the adrenal medulla. This stimulates the activation of adenyl cyclase which eventually leads via a cascade process (biological amplification) to the breakdown of glycogen to glucose-1-phosphate (fig. 8.1). The net result is a rapid increase in the availability of glucose to the muscles which allows the person to run away if so desired!

The simplest form of regulatory mechanism is the self regulation by a metabolite of an enzyme which is essential for its own biosynthesis.

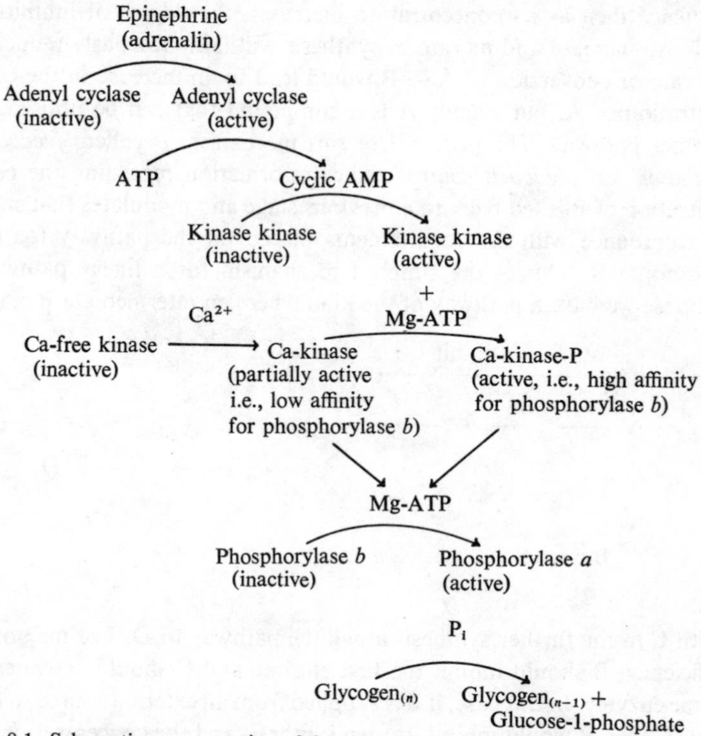

Fig. 8.1. Schematic representation of the cascade process resulting in the breakdown of glycogen to glucose-1-phosphate. (Adapted from Fischer, Pocker & Saari (1970) *Essays Biochem.* **6**, 23–68.)

Consider the following pathway in which a compound B is synthesised via a series of steps from a precursor A.

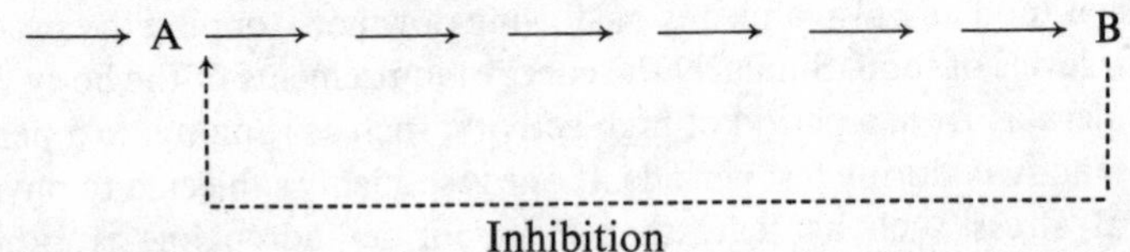

If there is no inhibition of the enzyme catalysing the degradation of A then after a certain time the concentration of B (and all the other components in the pathway) will reach a steady state depending on its rate of synthesis and utilisation. If conditions change, so that the requirements for B are lessened or if B is supplied from some external source, then the continued synthesis of B at the same rate would be unnecessary and its concentration could in fact rise to a level which might be toxic to the organism. In contrast, if B is an inhibitor of the first enzyme in the sequence then as its concentration increases, the degree of inhibition will also increase and its rate of synthesis will fall. Obviously reducing the rate of conversion of A $\rightarrow$ B would lead to an increase in the concentration of A, but usually A is a compound that can be utilised by another pathway. The principle of this mechanism is called '*feedback inhibition*' or '*feedback control*', since information regarding the concentration of B is fed back to a previous stage and modulates that stage in accordance with the requirements placed on the pathway for the metabolite B. This is the simplest mechanism for a linear pathway. Suppose we have a pathway of the kind where an intermediate B reacts

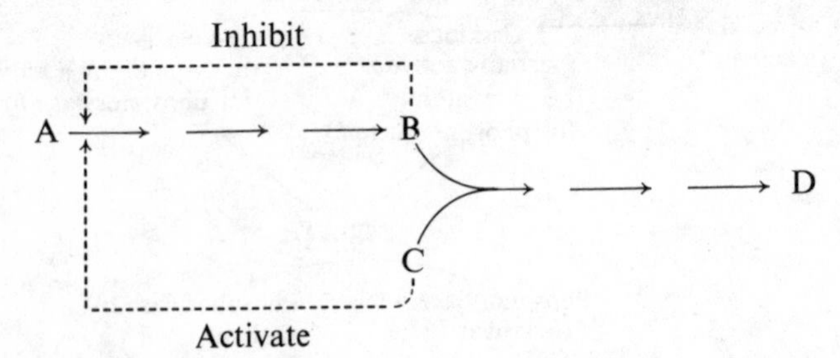

with C in the further synthesis along the pathway to D. For maximum efficiency, B should inhibit the first enzyme and C should activate the same enzyme. In this case, if B is supplied from an external source so that $B \gg C$ then B would inhibit its own synthesis and the concentrations of B and C would tend to equalise. Alternatively, if $C \gg B$ and C activates the production of B this will again equalise the concentrations of B and

C. The 'activation' by C is usually caused by C competing for the same binding site as B on the first enzyme and hence reducing the inhibition by B. The first enzyme of pyrimidine synthesis (aspartate transcarbamylase E.C. 2.1.3.2) is an example of this type of control mechanism. In this case, B is CTP and C is ATP and the two are used together in the further synthesis of nucleic acids. This example is one of converging pathways; however it is more usual for biosynthetic pathways to diverge as in the example below:

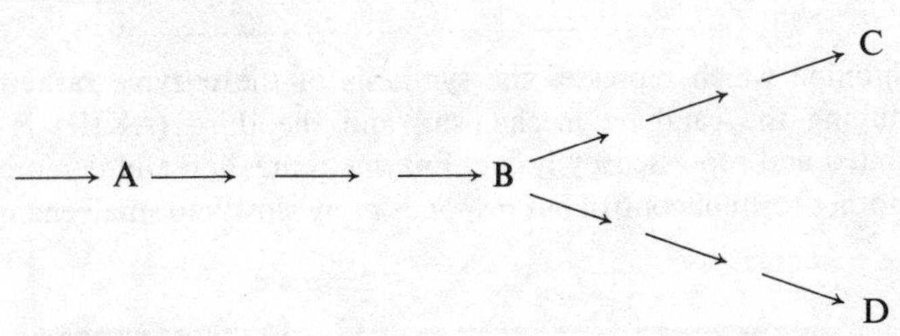

For optimal control of this pathway it should be possible to regulate the levels of C and D as well as B. There are a number of ways that this can be achieved. The first is called *isoenzymic* control. In this case the first enzyme exists as two isoenzymes, one which is inhibited by C and the other by D.

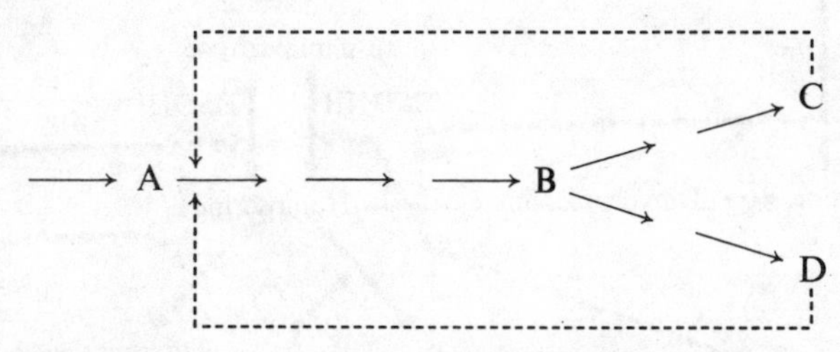

Obviously if these were the only controls that operated and if, say, C were at a high level so that the C isoenzyme were completely inhibited, C could still be synthesised from B via the D isoenzyme. More effective control would be achieved if C and D also inhibited their respective enzymes at the branch point B.

An example of this sort of mechanism is found in the aspartate pathway (fig. 8.2). The first enzyme (aspartokinase) has been shown to exist in multiple forms in both *E. coli* (Cohen, 1969) and *Salmonella typhimurium* (Cafferata & Freundlich, 1969). In the case of *E. coli*, Stadtman, Cohen and their collaborators have isolated three aspartokinases; one is sensitive to threonine (AKI), the second (AKII) is sensitive to

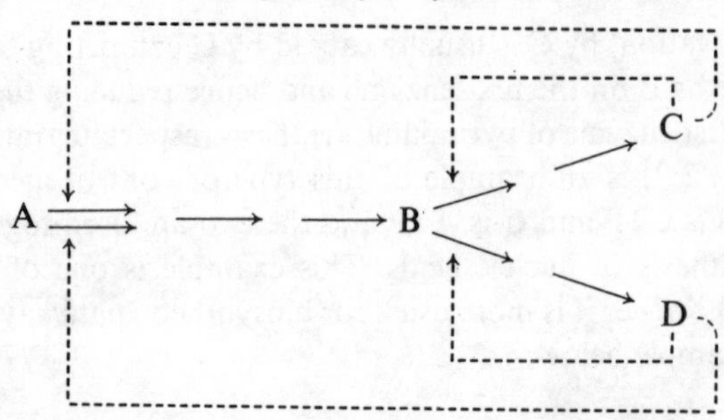

methionine which represses the synthesis of the enzyme rather than modifying the catalytic mechanism, and the third (AKIII) is both inhibited and repressed by lysine. Enzyme repression and derepression is another form of control but responds more slowly to small changes in

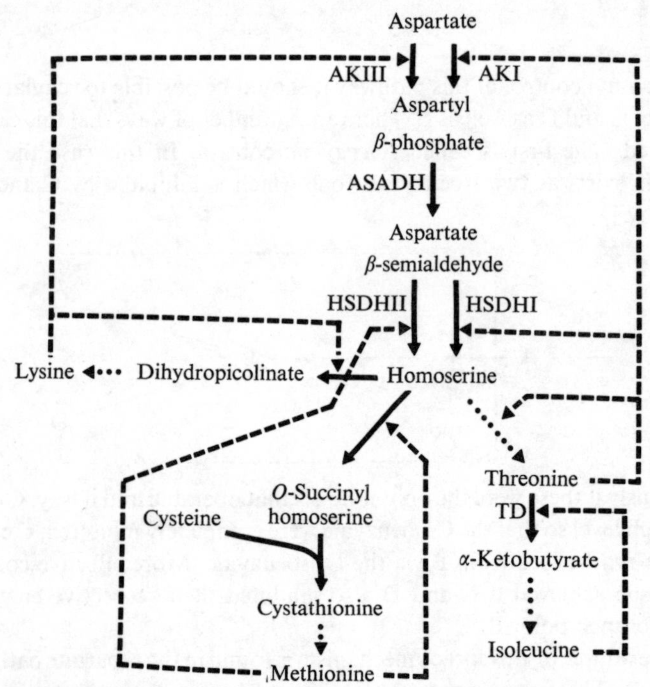

Fig. 8.2. Schematic representation of the aspartate pathway indicating the various feedback control paths (dashed lines) that operate. Dotted lines indicate more than one step in the pathway. The abbreviations are: AKI, AKIII, isoenzymes of aspartokinase; ASADH, aspartate β-semialdehyde dehydrogenase; HSDHI, HSDHII, isoenzymes of homoserine dehydrogenase; TD, threonine deaminase. (From Datta (1969) *Science*, **165**, 556–62. © 1969 American Association for the Advancement of Science.)

concentration of the controlling metabolite. In this pathway, there are also two species of homoserine dehydrogenase, one sensitive to methionine and the other affected by threonine. There is also evidence that the aspartokinase and homoserine dehydrogenase controlled by threonine are the same protein or protein–protein complex, and similarly for the two enzymes controlled by methionine. Threonine is also the precursor for the biosynthesis of isoleucine and the latter controls the first enzyme in the sequence – threonine deaminase. In this particular example the control mechanism is complex but quite specific for each end-product of the pathway.

A different mechanism operates in the aspartate pathway in *Rhodopseudomonas spheroides*. It can be called *sequential* feedback inhibition and can be represented by the scheme:

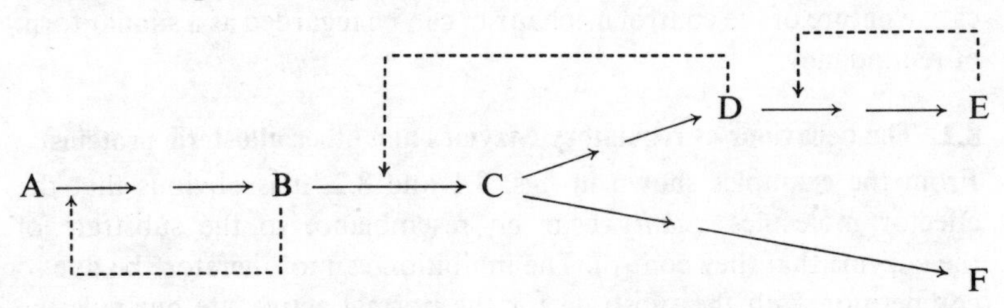

In this example, overproduction of isoleucine (E) inhibits threonine deaminase and the consequent rise in threonine (D) concentration reduces the rate of homoserine (C) synthesis by inhibition of homoserine dehydrogenase. This negative cascade process leads in turn to an increase in the concentration of aspartate-β-semialdehyde (B) which inhibits aspartokinase.

Another type of feedback mechanism is observed in the same pathway in *Rhodopseudomonas capsulata*. In this case, the enzyme aspartokinase is not inhibited by any of the end-products lysine, threonine or methionine *alone* but is strongly inhibited by threonine and lysine *together* (Datta & Gest, 1964). This form of feedback mechanism is called *concerted* feedback inhibition. In *Rhodospirillum rubrum* the concerted feedback inhibition of aspartokinase by threonine and lysine can be reversed by isoleucine and methionine acting together.

Apart from feedback inhibition and repression of enzyme synthesis, another form of control is shown in fig. 8.1. This is the control of a pathway by 'molecular conversion' (Monod & Jacob, 1961) which in this particular case is the ultimate conversion of phosphorylase *b* to phosphorylase *a* via a cascade process involving the conversion of inactive enzyme species to active enzymes by enzymes that are themselves activated by effector molecules, e.g. epinephrine or cyclic AMP.

The complex nature of the overall control mechanism of a particular metabolic pathway is probably another example of redundancy as a safeguard against harmful mutations. The structures of enzymes are highly redundant in that only perhaps ten amino acid residues out of say 200 are directly involved in the catalytic mechanism. The remaining amino acid residues provide a suitable environment and support structure for these essential amino acids. Thus a comparison of the homologies between the serine proteinases from different species shows that the important residues in the active site – histidine, serine and aspartic acid – are all conserved but that many modifications, insertions and deletions are found in the remaining parts of the molecules. The duplicative nature of the control mechanism can be regarded as a similar form of redundancy.

8.2. The behaviour of regulatory enzymes and other allosteric proteins

From the examples shown in figs. 8.1 and 8.2, it is obvious that the effector molecules *usually* bear no resemblance to the substrate of the enzyme that they control. The inhibition cannot therefore be due to competition with the substrate for the normal active site but must be due to the binding of the effector molecule at another site (allosteric site) which causes a change in the 'affinity' of the enzyme for the substrate. Regulatory enzymes are usually identified by their deviation from Michaelis–Menten behaviour. Typical velocity versus substrate plots that may be observed are shown in fig. 8.3(*a*) with the corresponding reciprocal plots in fig. 8.3(*b*). A control enzyme may exhibit curves that are sigmoidal (curve 1), rectangular hyperbolic (curve 2) or appear to be hyperbolic (curve 3) (fig. 8.3*a*). If these curves are replotted in the reciprocal form (fig. 8.3*b*), then it is more apparent that 3 is no longer a *rectangular hyperbola*. Koshland, Neméthy & Filmer (1966) have proposed that a useful parameter for comparing control enzymes and other allosteric proteins is the ratio

$$R_S = \frac{\text{substrate (or ligand) concentration at 0.9 saturation}}{\text{substrate (or ligand) concentration at 0.1 saturation}}.$$

For an enzyme that follows Michaelis–Menten kinetics $R_S = 81$. For a sigmoidal curve (Fig. 8.3*a*, curve 1), $R_S < 81$ and the enzyme is said to exhibit positive co-operativity. Positive co-operativity means, therefore, that the binding of the substrate becomes easier as the enzyme becomes

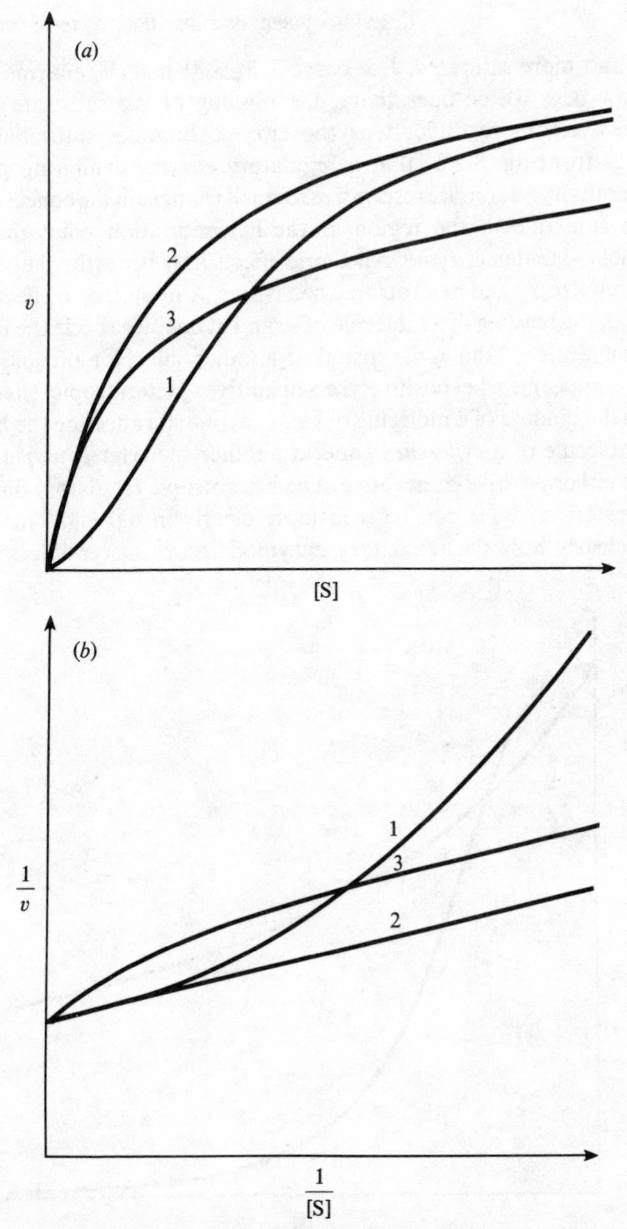

Fig. 8.3. (*a*) Possible kinetic behaviour that may be observed with regulatory enzymes plotted as v against [S]. 1, sigmoidal; 2, hyperbolic; 3, apparently hyperbolic. (*b*) The same data plotted in reciprocal form of $1/v$ against $1/[S]$. Only example 2 is linear and therefore hyperbolic, the other two examples, 1 and 3, are both non-linear.

more and more saturated. For curve 3 $R_S > 81$ and the enzyme is said to show negative co-operativity, the binding of the substrate getting progressively more difficult as the enzyme becomes saturated. It is obvious from fig. 8.3(*a*) that a regulatory enzyme exhibiting positive co-operativity is far more responsive to small changes in the concentration of the substrate in the region of the half-saturation point than is a Michaelis–Menten enzyme. Allosteric effects may be further subdivided into homotropic and heterotropic behaviour. A homotropic effect is one in which the binding of a molecule of ligand at one site affects the binding of a molecule of the *same* ligand at another site. Homotropic effects usually appear to be positively co-operative. Heterotropic effects are due to the binding of a molecule of ligand at one site affecting the binding of a molecule of a *different* ligand at another site. Heterotropic effects can be either positive or negative. The heterotropic regulatory nature of an allosteric enzyme can be seen more clearly in fig. 8.4. This shows quite clearly how the regulatory enzyme is more susceptible to small

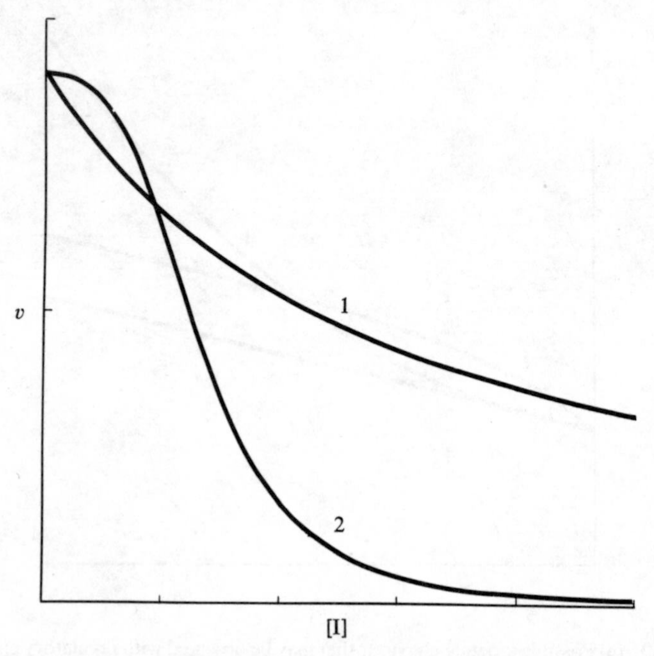

Fig. 8.4. A typical plot of v against [I] for: 1, a Michaelis–Menten enzyme; and 2, a sigmoidal enzyme. Note that the regulatory enzyme is controlled over a narrower range of inhibitor concentration than the Michaelis–Menten enzyme.

changes in the concentration of the effector molecule. This is exactly what is required of course for an enzyme that is controlling the overall flux of a metabolic pathway.

Suppose we were presented with a series of curves for a particular enzyme that were sigmoidal. Clearly, two major questions require answering before a mechanism can be proposed for the regulatory enzyme. Firstly, can we derive a model in terms of kinetic or binding equations that will accurately predict the form of the curves that are obtained experimentally? Secondly, can we explain on a molecular level the detailed processes that occur in the active site and the rest of the molecule during the catalytic process? The first 'mechanism' only requires specifying a minimal number of intermediates, binding constants and possible interconversions between states of the enzyme. The second and true mechanism requires a detailed knowledge of the physical and chemical structure of the enzyme at *all* steps of the reaction. The latter is the ultimate goal in understanding enzyme reactions on a molecular level but as yet it has not been possible to explain completely the processes that occur with enzymes that obey Michaelis–Menten kinetics, much less the more complicated control enzymes.

Let us look briefly at some of the facts that are known about two regulatory enzymes, aspartate transcarbamylase and glutamate dehydrogenase, as this will assist us in formulating models that describe their kinetic behaviour. Aspartate transcarbamylase from *E. coli* catalyses an early step in the pyrimidine biosynthesis pathway:

$$\text{L-aspartate} + \text{carbamyl phosphate} \rightarrow \text{carbamyl-L-aspartate} + P_i.$$

The discovery that aspartate transcarbamylase in crude extracts of *E. coli* is regulated by cytidine nucleotides, the end-products of the pyrimidine pathway, was made by Yates & Pardee (1956). Similar work by Umbarger (1956) on threonine deaminase led to the concept of feedback inhibition. The inhibition of the activity of aspartate transcarbamylase is an efficient way of controlling the overall synthesis of UTP and CTP but it is only part of a more complex regulatory scheme involving other enzymes in the pathway (Gerhart, 1970). Although the enzyme is not affected by UTP, it is activated by ATP which competes for the same binding site as CTP. The saturation of the enzyme by L-aspartate at high carbamyl phosphate concentrations and vice versa, is sigmoidal, indicating a positive co-operative binding mechanism (Gerhart & Pardee, 1962; Bethell, Smith, White & Jones, 1968). The effectors ATP and CTP do not alter the maximum velocity of the enzymic reaction but alter the shape of the binding curve. Thus the effect of CTP is to make the curve more sigmoidal whereas ATP makes the curve more hyperbolic (fig. 8.5). Treatment of the enzyme with mercurials produces two interesting effects. Firstly, the enzyme is no longer sensitive to ATP or

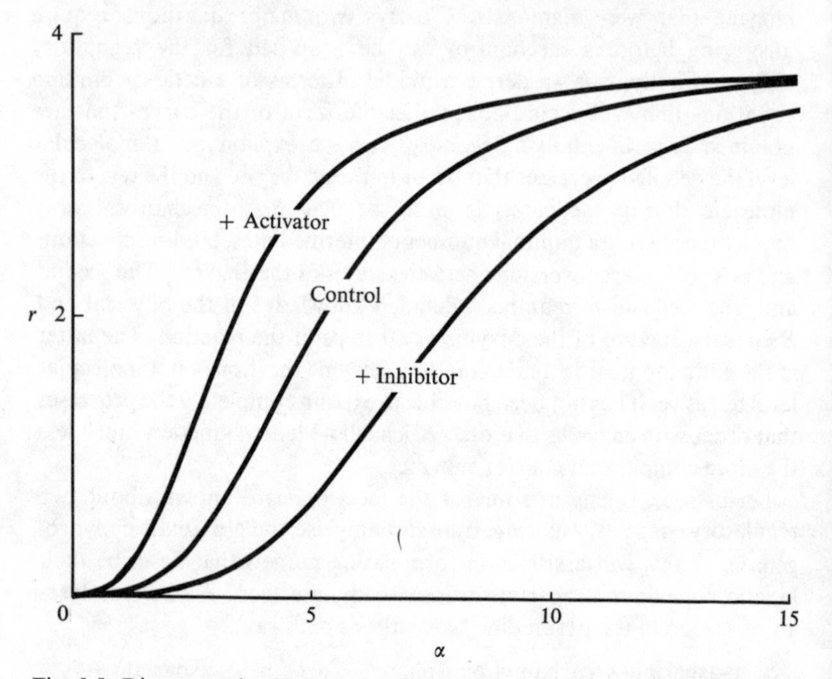

Fig. 8.5. Diagrammatic representation of the effect of ATP (activator) and CTP (inhibitor) on the activity of aspartate transcarbamylase. The effect of CTP is to make the curve more sigmoidal whereas ATP makes the curve more hyperbolic. (The meanings of r and α are explained in section 8.2.1, especially in relation to the derivation of equation (8.10).)

CTP and positive co-operativity due to the substrate alone also disappears (Gerhart & Pardee, 1962); secondly, the enzyme dissociates into two kinds of subunits (Gerhart & Schachman, 1965). The larger subunit is catalytically active towards the normal substrates, L-aspartate and carbamyl phosphate, and it behaves as a typical Michaelis–Menten enzyme. The smaller subunit binds the effector molecules CTP and ATP (Changeux, Gerhart & Schachman, 1968) and is catalytically inactive. The dissociated enzyme spontaneously re-forms after removal of the mercurials. Chemical and physicochemical studies by Weber (1968*a*,*b*) showed that the molecule consists of twelve polypeptide chains of two kinds, C (catalytic) and R (regulatory). The catalytic subunit has been shown to be a stable trimer of polypeptide chains (Meighen, Pigiet & Schachman, 1970; Davies & Stark, 1970; Rosenbusch & Weber, 1971) which shows no tendency to dissociate into individual polypeptide chains or to associate into larger species (Kirschner & Schachman, 1971). There are two such trimers per enzyme molecule.

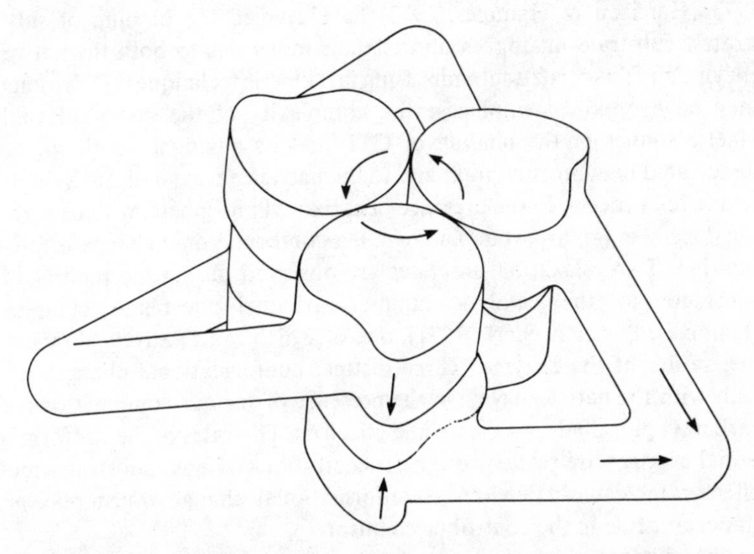

Fig. 8.6. Model of the possible arrangement of the regulatory and catalytic subunits of aspartate transcarbamylase. The six asymmetrically shaped catalytic polypeptide chains are arranged in the form of two trimer subunits. The regulatory subunits consist of a pair of polypeptide chains which are orientated to one another at an acute angle. Each regulatory subunit is bonded to an upper and lower catalytic polypeptide chain. (Adapted after J. A. Cohlberg *et al.* (1972).)

In the absence of zinc ions, the regulatory subunit exists in a monomer–dimer equilibrium. The effect of zinc ions is to stabilise the dimer (Cohlberg, Pigiet & Schachman, 1972). The complete enzyme structure is formed in high yield when the regulatory subunits are mixed with the catalytic subunits. Since the enzyme can be reconstituted by mixing the catalytic subunits with regulatory subunits which have been covalently cross-linked, it seems likely that the dimer form of the regulatory subunit is a structural entity of the enzyme. The structure of the enzyme proposed by Cohlberg *et al.* (1972) is shown in fig. 8.6. Electron micrographs suggest that the two catalytic trimers are not in direct contact but are separated by the three regulatory dimers. Gerhart (1970) suggested that the molecule can exist in two conformations, a totally eclipsed structure in which the catalytic trimers lie directly above one another and a partially eclipsed structure in which the trimers are rotated relative to each other. Additional evidence is required, especially from crystallographic studies, before the enzyme structure can be finalised. Hammes and his group (Eckfeldt, Hammes, Mohr & Wu, 1970; Wu & Hammes, 1973; Harrison & Hammes, 1973) have studied the binding of substrates, substrate analogues and effector molecules to both the native enzyme and dissociated subunits using fast kinetic techniques. The results they have obtained emphasise the complexity of the system. Rapid kinetic studies of the binding of CTP and its analogues both to the dissociated regulatory subunit and to the native enzyme indicate a single relaxation process. In the presence of carbamyl phosphate, with or without succinate (an aspartate analogue), a conformational change is rate-limiting. Two relaxation processes are observed during the binding of succinate to the catalytic subunit–carbamyl phosphate complex (Hammes, Porter & Stark, 1971), one of which is a relatively slow isomerisation of the enzyme. Three distinct conformational changes are seen with the native enzyme in the presence of various combinations of carbamyl phosphate, succinate and effectors. The rates of the conformational changes are related to the concentrations of both substrates and effector molecules. Obviously conformational changes must play an important role in the control mechanism.

The glutamate dehydrogenases, as a class of enzymes, catalyse the oxidative deamination of L-glutamic acid to α-ketoglutaric acid and ammonia. The enzymes from different sources vary considerably in terms of their kinetic characteristics, metabolic function and molecular properties. The animal enzymes are sensitive to the concentration of purine nucleotides and can catalyse the reaction using either NAD or NADP. They also frequently undergo a reversible polymerisation reaction which gives rise to molecules of very high molecular weight. The enzymes from non-animal sources are specific for either NAD or NADP but are not affected by the purine nucleotides and do not undergo polymerisation. These differences are undoubtedly related to the different metabolic role of the enzymes in animal species where they link the Krebs cycle with the pathways of amino acid synthesis. Although a number of glutamate dehydrogenases from different sources have been studied, by far the most information concerning the molecular properties of the enzyme has been obtained using the bovine liver enzyme. The weight-average molecular weight of the enzyme is concentration-dependent, which is compatible with polymerisation of the enzyme. Polymerisation can become quite pronounced at concentrations greater than 1 mg ml^{-1}; a hexameric form predominates at 5 mg ml^{-1}. In addition the weight-average molecular weight can be drastically altered by the addition of NADH in conjunction with an additional regulatory reagent. For instance the addition of GTP and NADH can lead to dissociation to the 'monomeric' or protomeric form (Frieden, 1959; Yielding & Tom-

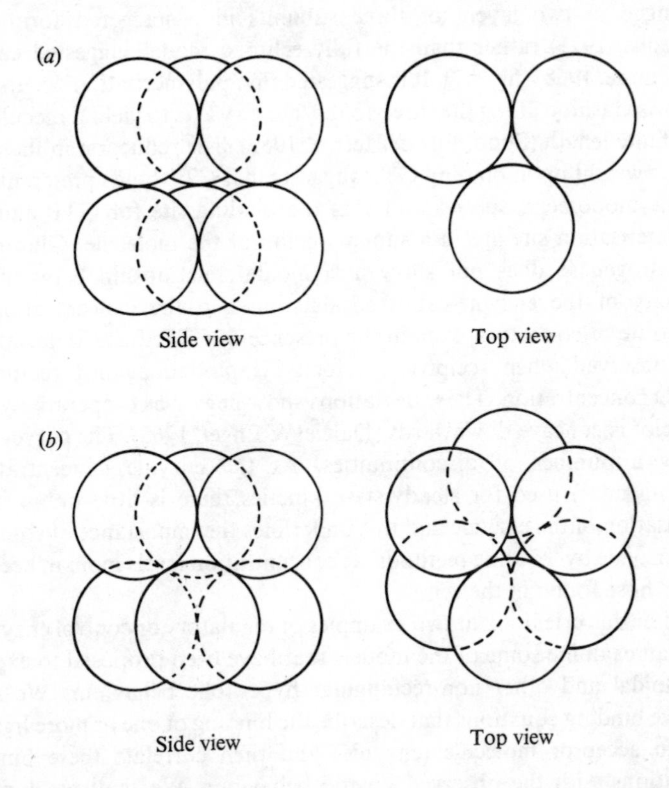

Fig. 8.7. Possible arrangements of the subunits of glutamate dehydrogenase. The basic subunit appears to be a trimer of polypeptide chains two of which are stacked together to form the normal molecule of the enzyme. Two possible stacking arrangements have been suggested: (*a*) an eclipsed conformation (Valentine, 1968; Eisenberg & Reisler, 1970) and (*b*) a staggered conformation (Josephs, 1971). Polymerisation of the molecules can take place along the three-fold axis to give molecules of indefinite length.

kins, 1961). This effect is dramatic and visible to the naked eye. A solution of glutamate dehydrogenase containing 5–10 mg ml^{-1} is quite turbid due to the light scattered by the high molecular weight polymeric material. If GTP and NADH are added to the solution to a final concentration of 10^{-3} M the mixture instantly becomes transparent. The polymerisation/depolymerisation process is therefore quite rapid. The fully dissociated enzyme or 'monomer' has a molecular weight of 316 000 $\pm$ 10 000. The 'monomer' is in fact composed of six subunits each with a molecular weight of 53000 (Eisenberg & Tomkins, 1968). The six subunits are arranged in two layers of three subunits in a staggered formation (Josephs, 1971) rather than the fully eclipsed model suggested earlier (Valentine, 1968) (fig. 8.7). It is suggested that polymerisation occurs in a stepwise fashion along the three-fold symmetry axis to yield molecules of indefinite length (Sund, Pilz & Herbst, 1969). The reduction in the molecular weight upon binding GTP suggests that GTP binds preferentially to the monomeric species and that the binding site for GTP and the polymerisation site are in a similar region of the molecule. Glutamate dehydrogenase does not show a sigmoidal relationship between the velocity of the enzyme-catalysed reaction and the concentration of substrate or co-factors, even in the presence of GTP. Instead deviations are observed when reciprocal velocity is plotted against reciprocal NAD concentration. These deviations show negative co-operativity, i.e. the plot is concave downwards (Dalziel & Engel, 1968). The curves also show a number of discontinuities. At the enzyme concentrations (~1 μg ml^{-1}) used for steady-state kinetics there is little or no polymerisation of the enzyme and this underlines the importance of studying the enzyme by 'binding methods' at concentration levels more in keeping with those found in the cell.

After this brief look at two examples of regulatory or control enzymes, we can examine some of the models that have been proposed to explain sigmoidal and other non-rectangular hyperbolic behaviour. We shall derive binding equations that describe the binding of one or more ligands to an acceptor molecule (enzyme) and then correlate these binding equations with the observed kinetic behaviour. We shall use binding equations in the first instance, since a lot of the information that has been obtained about regulatory enzymes has been obtained from the binding studies of substrate analogues and effector molecules at enzyme concentrations that would normally prohibit the use of steady-state kinetic methods. With these points in mind let us consider the first model.

8.2.1. The Monod, Wyman & Changeux model. The terminology used by Monod *et al.* (1965) is as follows:

(1) An oligomer is a polymeric protein containing a finite and relatively small number of identical subunits.
(2) The identical subunits are called protomers.
(3) The term monomer refers to the fully dissociated protomer.

The assumptions made by Monod *et al.* in formulating their first model are as follows:

(1) The protomers are associated in such a way that they all occupy equivalent positions. The molecule must therefore possess at least one axis of symmetry.
(2) Each ligand has only one binding site on each protomer or subunit.
(3) Two (at least two) states are reversibly accessible to the allosteric oligomers. These states differ in their affinity for the ligand.
(4) The affinity of the stereospecific site is altered when a transition from one state to the other occurs.

We will first derive the equation that describes the binding of a single ligand to the protein (e.g. homotropic effects). In this model the two states are defined as the R-state for the one with the highest affinity for the ligand and the T-state for the one with the lowest affinity for the ligand. In order to make the model more compatible with the other models we shall define the intrinsic *association* constant for the binding of the ligand to the R-state as K_R and similarly the intrinsic binding constant for the ligand to the T-state as K_T. (Monod *et al.* defined K_R and K_T as intrinsic *dissociation* constants for the ligand–R and ligand–T complexes.) L is the equilibrium or isomerisation constant for the conversion of $R_0 \rightleftharpoons T_0$ in the absence of ligand. Assuming there are n protomers (i.e. n identical subunits) we can write the following equations for the binding of successive ligand molecules, S, to the R- and T-states. $R_0, RS_1, \cdots, RS_n$ and $T_0, TS_1, \cdots, TS_n$ refer to the complexes involving $0, 1, \cdots, n$ molecules of ligand.

$$R_0 \rightleftharpoons T_0$$

$$\begin{array}{ll} R_0 + S \rightleftharpoons RS_1 & \qquad T_0 + S \rightleftharpoons TS_1 \\ RS_1 + S \rightleftharpoons RS_2 & \qquad TS_1 + S \rightleftharpoons TS_2 \\ \vdots & \qquad \vdots \\ RS_{n-1} + S \rightleftharpoons RS_n & \qquad TS_{n-1} + S \rightleftharpoons TS_n \end{array}$$

Consider the binding of the first molecule of ligand to R_0. Since there are n vacant sites there are n ways in which this first molecule can be bound. There will therefore be n equivalent forms of RS_1, i.e. ${}_1RS_1$, ${}_2RS_1$, ${}_3RS_1$, $\cdots$, ${}_nRS_1$ where the left subscript refers to the site that has bound that particular molecule of ligand. We can therefore write the equilibrium equation for the first equation, $R_0 + S \rightleftharpoons RS_1$, as

$$\frac{[RS_1]}{[R_0][S]} = K_1 = \frac{[{}_1RS_1]}{[R_0][S]} + \frac{[{}_2RS_1]}{[R_0][S]} + \cdots + \frac{[{}_nRS_1]}{[R_0][S]}.$$

Since each site is assumed to have an identical intrinsic binding constant K_R, then $K_1 = K_R + K_R + \cdots + K_R = nK_R$. After $m-1$ sites have been occupied there are then $n!/m!(n-m)!$ ways in which the mth molecule of ligand can be bound. For $m = 2$, therefore, there are $n(n-1)/2$ ways in which the second molecule can be bound. Since the association constants are equal for each site, if ${}_1RS_2$ is one form of RS_2 then

$$\frac{[RS_2]}{[RS_1][S]} = K_2 = \frac{\{n(n-1)/2\}[{}_1RS_2]}{n[{}_1RS_1][S]} = \frac{(n-1)}{2} K_R.$$

More generally, each form of RS_m (say ${}_jRS_m$) will be present in equal concentrations and there will be $n!/m!(n-m)!$ forms of RS_m. Similarly for RS_{m-1} (say ${}_iRS_{m-1}$) there will be $n!/(m-1)!(n-m+1)!$ forms of RS_{m-1}.

Hence

$$\frac{[RS_m]}{[RS_{m-1}][S]} = K_m = \frac{\{n!/m!(n-m)!\}[{}_jRS_m]}{\{n!/(m-1)!(n-m+1)!\}({}_iRS_{m-1})[S]}$$

and therefore

$$K_m = \left(\frac{n-m+1}{m}\right) K_R.$$

The concentration of any RS_i species is therefore

$$[RS_i] = \left(\frac{n-i+1}{i}\right) K_R[RS_{i-1}][S] \qquad \text{for } i = 1, \cdots, n.$$

By an exactly similar argument we can show that

$$[TS_i] = \left(\frac{n-i+1}{i}\right) K_T[TS_{i-1}][S].$$

The total concentration of protein is the sum of all its forms:

$$[P_0] = ([R_0] + [RS_1] + \cdots + [RS_n]) + ([T_0] + [TS_1] + \cdots + [TS_n]) \tag{8.1}$$

and thus the fraction of protein in the R-state is

$$\bar{R} = \frac{([R_0] + [RS_1] + \cdots + [RS_n])}{[P_0]}.$$

A convenient measure of the extent of combination of ligand with the protein is the parameter r, where

$$r = \frac{\text{moles of bound ligand}}{\text{moles of total protein}} = \frac{([S_0] - [S])}{[P_0]} \tag{8.2}$$

where $[S_0]$ is the total concentration of ligand and $[S]$ is the concentration of unbound ligand. The number of moles of ligand bound to the R- and T-states are respectively equal to:

$$[RS_1] + 2[RS_2] + 3[RS_3] + \cdots + n[RS_n] \tag{8.3}$$

and $[TS_1] + 2[TS_2] + 3[TS_3] + \cdots + n[TS_n]$.

If we take the equation (8.3) for the number of moles of ligand bound to the R-state and systematically substitute for the $[RS_i]$ for $i = 1, \cdots, n$ using the equilibrium relationship that

$$[RS_i] = K_i\,[RS_{i-1}]\,[S] = K_i\,K_{i-1}[RS_{i-2}]\,[S]^2 \cdots$$

we obtain

$$K_1[R_0]\,[S] + 2K_1\,K_2[R_0]\,[S]^2 + 3K_1\,K_2\,K_3[R_0]\,[S]^3 + \cdots$$
$$+ nK_1\,K_2\,K_3 \cdots K_n[R_0]\,[S]^n$$

and using the further relationship that

$$K_i = \left(\frac{n-i+1}{i}\right) K_R$$

this becomes

$$[R_0]\left\{nK_R[S] + 2\,\frac{n(n-1)}{2!}\,K_R^2[S]^2 + \cdots \right.$$
$$\left. + m\left(\frac{n(n-1)\cdots(n-m+1)}{m!}\right) K_R^m[S]^m + \cdots + n\frac{n!}{n!}K_R^n[S]^n\right\} \tag{8.4}$$

with an exactly analogous equation for the T-state in terms of $[T_0]$ and K_T. By the same procedure the equation (8.1) for the total protein becomes

$$[P_0] = [R_0]\left(1 + nK_R[S] + \frac{n(n-1)}{2!}K_R^2[S]^2 + \cdots + \frac{n(n-1)\cdots(n-m+1)}{m!}K_R^m[S]^m + \cdots + \frac{n!}{n!}K_R^n[S]^n\right) + [T_0]\left(1 + nK_T[S] + \frac{n(n-1)}{2!}K_T^2[S]^2 + \cdots + \frac{n(n-1)\cdots(n-m-1)}{m!}K_T^m[S]^m + \cdots + \frac{n!}{n!}K_T^n[S]^n\right). \quad (8.$$

This equation (8.5) is the binomial expansion of

$$[P_0] = [R_0](1 + K_R[S])^n + [T_0](1 + K_T[S])^n$$

and this is the denominator term in equation (8.2). The total number of moles of bound ligand is the sum of equation (8.4) and the analogous one in $[T_0]$ and this will be the numerator in equation (8.2). If we factorise out [S] from the numerator it can be shown by term-by-term differentiation that the numerator is equal to $\frac{[S]\,d\,(\text{denominator})}{d[S]}$ and therefore

$$\text{numerator} = \frac{[S]\,d\{[R_0](1 + K_R[S])^n + [T_0](1 + K_T[S])^n\}}{d[S]}$$
$$= n[S]\{K_R[R_0](1 + K_R[S])^{n-1} + K_T[T_0](1 + K_T[S])^{n-1}\}.$$

The expression for r now becomes

$$r = \frac{n[S]\{K_R[R_0](1 + K_R[S])^{n-1} + K_T[T_0](1 + K_T[S])^{n-1}\}}{[R_0](1 + K_R[S])^n + [T_0](1 + K_T[S])^n}.$$

If we set $K_R[S] = \alpha$ and $K_T/K_R = c$ and use the relationship $L = [T_0]/[R_0]$ then this equation becomes

$$r = \frac{n\{\alpha(1 + \alpha)^{n-1} + Lc\alpha(1 + c\alpha)^{n-1}\}}{(1 + \alpha)^n + L(1 + c\alpha)^n}. \quad (8.6)$$

Note that the definition of r in equation (8.2) in terms of moles is different from that used by Monod *et al.* (1965) so that $r = n\bar{Y}_F$ where $\bar{Y}_F$ is

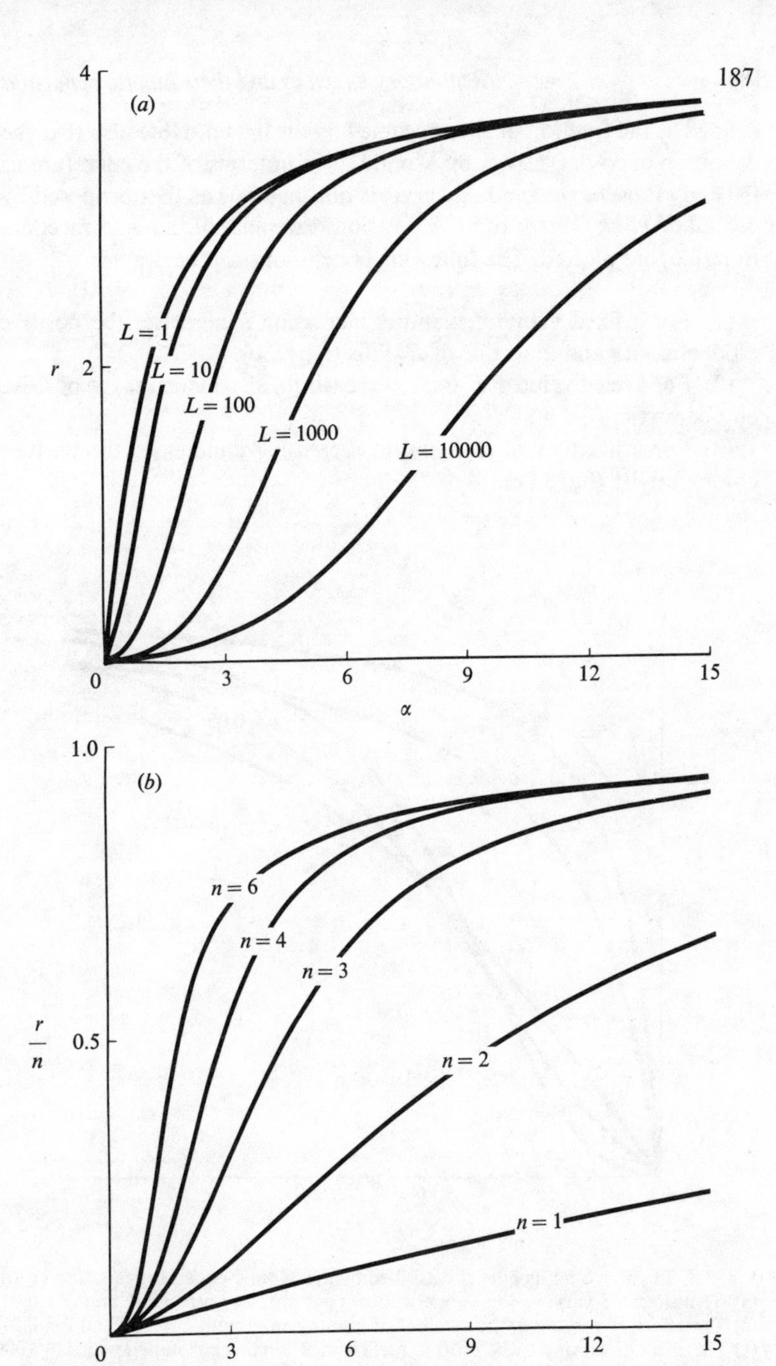
(a)
4
r
2
0
3
6
9
12
15
α
L = 1
L = 10
L = 100
L = 1000
L = 10000
(b)
1.0
0.5
r/n
0
3
6
9
12
15
α
n = 6
n = 4
n = 3
n = 2
n = 1

defined as the fraction of sites occupied by the ligand. Note also that the definition of co-operativity by Monod *et al.* in terms of the curvature of the early stage of the binding curve, is not the same as that proposed by Koshland *et al.* (1966). In fig. 8.8 various examples of the binding equation (8.6) are plotted. The following conclusions can be drawn:

(1) For a fixed value of n and c, increasing L increases the positive co-operativity and the value of R_S falls (fig. 8.8*a*).

(2) For a fixed value of L and c, increasing n also increases the positive co-operativity (fig. 8.8*b*).

(3) For a fixed value of L and n, *decreasing* c increases the positive co-operativity (fig. 8.8*c*).

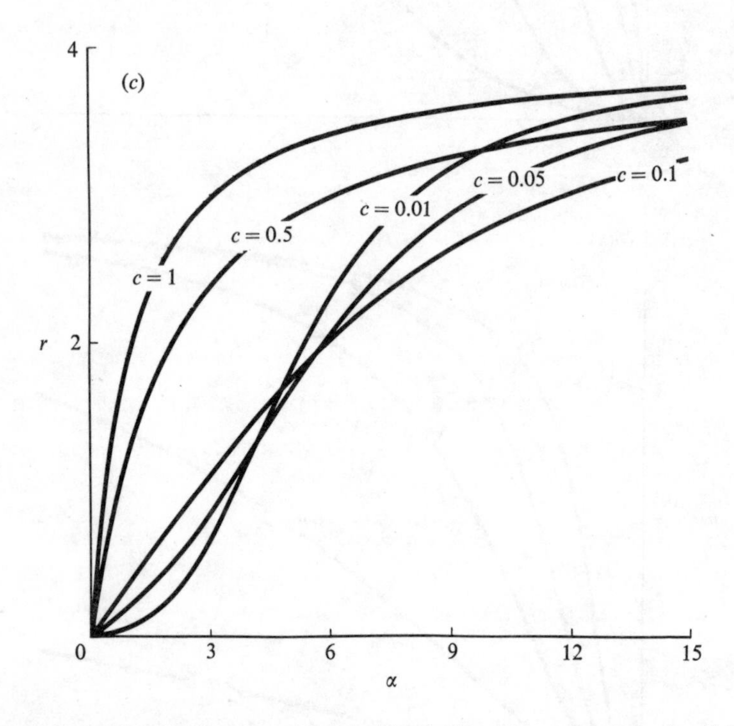

Fig. 8.8. Possible binding curves described by the Monod equation (equation (8.6)). (*a*) The effect of varying the isomerisation constant, L, with $n = 4$ and $c = 0.01$. (*b*) The effect of varying n, the number of binding sites, with $L = 100$ and $c = 0.01$. (*c*) The effect of varying c, the ratio of the T and R binding constants, with $L = 1000$ and $n = 4$.

Note that for the two cases $L = 0$ or $c = 1$ equation (8.6) becomes

$$r = \frac{n\alpha}{1 + \alpha}$$

which is a normal rectangular hyperbola (Michaelis–Menten curve) (Klotz, 1946).

So far this model only accounts for homotropic positive co-operative effects. It does not explain the heterotropic effects which are frequently observed with regulatory enzymes. Obviously, to explain heterotropic effects we have to introduce the binding of a second ligand to the protein. To simplify the model let us assume that the substrate ligand binds only to the R-state. For the binding of another ligand to bring about activation of the enzyme it must promote a shift in the equilibrium between the R- and T-states in favour of the R-state. An activator molecule A must therefore bind to the R-state. By the same reasoning, an inhibitor molecule I must bind to the T-state and cause a shift in the equilibrium in favour of the T-state. Let us assume that there are n sites on the R-state for the activator and also n sites on the T-state for the inhibitor and that the intrinsic binding constant (association) for A is K_A, for I is K_I and for S is K_S.

The amount of protein in the T-state is

$$[T_0] + [TI_1] + [TI_2] + \cdots + [TI_n]$$

which by the same arguments outlined in the earlier model is

$$[T_0](1 + K_I[I])^n. \tag{8.7}$$

The amount of protein in the R-state is slightly more difficult to visualise. If R_0, RS_1, RS_2, $\cdots$, RS_n are the possible states containing bound substrate then for each $[RS_i]$, $i = 1,n$ and also for R_0, there can be bound $[A_j]$, $j = 1,n$ molecules of A so we have the following array of possible R-states:

$$\begin{array}{l} R_0 + R_0A_1 + R_0A_2 + \cdots + R_0A_n \\ RS_1 + RS_1A_1 + RS_1A_2 + \cdots + RS_1A_n \\ RS_2 + RS_2A_1 + RS_2A_2 + \cdots + RS_2A_n \\ \vdots \quad\quad \vdots \quad\quad \vdots \quad\quad\quad\quad \vdots \\ RS_n + RS_nA_1 + RS_nA_2 + \cdots + RS_nA_n \end{array} \tag{8.8}$$

where RS_jA_k signifies that there are j molecules of substrate and k molecules of A bound to the enzyme in the R-state. The total protein concentration is obtained by the summation of the concentrations of all the R-states in the scheme (8.8) and the T-states given in scheme (8.7). For each row in equation (8.8), we have the following equilibrium for the binding of A molecules,

$$\frac{[RS_i\ A_j]}{[RS_i\,A_{j-1}]\,[A]} = K_{jA}.$$

The first row ($i = 0$) can be expressed as

$$[R_0]\,(1 + K_{1A}[A] + K_{1A}\,K_{2A}[A]^2 + \cdots + K_{1A}\,K_{2A}\cdots K_{nA}[A]^n). \quad (8.9)$$

Using the same statistical reasoning as in the earlier example we can express each apparent binding constant K_{jA} in terms of the intrinsic binding constant K_A by

$$K_{jA} = \frac{(n-j+1)}{j}\cdot K_A.$$

Substitution for K_{jA} from this expression in equation (8.9) we obtain an equation for the first row:

$$[R_0]\,(1 + K_A[A])^n.$$

By a similar argument each row can be written as

$$[RS_i]\,(1 + K_A[A])^n, \qquad i = 1, n$$

so that the total protein in the R-state is

$$(1 + K_A[A])^n\,([R_0] + [RS_1] + [RS_2] + \cdots + [RS_n])$$

which can be written as

$$[R_0]\,(1 + K_A[A])^n\,(1 + K_S[S])^n.$$

The total protein concentration in the R- and T-states is therefore

$$[P_0] = [R_0]\,(1 + K_A[A])^n\,(1 + K_S[S])^n + [T_0]\,(1 + K_I\,[I])^n.$$

The total amount of bound substrate is the summation of all the R-states in the scheme (8.8) except those containing R_0. The number of moles of bound substrate is therefore

$$n[S]\,K_S[R_0]\,(1 + K_A[A])^n\,(1 + K_S[S])^{n-1}.$$

We can now express the binding function, r, as

$$r = \frac{nK_S[\mathrm{S}](1 + K_A[\mathrm{A}])^n (1 + K_S[\mathrm{S}])^{n-1}}{(1 + K_A[\mathrm{A}])^n (1 + K_S[\mathrm{S}])^n + L(1 + K_I[\mathrm{I}])^n}$$

or $$r = \frac{n\alpha(1+\alpha)^{n-1}}{(1+\alpha)^n + \{L(1+\beta)^n/(1+\gamma)^n\}} \tag{8.10}$$

where $\alpha = K_S[\mathrm{S}]$, $\beta = K_I[\mathrm{I}]$, $\gamma = K_A[\mathrm{A}]$ and $L = [\mathrm{T_O}]/[\mathrm{R_O}]$. The term $L(1+\beta)^n/(1+\gamma)^n$ can be regarded as an 'allosteric constant'. If this term is zero then equation (8.10) reverts to a normal hyperbolic binding function. Obviously an increase in the concentration of the activator (γ-term) will cause the binding function to become more hyperbolic whereas an increase in the concentration of the inhibition (β-term) will cause the binding function to become more sigmoidal. This is shown more clearly in fig. 8.9.

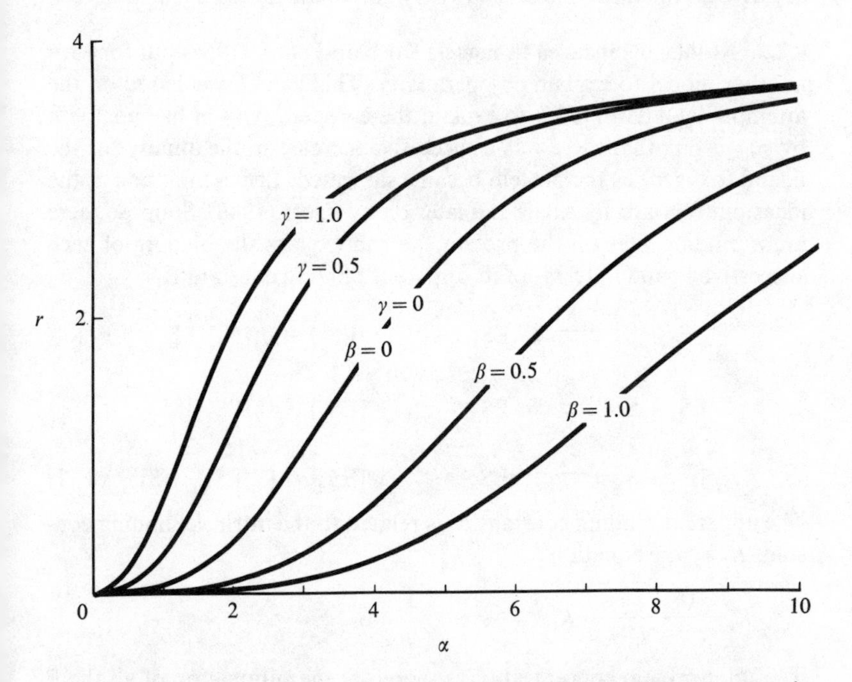

Fig. 8.9. Diagrammatic representation of the behaviour of an allosteric enzyme in the presence of effector molecules. An activator, γ-term in equation (8.10), causes the curve to become more hyperbolic whereas an inhibitor, β-term in equation (8.10), causes the curve to become more sigmoidal. ($L = 1000$, $n = 4$.)

The curves in fig. 8.9 show that the heterotropic effect of an allosteric ligand on the binding of another allosteric ligand is to modify the homotropic effect of the latter. This results in a change in shape of the binding function. An activator which binds to the R-state causes a shift in the R–T equilibrium in favour of the catalytically active R-state and consequently the binding function for the substrate becomes more hyperbolic and less co-operative. The converse is true for an inhibitor that binds to the T-state. Each ligand will exhibit homotropic co-operative effects. An earlier figure (fig. 8.4) shows a comparison of the effect of an inhibitor on a Michaelis–Menten enzyme and on an allosteric enzyme. It is clear that the allosteric enzyme is more susceptible to small changes in the concentration of the effector molecule. The Monod model only explains homotropic and heterotropic positive co-operative effects. As it stands there is no way that the model can produce reciprocal $1/r$ against $1/\alpha$ plots which are concave downwards; hence, it can not explain the negative co-operative effects shown by glutamate dehydrogenase.

8.2.2. Koshland's induced fit model. Koshland *et al.* (1966) put forward another model to explain co-operativity. This model was based on the attempts by Adair (1925) to explain the co-operativity of haemoglobin by suggesting that there was a successive increase in the affinity for the ligand (oxygen) as the protein became saturated. Let us look first at the ideas put forward by Adair and later on by Klotz (1946). Suppose there are n binding sites on the protein, we can express the binding of each successive ligand in terms of its apparent binding constant K_i'

$$
\begin{array}{lclcl}
E_0 + S & \overset{K_1'}{\rightleftharpoons} & ES_1 & \qquad & [ES_1] = K_1'[E_0][S] \\
ES_1 + S & \overset{K_2'}{\rightleftharpoons} & ES_2 & & [ES_2] = K_2'[ES_1][S] \\
\vdots & & \vdots & & \vdots \qquad \vdots \\
ES_{n-1} + S & \overset{K_n'}{\rightleftharpoons} & ES_n & & [ES_n] = K_n'[ES_{n-1}][S]
\end{array}
\qquad (8.11)
$$

The apparent binding constant K_i' is related to the intrinsic binding constant K_i by the equation

$$K_i' = \frac{(n-i+1)}{i} K_i.$$

The total protein concentration is therefore the summation of all the E complexes

$$[P_0] = [E_0] + [ES_1] + \cdots + [ES_n]$$

which can be written as

$$[P_0] = [E_0](1 + K_1'[S] + K_1' K_2'[S]^2 + \cdots + K_1' K_2' \cdots K_n'[S]^n)$$

or in terms of the intrinsic binding constants as

$$[P_0] = [E_0]\left(1 + nK_1[S] + n\frac{(n-1)}{2!}K_1 K_2[S]^2 + \cdots + \frac{n!}{n!}K_1 K_2 \cdots K_n[S]^n\right)$$

and hence

$$[P_0] = [E_0]\left[1 + \sum_{i=1}^{n} [S]^i \frac{n!}{(n-i)!\,i!} \prod_{j=1}^{i} K_j\right].$$

The total ligand bound is

$$[ES_1] + 2[ES_2] + 3[ES_3] + \cdots + n[ES_n]$$

which eventually can be written as

$$[E_0]\left[\sum_{i=1}^{n} i[S]^i \frac{n!}{(n-i)!\,i!} \prod_{j=1}^{i} K_j\right].$$

The binding function, r, is therefore

$$r = \frac{\sum_{i=1}^{n}\left[i[S]^i \frac{n!}{(n-i)!i!} \prod_{j=1}^{i} K_j\right]}{1 + \sum_{i=1}^{n}\left[[S]^i \frac{n!}{(n-i)!i!} \prod_{j=1}^{i} K_j\right]}. \tag{8.12}$$

Since at this stage we have placed no restrictions on the possible values of K_i, this model is therefore more flexible than that due to Monod. Because there are more variable parameters, using it to fit a curve to real data will be much more difficult. Let us consider four cases: (1) all K_i are identical, (2) K_i increase sequentially, (3) K_i decrease sequentially, (4) K_i can take any value. Suppose $n = 4$ for a tetrameric protein, then from equation (8.12)

$$r = \frac{4K_1[S] + 12K_1 K_2[S]^2 + 12K_1 K_2 K_3[S]^3 + 4K_1 K_2 K_3 K_4[S]^4}{1 + 4K_1[S] + 6K_1 K_2[S]^2 + 4K_1 K_2 K_3[S]^3 + K_1 K_2 K_3 K_4[S]^4}.$$

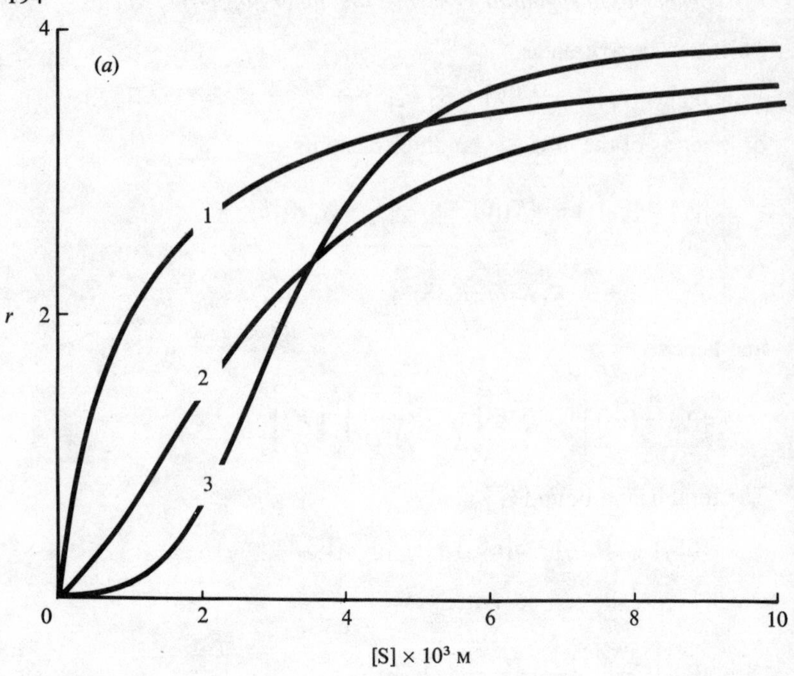
(a)
1
2
3
r
4
2
0
2
4
6
8
10
[S] × 10³ M

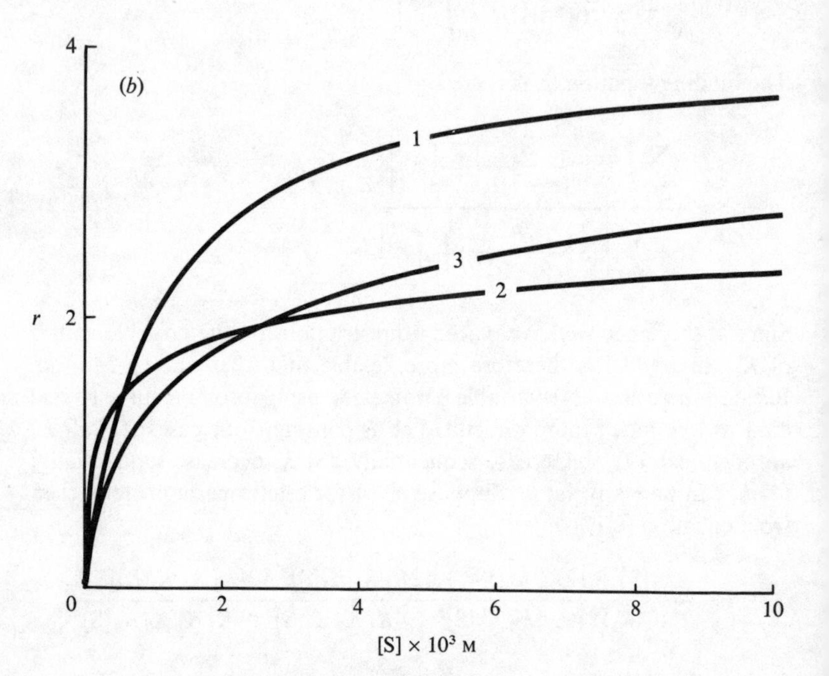
(b)
1
3
2
r
4
2
0
2
4
6
8
10
[S] × 10³ M

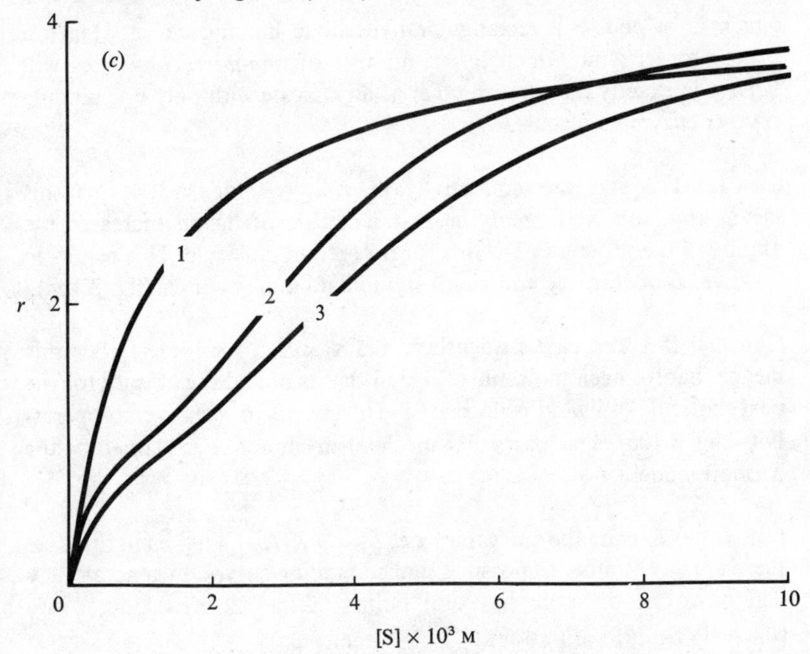

Fig. 8.10. Typical binding curves obtained using the Adair equation (equation (8.12)) for $n = 4$.

(*a*) 1, $K_1 = 10^3$, $K_2 = 10^3$, $K_3 = 10^3$, $K_4 = 10^3$.
2, $K_1 = 1.25 \times 10^2$, $K_2 = 2.5 \times 10^2$, $K_3 = 5.0 \times 10^2$, $K_4 = 10^3$.
3, $K_1 = 10^1$, $K_2 = 10^2$, $K_3 = 10^3$, $K_4 = 10^4$.

(*b*) 1, $K_1 = 10^3$, $K_2 = 10^3$, $K_3 = 10^3$, $K_4 = 10^3$.
2, $K_1 = 10^4$, $K_2 = 10^3$, $K_3 = 10^2$, $K_4 = 10$.
3, $K_1 = 10^3$, $K_2 = 5 \times 10^2$, $K_3 = 2.5 \times 10^2$, $K_4 = 1.25 \times 10^2$.

(*c*) 1, $K_1 = 10^3$, $K_2 = 10^3$, $K_3 = 10^3$, $K_4 = 10^3$.
2, $K_1 = 10^3$, $K_2 = 10^2$, $K_3 = 10^2$, $K_4 = 10^4$.
3, $K_1 = 10^3$, $K_2 = 10^2$, $K_3 = 10$, $K_4 = 10^4$.

Case (1). If all K_i are identical, the binding function becomes

$$r = \frac{4K_1[S] + 12K_1^2[S]^2 + 12K_1^3[S]^3 + 4K_1^4[S]^4}{1 + 4K_1[S] + 6K_1^2[S]^2 + 4K_1^3[S]^3 + K_1^4[S]^4}.$$

The denominator is $(1 + K_1[S])^4$ and the numerator after factorising out [S] is the derivative of the denominator. Hence

$$r = \frac{[S]4K_1(1 + K_1[S])^3}{(1 + K_1[S])^4}$$

$$r = \frac{4K_1[S]}{1 + K_1[S]},$$

which is of course a rectangular hyperbolic binding curve. Hence a protein or enzyme which has a number of *non-interacting* sites will behave in exactly the same manner as an enzyme with only one binding site per enzyme molecule.

Case (*2*). If K_i increase sequentially, i.e. $K_{i-1} < K_i$ for $i = 1, 4$, then this means that the binding of the first molecule of ligand increases the affinity of the protein or enzyme for the second molecule. This results in positive co-operativity and yields sigmoidal binding curves (fig. 8.10*a*).

Case (*3*). If K_i decrease sequentially, i.e. $K_i < K_{i-1}$ for $i = 1, 4$, then this means that for each molecule of ligand that is bound the affinity for the next molecule of ligand is decreased. This results in negative co-operativity (fig. 8.10*b*). This particular mechanism cannot be explained by the Monod model.

Case (*4*). If K_i can take any values, i.e. $K_{i-1} \neq K_i$ for all i, then in this case there are any number of possible shaped binding curves. In general, they will have one or more inflexion points (Teipel & Koshland, 1969) (fig. 8.10*c*).

Typical plots are shown in fig. 8.10(*a*),(*b*) and (*c*). The first curve in each case is a rectangular hyperbola (case (1), all K_i equal). In this scheme, based on Adair's model, we have made no assumptions about possible conformational changes or on subunit interactions that may be disallowed. For instance, in a tetrameric protein, with a square arrangement, diagonal interactions may not be allowed. In this scheme, all interactions are allowed simply by choosing the K_i values freely. It explains Michaelis–Menten behaviour, and also positive and negative co-operativity.

In Koshland's 'induced fit' model the protein is assumed to exist in two states, A and B. In the absence of a ligand the protein remains in the A conformation. Any subunit that binds a ligand changes its conformation to the B-state. Depending on the number of ligand molecules bound there will be three possible interaction constants between the subunits, K_{AA}, K_{AB} and K_{BB}. The actual interactions will depend on the arrangement of the subunits. Koshland proposed four models – tetrahedral, square, linear and concerted. It is not suggested that these are the physical arrangements of the subunits in the molecule but they are used solely to calculate the possible A–A, A–B and B–B interactions.

To describe the model we define the following expressions.

$$K_S = \frac{[\mathrm{BS}]}{[\mathrm{B}]\,[\mathrm{S}]}$$

where K_S is the intrinsic binding constant for the binding of ligand S to the subunit.

K_t represents the equilibrium constant for the conformational change $\mathrm{A} \rightleftharpoons \mathrm{B}$, hence

$$K_t = \frac{[\mathrm{B}]}{[\mathrm{A}]}.$$

The constant K_{AB} for the interaction between A and B subunits is defined as

$$K_{AB} = \frac{[\mathrm{AB}]\,[\mathrm{A}]}{[\mathrm{AA}]\,[\mathrm{B}]}$$

where [AB] and [AA] refer to the interacting subunits. Similarly

$$K_{BB} = \frac{[\mathrm{BB}]\,[\mathrm{A}]\,[\mathrm{A}]}{[\mathrm{AA}]\,[\mathrm{B}]\,[\mathrm{B}]}.$$

Consider the simplest tetrahedral model:

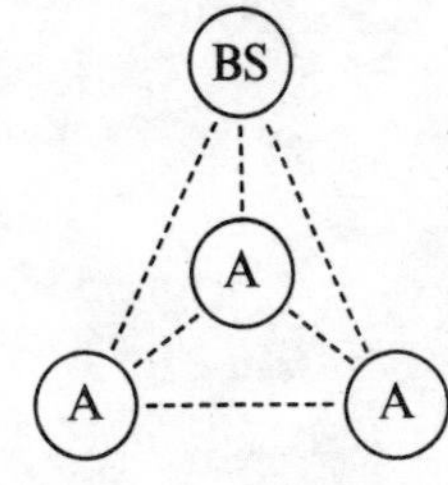

For this species A_3BS, there are four equivalent ways of binding S, there are three A–B interactions and three A–A interactions.

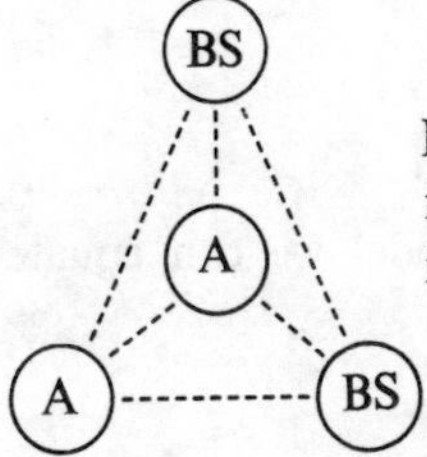

For this species $A_2B_2S_2$, there are six ways of binding S, there are four A–B interactions, one B–B interaction and one A–A interaction.

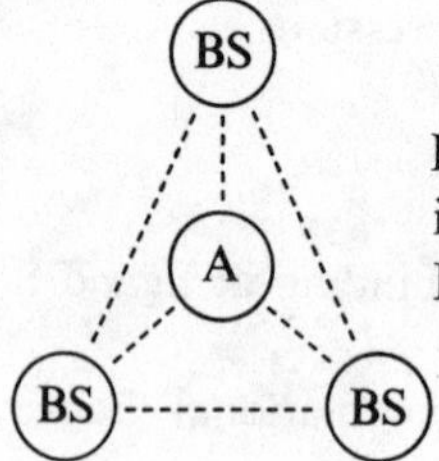

For this species AB_3S_3, there are four ways of binding S, there are three A–B interactions and three B–B interactions.

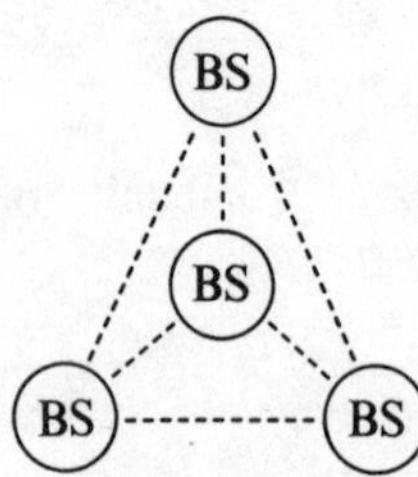

For species B_4S_4 there is only one way of binding all four molecules of S, and there are six B–B interactions.

Using the equations (8.11) with $n = 4$, we can write

$$[ES_1] = A_3\,BS = 4K_{AB}^3(K_S\,K_t[S])\,(A_4),$$

$$[ES_2] = A_2\,B_2\,S_2 = 6K_{AB}^4\,K_{BB}(K_S\,K_t[S])^2\,(A_4),$$

$$[ES_3] = AB_3\,S_3 = 4K_{AB}^3\,K_{BB}^3(K_S\,K_t[S])^3\,(A_4),$$

$$[ES_4] = B_4\,S_4 = K_{BB}^6(K_S\,K_t[S])^4\,(A_4).$$

We can write the binding function as

$$r = \frac{\sum_{i=1}^{4} i[ES_i]}{\sum_{i=0}^{4} [ES_i]}$$

and hence

$$r = \frac{\{4K_{AB}^3(K_S\,K_t[S]) + 12K_{AB}^4\,K_{BB}\,(K_S\,K_t[S])^2 + 12K_{AB}^3\,K_{BB}^3\,(K_S\,K_t[S])^3 + 4K_{BB}^6(K_S\,K_t[S])^4\}}{\{1 + 4K_{AB}^3(K_S\,K_t[S]) + 6K_{AB}^4\,K_{BB}(K_S\,K_t[S])^2 + 4K_{AB}^3\,K_{BB}^3(K_S\,K_t[S])^3 + K_{BB}^6(K_S\,K_t\,[S])^4\}}$$

Comparing this with the more general Adair model we can equate coefficients in [S] and hence

$$K_1 = K_{AB}^3\,K_S\,K_t,$$

$$K_1\,K_2 = K_{AB}^4\,K_{BB}\,K_S^2\,K_t^2 \;\therefore\; K_2 = K_{AB}\,K_{BB}\,K_S\,K_t,$$

$$K_1 K_2 K_3 = K_{AB}^3 K_{BB} K_S^3 K_t^3 \therefore K_3 = \frac{K_{BB}^2}{K_{AB}} K_S K_t,$$

$$K_1 K_2 K_3 K_4 = K_{BB}^6 K_S^4 K_t^4 \therefore K_4 = \frac{K_{BB}^3}{K_{AB}^3} K_S K_t.$$

In contrast to the Adair model where the values of K_i could be chosen freely, the tetrahedral model of Koshland places certain limitations on the values of K_1–K_4 due to the appearance of K_{AB}, K_S and K_t in each term and K_{BB} in three terms. Therefore it is no longer possible to describe

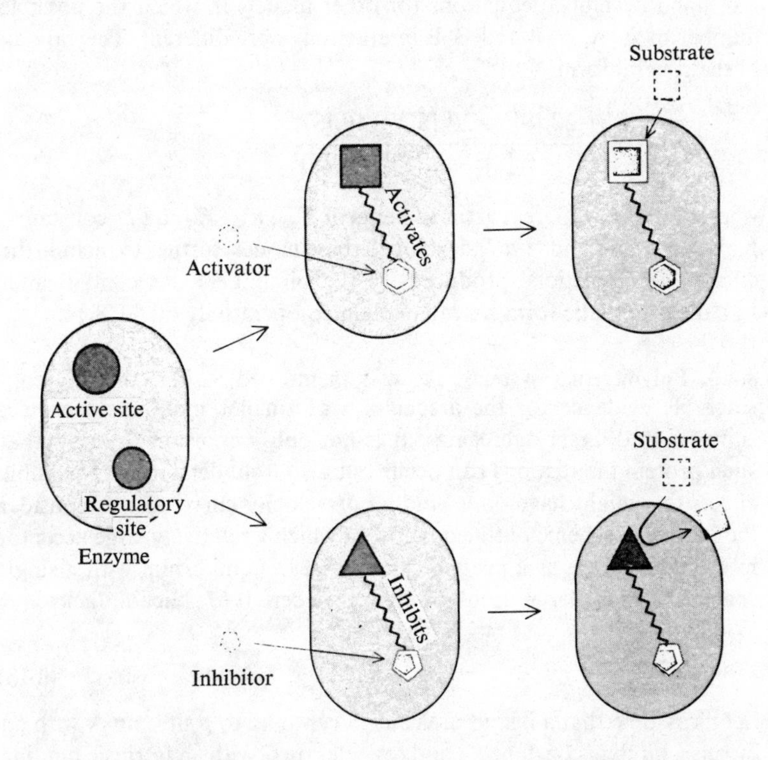

Fig. 8.11. Diagrammatic representation of the controlling effect of small regulatory molecules on the catalytic activity of an enzyme. An activator, binding to the regulatory site, induces a conformational change which favours the binding of the substrate. Conversely, an inhibitor induces a conformational change which inhibits the binding of the substrate. Positive co-operativity, in a multi-site enzyme is explained in a like manner by the binding of a substrate molecule at one active site inducing a conformational change which favours the binding of another substrate at a second active site. (Adapted after Koshland (1973).)

mixed co-operativity with this particular model. If $K_{AB} = 1$ and $K_{BB} = 1$ then this model shows a hyperbolic saturation curve. If $K_{AB} > 1$, $K_{AB} > K_{BB}$, then $K_{i-1} > K_i$, and the model shows negative co-operativity whereas if $K_{AB} < 1$ and $K_{AB} > K_{BB}$, both positive and negative co-operativity can be shown depending on the exact values of K_{AB} and K_{BB}. Each constant K_i is related to the previous one K_{i-1} by the equation

$$K_{i-1} = K_i \frac{K_{AB}^2}{K_{BB}}.$$

Koshland developed equations for other models in which the possible number of A–A, A–B and B–B interactions were different. They are all of the general form

$$r = \frac{\theta_1[S] + 2\theta_2[S]^2 + 3\theta_3[S]^3 + 4\theta_4[S]^4}{1 + \theta_1[S] + \theta_2[S]^2 + \theta_3[S]^3 + \theta_4[S]^4}$$

where the θ_i are different arrangements of K_{AB}, K_{BB}, K_S and K_t constants. Kirtley & Koshland (1967) developed these models further to include the effects of interactions produced by the binding of a second ligand. In a diagrammatic form we can explain co-operativity by fig. 8.11.

8.2.3. Polymerising systems. As was mentioned earlier, there is considerable evidence for the association of smaller monomeric protein subunits into larger aggregates. It is not only important to know that such protein interactions can occur but also to understand the possible effects this might have on a binding or velocity curve. Let us consider the following scheme outlined below in which a relatively large acceptor molecule, A, such as a protein can coexist in equilibrium with a single species, C, of higher molecular weight (Frieden, 1967; Nichol, Jackson & Winzor, 1967):

$$nA \rightleftharpoons C. \qquad (8.13)$$

Let us assume that a ligand molecule S can bind to p sites on A with an intrinsic binding constant K_A and to q sites on C with an intrinsic binding constant L_C. At equilibrium there will coexist in solution a series of complexes AS_i and CS_j where $i = 1,p$ and $j = 1,q$. For each species AS_i and CS_j, we can write an equilibrium equation:

$$K_i = \frac{[AS_i]}{[AS_{i-1}][S]}, \qquad L_j = \frac{[CS_j]}{[CS_{j-1}][S]}.$$

Each K_i and L_j are related to their respective intrinsic binding constants by the equations

$$K_i = \left(\frac{p-i+1}{i}\right) K_A, \qquad L_j = \left(\frac{q-j+1}{j}\right) L_C.$$

The total protein concentration is given by

$$[\mathrm{P_0}] = \sum_{i=0}^{p} [\mathrm{AS}_i] + n \sum_{j=0}^{q} [\mathrm{CS}_j].$$

By a similar argument to that outlined in the Monod model (section 8.2.1) this becomes

$$[\mathrm{P_0}] = [\mathrm{A_0}](1 + K_A[\mathrm{S}])^p + n[\mathrm{C_0}](1 + L_C[\mathrm{S}])^q \tag{8.14}$$

where $[\mathrm{A_0}]$ and $[\mathrm{C_0}]$ are the concentrations of free A and free C. The total concentration of substrate bound is given by the summation of the terms $[\mathrm{AS}_i]$ and $[\mathrm{CS}_j]$ so that

$$[\mathrm{S}]_{\mathrm{Bound}} = [\mathrm{A_0}]pK_A[\mathrm{S}](1 + K_A[\mathrm{S}])^{p-1} + [\mathrm{C_0}]qL_C[\mathrm{S}](1 + L_C[\mathrm{S}])^{q-1}.$$

The binding function r is therefore

$$r = \frac{[\mathrm{A_0}]pK_A[\mathrm{S}](1 + K_A[\mathrm{S}])^{p-1} + [\mathrm{C_0}]qL_C[\mathrm{S}](1 + L_C[\mathrm{S}])^{q-1}}{[\mathrm{A_0}](1 + K_A[\mathrm{S}])^p + n[\mathrm{C_0}](1 + L_C[\mathrm{S}])^q}.$$

If we use the transforms $\alpha = K_A[\mathrm{S}]$, $\beta = L_C/K_A$, this becomes

$$r = \frac{p[\mathrm{A_0}]\alpha(1 + \alpha)^{p-1} + q[\mathrm{C_0}]\beta\alpha(1 + \beta\alpha)^{q-1}}{[\mathrm{A_0}](1 + \alpha)^p + n[\mathrm{C_0}](1 + \beta\alpha)^q}. \tag{8.15}$$

We can now apply this general binding equation to more specific cases.

A single molecular species. By putting $[\mathrm{A_0}]$ or $[\mathrm{C_0}]$ equal to zero the physical situation described is that of a single macromolecule binding a ligand to p (or q) equivalent sites. A plot of r versus α is a hyperbola described by the equation:

$$r = \frac{p\alpha}{1+\alpha} \quad ([\mathrm{C_0}] = 0) \text{ (Klotz, 1946)}. \tag{8.16}$$

An isomerising system. By substituting $n = 1$ in equation (8.13), the model simplifies to an isomerising system with an equilibrium (or isomerisation) constant $L = [C_0]/[A_0]$. Equation (8.15) can therefore be written as

$$r = \frac{p\alpha(1+\alpha)^{p-1} + qL\beta\alpha(1+\beta\alpha)^{q-1}}{(1+\alpha)^p + L(1+\beta\alpha)^q}.$$

This is analogous to equation (8.6) produced by Monod *et al.* (1965) except that their model assumed that there were no binding sites lost or gained on isomerisation, i.e. $p = q = n$ in equation (8.6). The binding curves for this model with $p = q$ and $n = 1$ have already been described (fig. 8.3).

A polymerising system ($n > 1$). If there is no loss of binding sites on polymerisation, $q = np$; if the intrinsic binding constants are the same for the monomer and polymer ($\beta = 1$) then equation (8.15) reverts to equation (8.16) and the binding curve is a rectangular hyperbola for all values of n. If the intrinsic binding constants are different then we can consider two possible cases.

(*a*) The binding and polymerisation sites are identical or nearly so and hence polymerisation leads to a loss of binding sites and vice versa. Binding of the ligand and polymerisation are thus *competitive*. Polymerisation can therefore only occur between A_0 and C_0 and not between any species AS_i or CS_j. The polymerisation process can be expressed in terms of an equilibrium constant L where

$$L = \frac{[C_0]}{[A_0]^n}. \tag{8.17}$$

Using this equation we can substitute for $[C_0]$ in equation (8.15) and obtain the equation

$$r = \frac{p\alpha(1+\alpha)^{p-1} + L[A_0]^{n-1}q\beta\alpha(1+\beta\alpha)^{q-1}}{(1+\alpha)^p + nL[A_0]^{n-1}(1+\beta\alpha)^q}. \tag{8.18}$$

Note that the binding equation is now dependent on the concentration of *free* acceptor. The solution for $[A_0]$ can be obtained by substituting equation (8.17) in equation (8.14) and solving the polynomial in $[A_0]$:

$$L(1+\beta\alpha)^q[A_0]^n + (1+\alpha)^p[A_0] - [P_0] = 0. \tag{8.19}$$

For $n = 2$, i.e. a monomer–dimer system, equation (8.19) is a quadratic with one positive root for each value of α. A plot of r against α for different values of n is shown in fig. 8.12. As n increases, the binding curves

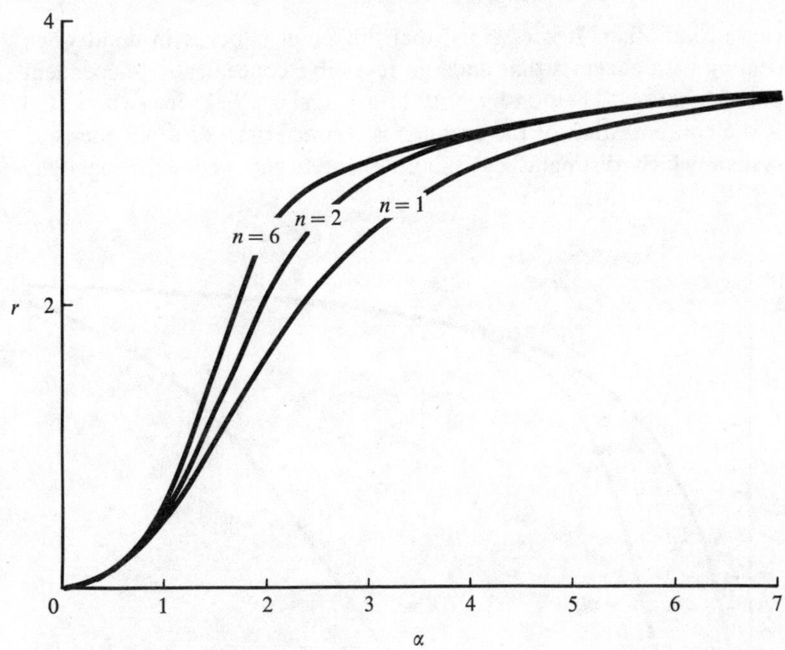

Fig. 8.12. Plots of r against α illustrating the effect of changing n, the polymerisation number. In each case, $p = 4, q = 0, \beta = 0$ and $[P_0] = 10^{-6}$ M (base molar concentration). Values of L were chosen so that the molar ratio of non-binding polymer to monomer, in the absence of substrate, was 50:1.

exhibit increasing positive co-operativity and hence polymerisation is a more effective control mechanism for a regulatory enzyme than isomerisation. Fig. 8.13 shows the effect of increasing the polymerisation constant, L, for a monomer–dimer system ($n = 2$). Sigmoidality is more pronounced for high values of L which favour the formation of the non-binding ($q = 0$) dimeric form of the acceptor molecule. One important aspect of a binding curve for a polymerising system that needs to be remembered when applying equation (8.18) to either kinetic or binding studies stems from the presence of $[A_0]^{n-1}$ in the equation. If one carried out a series of experiments at an enzyme concentration $[P_0]$ and then for some reason (possibly due to limitations of the experimental apparatus) changed the concentration of the enzyme to a different value $[P_0]'$, then this would affect the polymerisation equilibrium. Since the shape of the binding curve (fig. 8.14) depends on the total concentration of enzyme, changing this would result in a jump from one v against [S] curve to another. It is essential that this point is borne in mind when dealing with enzymes that undergo reversible concentration-dependent polymerisation. The non-linearity of a v against $[E_0]$ plot (where $[E_0]$ is the concentration of the enzyme) is characteristic of a polymerising system which distinguishes it both theoretically and experimentally

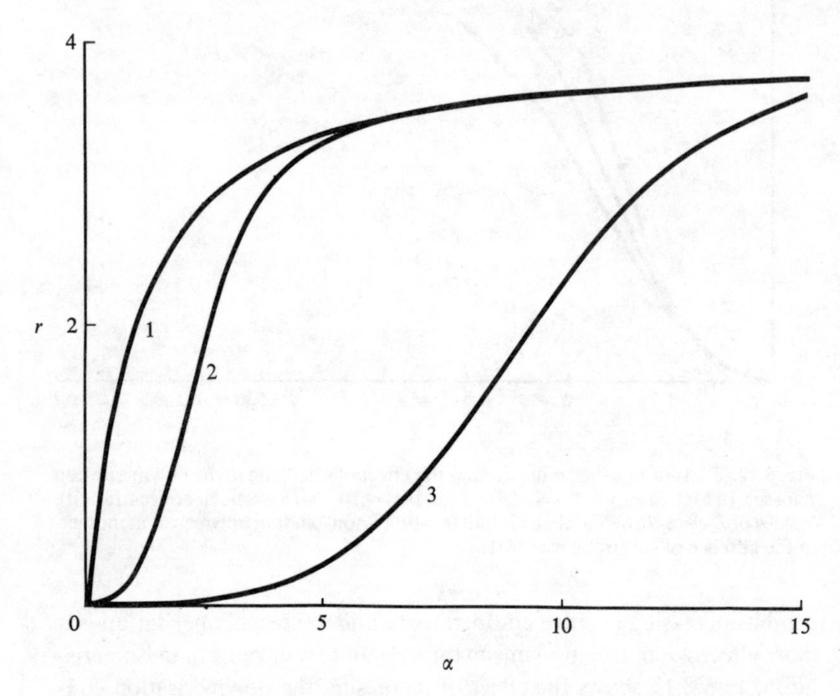

Fig. 8.13. The effect of changing the polymerisation constant (L) upon the binding curves for the scheme described by equation (8.18). In each case $n = 2$, $p = 4$, $q = 0$, $\beta = 0$ and $[P_0] = 10^{-6}$ M. Values of L were chosen as follows: 1, 10^6 M^{-1}; 2, 10^{10} M^{-1}; 3, 10^{14} M^{-1}.

from an isomerising system, from an enzyme obeying simple Michaelis–Menten kinetics and from one whose allosteric nature depends on conformational changes (Koshland's model).

(*b*) If the polymerisation and binding sites are entirely independent, polymerisation of the acceptor molecule can take place through any species, $AS_i (i = 1,p)$, to give a polymeric form CS_j where $j = 1,q$ and $q = np$. The polymerisation and binding processes are therefore *non-*

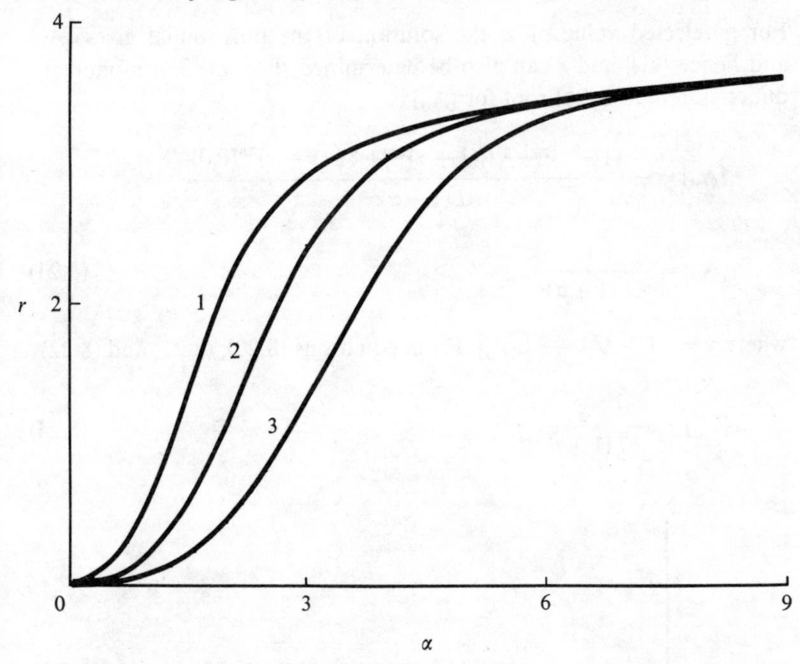

Fig. 8.14. The effect of changing the protein concentration upon the binding curves for a monomer–dimer system. In each case $L = 10^{10}$ M^{-1}, $p = 4$, $q = 0$, $\beta = 0$, $n = 2$. The base molar protein concentration is: 1, 10^{-7} M; 2, 10^{-6} M; and 3, 10^{-5} M.

competitive and we can write an equilibrium equation for the polymerisation in terms of the total concentration of A and C:

$$\bar{L} = \frac{[\bar{C}_0]}{[\bar{A}_0]^n} \tag{8.20}$$

where

$$[\bar{A}_0] = [A_0](1 + \alpha)^p \tag{8.21}$$

and $$[\bar{C}_0] = [C_0](1 + \beta\alpha)^{np}. \tag{8.22}$$

Combining equation (8.14) with these last three equations gives

$$n\bar{L}(1 + \alpha)^{np}[A_0]^n + (1 + \alpha)^p[A_0] - [P_0] = 0.$$

For a selected value of α, the solution of the polynomial gives $[A_0]$ and hence $[C_0]$ and r can also be determined. For $n = 2$, a monomer–dimer system, the real root for $[A_0]$ is

$$[A_0] = -\frac{(1+\alpha)^p + \sqrt{\{(1+\alpha)^{2p} + 8\bar{L}(1+\alpha)^{2p}[P_0]\}}}{4\bar{L}(1+\alpha)^{2p}}$$

$$[A_0] = \frac{y}{4\bar{L}(1+\alpha)^p} \tag{8.23}$$

where $y = -1 + \sqrt{1 + 8\bar{L}[P_0]}$. From equations (8.20), (8.21) and (8.22)

$$[C_0] = \frac{y^2}{16\bar{L}(1+\beta\alpha)^{2p}} \tag{8.24}$$

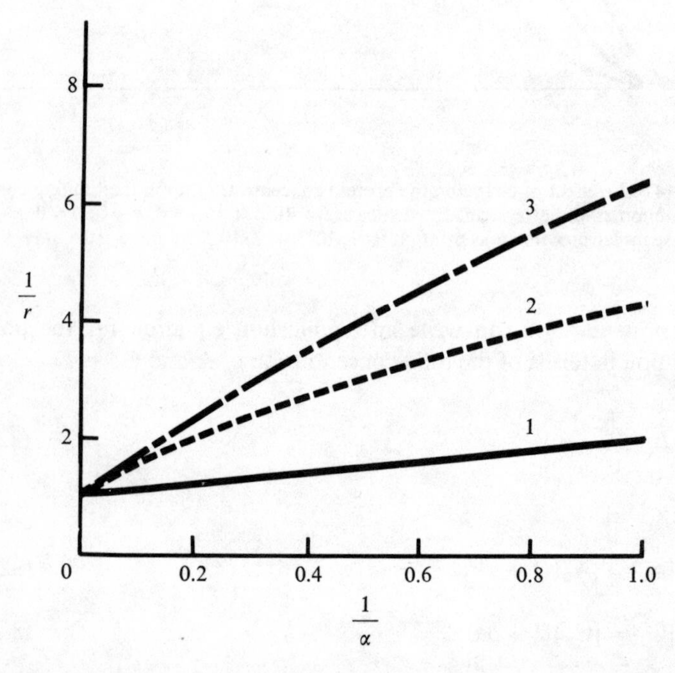

Fig. 8.15. Double reciprocal plots of binding curves for the non-competitive case where $n = 2$ (monomer–dimer system), $p = 1$ and $q = 2$. Line 1 refers to the case where $\beta = 1$ (hyperbolic), the broken lines to $\beta = 0.1$ (2, $L[P_0] = 6 \times 10^4$; 3, $L[P_0] = 6 \times 10^5$). (Adapted from Nichol, Jackson & Winzor (1967) *Biochem.* **6**, 2449–56. © American Chemical Society.)

and hence substitution for $[A_0]$ and $[C_0]$ from equations (8.23) and (8.24) in equation (8.13) gives a binding equation

$$r = \frac{p\alpha y}{4\bar{L}[P_0]}\left[\frac{1}{1+\alpha} + \frac{2y\beta}{4(1+\beta\alpha)}\right].$$

A plot of $1/r$ against $1/\alpha$ for this system is shown in fig. 8.15. The lines are concave downwards and hence this system can show negative co-operative effects.

So far we have only considered the binding of a single ligand to a polymerising system and obviously this model will only explain homotropic effects. An essential modification to our model would be to allow the binding of an effector ligand (E) to the monomer–polymer as outlined below (Nichol, O'Dea & Baghurst, 1972).

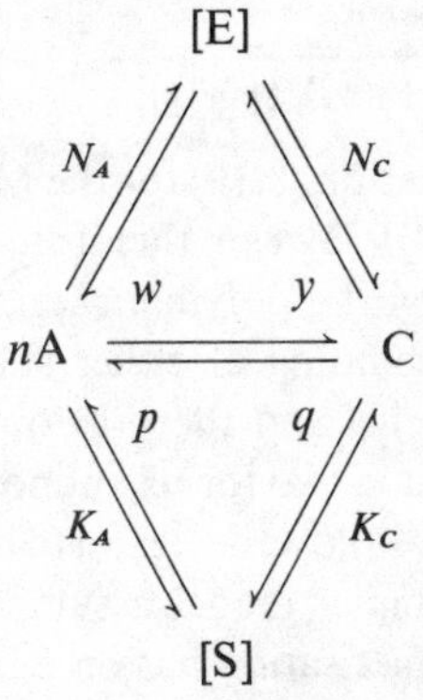

Let us assume that there are w sites on A for the effector molecule and these sites have an intrinsic binding constant N_A. Similarly there are y sites on C for E and these sites have an intrinsic binding constant N_C. There are p sites for S on A with a binding constant K_A and q sites for S on C with a binding constant K_C. In this case we can derive the following equation for total protein:

$$[P_0] = [A_0](1+K_A[S])^p(1+N_A[E])^w + n[C_0](1+K_C[S])^q(1+N_C[E])^y,$$

where $[A_0]$ and $[C_0]$ refer to free A and free C. The total amount of bound substrate is

$$[A_0]pK_A[S](1+K_A[S])^{p-1}(1+N_A[E])^w + q[C_0]K_C[S](1+K_C[S])^{q-1}(1+N_C[E])^y.$$

The derivation is similar to that outlined in the Monod model (equations (8.8) to (8.10)). The binding function for S is therefore.

$$r = \frac{\{[A_0]pK_A[S](1+K_A[S])^{p-1}(1+N_A[E])^w + q[C_0]K_C[S](1+K_C[S])^{q-1}(1+N_C[E])^y\}}{\{[A_0](1+K_A[S])^p(1+N_A[E])^w + n[C_0](1+K_C[S])^q(1+N_C[E])^y\}}$$

$$r = \frac{p\alpha(1+\alpha)^{p-1}(1+\gamma)^w + qL[A_0]^{n-1}\beta\alpha(1+\beta\alpha)^{q-1}(1+\gamma\delta)^y}{(1+\alpha)^p(1+\gamma)^w + nL[A_0]^{n-1}(1+\beta\alpha)^q(1+\gamma\delta)^y}, \tag{8.25}$$

where $\alpha = K_S[S]$, $\beta = K_C/K_A$, $\gamma = N_A[E]$, $\delta = N_C/N_A$ and $L = [C_0]/[A_0]^n$. For the case of $n = 1$ and $q = 0$ equation (8.25) becomes

$$r = \frac{p\alpha(1+\alpha)^{p-1}}{(1+\alpha)^p + \{L(1+\gamma\delta)^y/(1+\gamma)^w\}}$$

which is analogous to equation (8.10) for the Monod model. This second model of Nichol *et al.* (1972) can therefore explain the heterotropic allosteric effects caused by the polymerisation of the protein molecule which is induced by the binding of either substrate molecule or some other effector ligand. In this and the previous example, we have only allowed the protein to exist in two forms, monomer and *n*-mer; obviously it would be more realistic to allow the ligand to bind to all the intermediate polymeric species. The simpler model is sufficient, however, to demonstrate the effect that polymerisation has on a binding curve.

8.2.4. Active site competitive binding model. In the models that have been described, it is postulated that the effector molecules bind to sites on the enzyme which are distinct from the substrate binding (active) site, and with certain enzymes this is undoubtedly the case. Thus with aspartate transcarbamylase, it can be shown that the effector molecules ATP and CTP bind to a site that is on a different polypeptide chain from the catalytic binding site. With other enzymes, however, the similarity between the chemical structures of the substrate and effector molecules suggests that both may compete for the same binding sites on the enzyme molecules. The following model (Smith, Roberts & Kuchel, 1975) can be used to interpret the behaviour of allosteric enzymes in the presence of active-site-directed effector molecules, compounds that for simple Michaelis–Menten enzymes would be regarded as competitive inhibitors.

Consider an oligomeric protein E which exists in solution in two forms, E_1 and E_2, in equilibrium,

$$E_1 \rightleftharpoons E_2.$$

$X = [E_2]/[E_1]$, where X is the equilibrium constant. In solution there are two ligands, S and L, which compete for the same binding sites on the E_1 and E_2 forms of the protein. Ligand S binds to n_1 and n_2 sites on E_1 and E_2, with intrinsic association constants K_{1S} and K_{2S} respectively. Similarly, ligand L binds to the same n_1 sites on E_1 and the same n_2 sites on E_2 with respective association constants K_{1L} and K_{2L}. In a solution of E, S and L there will exist unbound S and L and a series of complexes $E_h S_i L_j$, where $0 \leqslant i+j \leqslant n_h$, $h = 1, 2$. The intrinsic association constants for each ligand can be expressed as

$$K_{hS} = [E_h S_i L_j]/[E_h S_{i-1} L_j][S]; \qquad h = 1, 2$$
$$K_{hL} = [E_h S_i L_j]/[E_h S_i L_{j-1}][L]; \qquad h = 1, 2$$

with a statistical weighting factor given by (Nichol, Smith & Ogston, 1969)

$$w_{ij} = n_h!/i!j!(n_h - i - j)!; \quad h = 1, 2.$$

The binding of each ligand can be expressed in terms of a molar binding function, thus

$$r_S = ([S_0] - [S])/[E_0]$$

and $r_L = ([L_0] - [L])/[E_0]$

where [S] and [L] are the molar concentrations of unbound S and L and the subscript 0 refers to the total molar concentration of the particular species. Using the same statistical reasoning outlined in the earlier models, it can be shown that

$$r_S = \frac{\{n_1 K_{1S}[S](1 + K_{1S}[S] + K_{1L}[L])^{n_1-1} + n_2 XK_{2S}[S](1 + K_{2S}[S] + K_{2L}[L])^{n_2-1}\}}{(1 + K_{1S}[S] + K_{1L}[L])^{n_1} + X(1 + K_{2S}[S] + K_{2L}[L])^{n_2}}. \tag{8.26}$$

It may be noted that with $X = 0$, equation (8.26) reduces to

$$r_S = \frac{n_1 K_{1S}[S]}{(1 + K_{1S}[S] + K_{1L}[L])},$$

which in terms of Michaelis constants (i.e. reciprocal binding constants) becomes

$$r_S = \frac{n_1[\mathrm{S}]}{[\mathrm{S}] + K_S\{1 + ([\mathrm{L}]/K_L)\}}, \tag{8.27}$$

where $K_S = 1/K_{1S}$ and $K_L = 1/K_{1L}$. Equation (8.27) is analogous to equation (3.1) describing competitive inhibition of a simple Michaelis–Menten type enzyme.

The various types of behaviour which may be described by equation (8.26) are best illustrated by numerical examples. When S binds preferentially to $E_1(K_{1S} > K_{2S})$ and E_2 predominates initially (large X), the

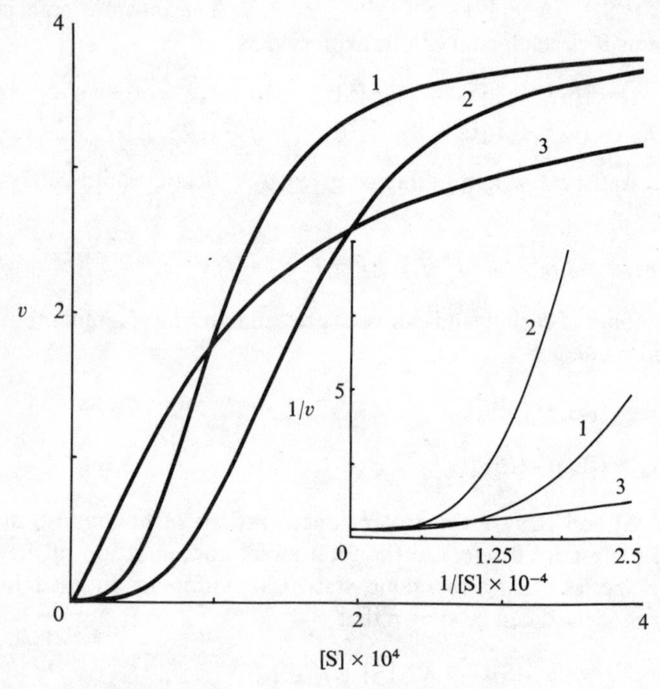

Fig. 8.16. Plots of initial velocity, v, against substrate concentration, [S], and corresponding double reciprocal plots (inset) calculated using equation (8.33), the kinetic version of the binding equation (8.26), which can, in this case, be compared directly with the binding equation by equating $r = v$ and expressing the kinetic constants (K_{1S}, K_{2S}, etc.) as *association* constants. The plots refer to an enzyme with four catalytic sites on each of two alternative conformations ($n_1 = n_2 = 4$) and with the following parameters; $K_{1S} = 5 \times 10^4$, $K_{2S} = 5 \times 10$, $X = 10^3$. The additional parameters for each curve are: Curve 1, $[\mathrm{L}] = 0$; Curve 2, $K_{1L} = 5 \times 10^2$, $K_{2L} = 5 \times 10^3$, $[\mathrm{L}] = 10^{-4}$ M; Curve 3, $K_{1L} = 5 \times 10^4$, $K_{2L} = 5 \times 10$, $[\mathrm{L}] = 8 \times 10^{-5}$ M. (From Smith *et al.* (1975) *Biochim. Biophys. Acta*, **377**, 197–202.)

plot of v against [S] obtained in the absence of L is sigmoidal in form as shown by curve 1 in fig. 8.16. If L also binds preferentially to $E_1(K_{1L} > K_{2L})$ then its addition results in behaviour as shown in curve 3, in which activation is observed at low [S] and inhibition at high [S]. The presence or absence of cross-over is dependent on the magnitudes of the various parameters; it can be shown mathematically that such a cross-over is never exhibited by the 'alternate effector site' model where the basic equation is of different form (e.g. equation (8.25)) and the observation of cross-over excludes the latter model. In contrast, if L were to bind preferentially to E_2, inhibition over the entire range of [S] would be observed and the curve would be more sigmoidal as shown in curve 2 of fig. 8.16. Both types of behaviour have been observed with the enzyme deoxycytidylate aminohydrolase (Scarano, Geraci & Rossi, 1967)

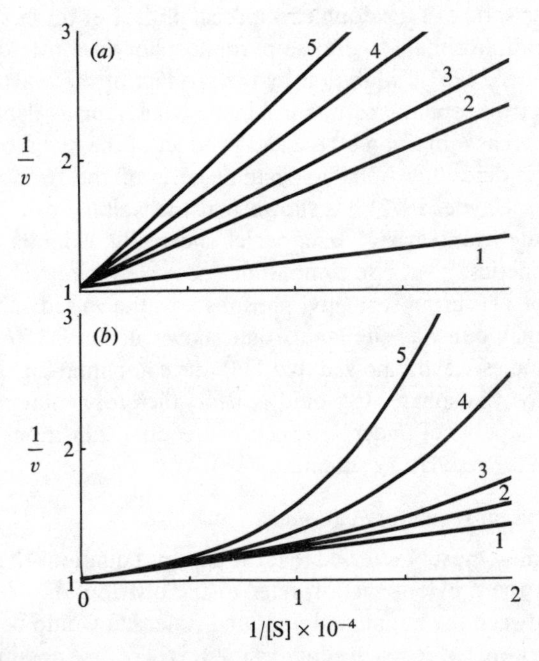

Fig. 8.17. Double reciprocal plots, again using equation (8.33), which can be compared directly with the binding equation by the transformation $v = Vr/n$ and expressing the kinetic constants (K_{1S}, K_{2S} etc.) as association constants. In this case the parameters used were: $n = 4$, $V = 0.967$, $X = 0.1$, $K_{1S} = 5.55 \times 10^4$, $K_{2S} = 1.66 \times 10^3$. For (*a*), $K_{1L} = 9.09 \times 10^4$, $K_{2L} = 2.72 \times 10^3$; and for (*b*), $K_{1L} = 2.72 \times 10^3$, $K_{2L} = 9.09 \times 10^4$. The molar concentrations of unbound effector, [L], are: 1, 0; 2, 3×10^{-5} M; 3, 4×10^{-5} M; 4, 6×10^{-5} M; and 5, 8×10^{-5} M. (From Smith *et al.* (1975) *Biochim. Biophys. Acta* **377**, 197–202.)

which is activated (at low [S]) by deoxycytidine triphosphate and inhibited by deoxythymidine triphosphate at all substrate concentrations. In view of the structural similarity of the effector molecules to the substrate (deoxycytidine monophosphate) and the kinetic similarity to that shown in fig. 8.16, the competitive binding model seems more likely than the alternate site model.

An alternative situation is where S binds preferentially to the predominant form E_1 (small X). In the absence of L the plot of v against [S] is essentially hyperbolic, i.e. essentially linear in double reciprocal form (curves 1 in fig. 8.17). If L also binds preferentially to E_1, the form of the plots resembles classical linear competitive inhibition (fig. 8.17*a*). Addition of L, however, when L binds preferentially to E_2, results in plots of v against [S] which become increasingly sigmoidal as the concentration of L is increased. This is illustrated in fig. 8.17(*b*) by the increasing curvature of the double reciprocal plots. For the enzyme deoxythymidine diphosphate-D-glucose pyrophosphorylase (Melo & Glaser, 1965; Frieden, 1967), inhibition by the product of the reaction (deoxythymidine triphosphate) results in kinetic behaviour as depicted in fig. 8.17(*a*) whereas inhibition by the end-product of the metabolic pathway (deoxythymidine diphosphate-L-rhamnose) is of the form depicted in fig. 8.17(*b*). Heyde (1973) has shown that the isolated catalytic subunit of aspartate transcarbamylase, which normally exhibits Michaelis–Menten kinetics, gives rise to non-linear double reciprocal plots in the presence of ITP using carbamyl phosphate as the varied substrate. The kinetic behaviour was similar to that shown in fig. 8.17(*b*) and from binding studies Heyde showed that ITP was competing for the same site as carbamyl phosphate. It would appear, therefore, that the catalytic subunit is capable of undergoing conformational transitions even in the absence of the regulatory subunit.

8.3. Kinetics of regulatory enzymes

At this point it must be stressed that the binding equations that have been derived in the previous section refer to the distribution of a ligand or ligands between the bound and unbound state that would be obtained if the system were left to reach equilibrium. It is of course possible to obtain the free energy changes for the various isomerisation or polymerisation processes from their respective equilibrium constants ($\Delta G° = -RT \ln K$) but this gives no indication of the rate at which the various processes occur. If the ligand is a substrate, then the binding function is simply the ratio of the substrate bound in the form of ES_i complexes to the total enzyme concentration. Obviously the rate of the reaction must depend on the amount of bound substrate but the binding equation as it stands cannot be converted into a kinetic equation without making some assumptions.

Firstly, we have to consider how fast the isomerisation, conformational or polymerisation changes take place. These changes may be (1) very fast, (2) of a similar rate or (3) much slower than the catalytic step in the mechanism. Let us consider the first case in which these changes are very rapid. Then at any instant in the enzyme reaction we can regard them as being in equilibrium with the rest of the system. The initial velocity, v_0, of a reaction catalysed by an enzyme with a single binding site is given by $v_0 = k[\mathrm{ES}]$ where k is the first order rate constant for the breakdown of ES into products. For the case of an enzyme with a number of binding sites the overall rate v_0 will be the sum of the individual rates of breakdown for each ES_i complex. Hence the rate v_0 can be expressed as

$$v_0 = \sum_{i=1}^{n} ik_i[\mathrm{ES}_i].$$

If it is assumed that the rate of equilibrium between enzyme and substrate is rapid relative to the overall rate of catalysis then the ES_i can be related to the substrate concentration by the various equilibrium expressions. In view of the known rate constants for the initial binding step, $\mathrm{E} + \mathrm{S} \rightarrow \mathrm{ES}$, for some enzymes ($10^7$–$10^9$ s^{-1}; pseudo first order rate constants), the initial binding steps are likely to equilibrate rapidly and this assumption is therefore reasonably justified. Let us consider the Monod model when $n = 4$; the various enzyme substrate complexes will be RS_i and TS_i for the R- and T-states respectively. We can replace the intrinsic association constants K_R for the R-state by the individual microscopic rate constants for the forward direction k_1 and for the reverse direction k_{-1}. Similarly for the T-state, K_T can be replaced by l_1 and l_{-1} for the forward and reverse directions. Each apparent binding constant K_i for the R-state is related to the intrinsic binding constant K_R by the statistical factor $(n - i + 1)/i$ and hence when $n = 4$

$$K_1 = 4K_R = 4k_1/k_{-1},$$

$$K_2 = \tfrac{3}{2}K_R = 3k_1/2k_{-1},$$

$$K_3 = \tfrac{2}{3}K_R = 2k_1/3k_{-1},$$

$$K_4 = K_R/4 = k_1/4k_{-1}.$$

There is a similar set of expressions for the T-state. We can therefore write the Monod model in the following kinetic form (Dalziel, 1968):

$$R_0 \rightleftharpoons T_0$$

$$R_0 + S \underset{k_{-1}}{\overset{4k_1}{\rightleftharpoons}} RS_1 \xrightarrow{k_2} P \qquad T_0 + S \underset{l_{-1}}{\overset{4l_1}{\rightleftharpoons}} TS_1 \xrightarrow{l_2} P$$

$$RS_1 + S \underset{2k_{-1}}{\overset{3k_1}{\rightleftharpoons}} RS_2 \xrightarrow{2k_2} P \qquad TS_1 + S \underset{2l_{-1}}{\overset{3l_1}{\rightleftharpoons}} TS_2 \xrightarrow{2l_2} P$$

$$RS_2 + S \underset{3k_{-1}}{\overset{2k_1}{\rightleftharpoons}} RS_3 \xrightarrow{3k_2} P \qquad TS_2 + S \underset{3l_{-1}}{\overset{2l_1}{\rightleftharpoons}} TS_3 \xrightarrow{3l_2} P$$

$$RS_3 + S \underset{4k_{-1}}{\overset{k_1}{\rightleftharpoons}} RS_4 \xrightarrow{4k_2} P \qquad TS_3 + S \underset{4l_{-1}}{\overset{l_1}{\rightleftharpoons}} TS_4 \xrightarrow{4l_2} P$$

The microscopic rate constants for the breakdown of the RS_i and TS_i species into products are ik_2 and il_2 respectively. It is also convenient to replace the binding constants by the Michaelis constants defined by the steady-state treatment of Briggs & Haldane as $K_m^R = (k_2 + k_{-1})/k_1$ for the R-state and $K_m^T = (l_2 + l_{-1})/l_1$ for the T-state. Since we have assumed that there is a rapid equilibrium between R_0 and T_0 then we can put $L = [T_0]/[R_0]$. The initial velocity will be the sum of the rates of breakdown into products of the complexes RS_i and TS_i and hence

$$v = k_2(RS_1 + 2RS_2 + 3RS_3 + 4RS_4) + l_2(TS_1 + 2TS_2 + 3TS_3 + 4TS_4).$$

By algebraic manipulation similar to that used to derive equation (8.6), we obtain

$$v = \frac{4[E_0]\left(\frac{k_2[S]}{K_m^R}\left\{1 + \frac{[S]}{K_m^R}\right\}^3 + L\frac{l_2[S]}{K_m^T}\left\{1 + \frac{[S]}{K_m^T}\right\}^3\right)}{\left(1 + \frac{[S]}{K_m^R}\right)^4 + L\left(1 + \frac{[S]}{K_m^T}\right)^4}.$$

For the general case of n sites and using the transforms $\alpha = [S]/K_m^R$ and $c = K_m^R/K_m^T$

$$v = \frac{n[E_0]\,(k_2\,\alpha(1+\alpha)^{n-1} + Ll_2\,c\alpha(1+c\alpha)^{n-1})}{(1+\alpha)^n + L(1+c\alpha)^n}. \tag{8.28}$$

The 'allosteric' constant L, which in this case applies simply to homotropic effects, can of course be replaced by $L' = L(1+\beta)^p/(1+\gamma)^p$ as in equation (8.10) where p refers to the number of binding sites for the effector molecules. The inclusion of different rate constants for the breakdown of the RS_i species (k_2) and the TS_i species (l_2) allows the *kinetic* model to be further classified into *V* and *K* systems. In a *V* system the substrate has the same affinity for the two states, so that $K_m^R = K_m^T$, but the microscopic rate constants are different, $k_2 \neq l_2$. The two states differ, therefore, in their catalytic activity. The addition of an effector molecule will change the 'allosteric' constant L' (above) and depending upon whether it has a maximum affinity for the active or inactive state it will behave as an activator (positive *V* system) or inhibitor (negative *V* system). With $K_m^R = K_m^T$ equation (8.28) reduces to

$$v = \frac{n[E_0]\,[S]\,(k_2 + L'\,l_2)/(1 + L')}{([S] + K_m^R)}$$

and hence $V = n[E_0](k_2 + L'\,l_2)/(1 + L')$.

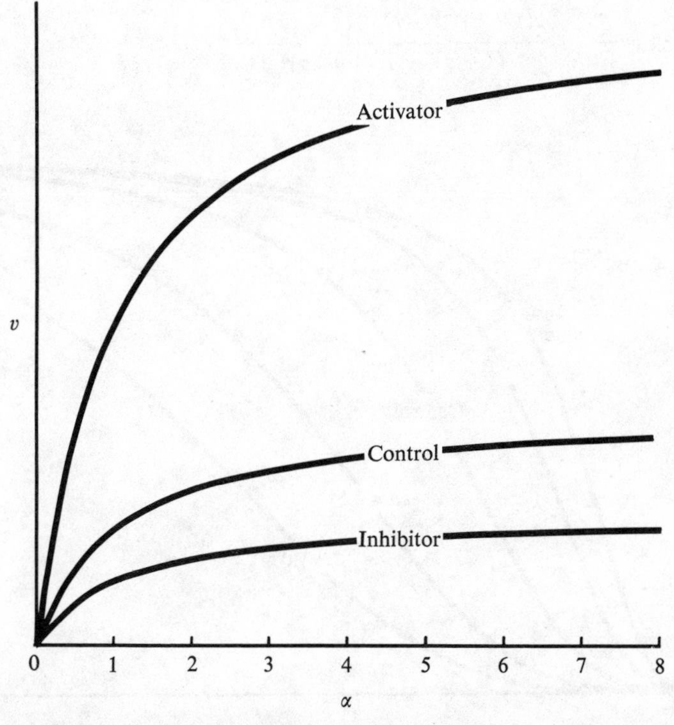

Fig. 8.18. Schematic plot of v versus α ($[S]/K_m^R$) illustrating the behaviour of a *V* system ($k_2 \neq l_2$, $K_m^R = K_m^T$). An activator (γ-term in L') causes the reaction to reach a higher maximum velocity by a reduction in the allosteric constant L' whereas an inhibitor (β-term in L') has the opposite effect. The parameters used were $n = 4$, $k_2 = 1$, $l_2 = 10$, $L = 0.1$, $p = 3$, $\beta = 1$, $\gamma = 4$.

For a particular set of parameters the curves are always hyperbolic. An example is plotted in fig. 8.18. An activator (γ-term in L') causes a reduction in L' and hence the curve tends towards a higher V; conversely an inhibitor (β-term in L') increases L' and hence a lower value of V is reached.

In a K system, both the effector molecule and the substrate have different affinities towards the R- and T-states. The presence of the effector molecule will therefore alter the binding of the substrate and vice versa. In a true K system, the catalytic rate constants k_2 and l_2 are the same and hence the presence of the effector molecule alters the shape of the v against [S] plot but not the maximum velocity that is ultimately reached. In this particular case we can relate the binding equation directly to the kinetic equation since

$$r = \frac{v}{nV} = \frac{\alpha(1+\alpha)^{n-1} + L'\,c\alpha(1+c\alpha)^{n-1}}{(1+\alpha)^n + L'(1+c\alpha)^n}$$

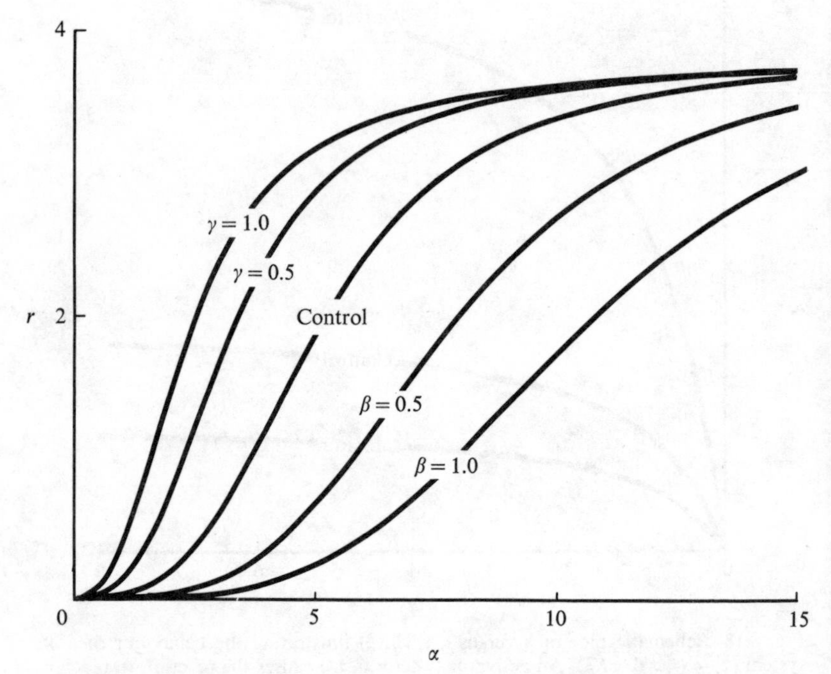

Fig. 8.19. Schematic plot of r against α illustrating the behaviour of a K system ($k_2 = l_2$, $K_m^R \neq K_m^T$). The parameters used were $L = 10^3$, $n = 4$, $p = 4$, $c = 0.01$, β and γ as indicated. For the control, $\beta = \gamma = 0$.

where $V = k_2[E_0](l_2 = k_2)$. An example of a K system is shown in fig. 8.19.

The Koshland or Adair model can be treated in a similar fashion. The rate of the reaction will be the sum of the breakdown of each ES_i complex at a rate determined by the respective intrinsic rate constant:

$$v = \sum_{i=1}^{n} ik_i[ES_i].$$

The numerator in equation (8.12) is $\sum_{i=1}^{n} i[ES_i]$ and the denominator is the total enzyme concentration $[E_0]$. We can therefore express the velocity of the reaction as

$$v = \frac{[E_0]\sum_{i=1}^{n} ik_i \frac{[S]^i n!}{(n-i)!\,i!}\prod_{j=1}^{i} K_j}{1 + \sum_{i=1}^{n} \frac{[S]^i n!}{(n-i)!\,i!}\prod_{j=1}^{n} K_j}. \tag{8.29}$$

For an enzyme possessing four sites this would yield

$$v = \frac{[E_0]\{4k_1 K_1[S] + 12k_2 K_1 K_2[S]^2 + 12k_3 K_1 K_2 K_3[S]^3 + 4k_4 K_1 K_2 K_3 K_4[S]^4\}}{(1 + 4K_1[S] + 6K_1 K_2[S]^2 + 4K_1 K_2 K_3[S]^3 + K_1 K_2 K_3 K_4[S]^4)} \tag{8.30}$$

The association constants could be replaced by their reciprocals or Michaelis constants but in this particular case it would make the equation much more cumbersome. In the event that all the catalytic rate constants are equal ($k_1 = k_2 = k_3 = k_4 = k$), equation (8.30) can be written in terms of the binding function

$$r = \frac{4K_1[S] + 12K_1 K_2[S]^2 + 12K_1 K_2 K_3[S]^3 + 4K_1 K_2 K_3 K_4[S]^4}{1 + 4K_1[S] + 6K_1 K_2[S]^2 + 4K_1 K_2 K_3[S]^3 + K_1 K_2 K_3 K_4[S]^4}$$

where $r = v/V$ and $V = k[E_0]$. In the case of a multi-site enzyme in which there are no interactions between sites (all K_i equal and all k_i equal) then equation (8.29) reduces to

$$r = \frac{v}{V} = \frac{nK[S]}{1 + K[S]}.$$

An enzyme with a number of *non-interacting* sites behaves like a single-site enzyme. It will be noticed that the maximum power to which the substrate concentration is raised is determined by the total number of binding sites. A simple mechanism proposed by Hill (1913) in which an enzyme with n sites can bind a ligand S to give only an ES_n complex (i.e. $E + nS \rightleftharpoons ES_n$) gives rise to a kinetic equation

$$\frac{v}{V} = \frac{1}{1 + (K/[S])^n}; \qquad \frac{v}{V - v} = \left(\frac{[S]}{K}\right)^n.$$

A plot of $\ln v/(V - v)$ against $\ln$ [S] would yield a straight line of slope n and hence give the number of binding sites in Hill's model. This type of plot, the Hill plot, however, has been applied quite commonly to multi-site enzymes in which Hill's restriction of only one ES_n complex is probably not valid. The value of n that is determined is regarded as an 'interaction' coefficient and an indication of the number of sites. Hill plots, however, being log–log, are rather insensitive, often curved and frequently give non-integer values for n. This is of course to be expected since the Hill model is a very restricted case of a multi-site enzyme and whether the time spent plotting the graph and the value of n that is obtained are worthwhile, is certainly open to question. It is probably more reasonable to fit the data direct to the Monod or Adair equation by computer and extract the parameter values from the best fit that can be achieved.

For the competitive binding model (equation (8.26)) the overall reaction velocity is given by

$$v_0 = n_1 k_1[E_1]\frac{[S]}{K_{1S}}\left(1 + \frac{[S]}{K_{1S}} + \frac{[L]}{K_{1L}}\right)^{n_1 - 1}$$
$$+ n_2 k_2[E_2]\frac{[S]}{K_{2S}}\left(1 + \frac{[S]}{K_{2S}} + \frac{[L]}{K_{2L}}\right)^{n_2 - 1} \qquad (8.31)$$

where k_1 and k_2 are the catalytic breakdown constants for the respective forms of the enzyme and K_{hS}, $h = 1,2$ and K_{hL}, $h = 1,2$ are now the Michaelis constants and reciprocal binding constants respectively. Hence equation (8.31) can be written as

$$v_0 = \frac{V_1 \dfrac{[S]}{K_{1S}}\left(1 + \dfrac{[S]}{K_{1S}} + \dfrac{[L]}{K_{1L}}\right)^{n_1-1} + XV_2 \dfrac{[S]}{K_{2S}}\left(1 + \dfrac{[S]}{K_{2S}} + \dfrac{[L]}{K_{2L}}\right)^{n_2-1}}{\left(1 + \dfrac{[S]}{K_{1S}} + \dfrac{[L]}{K_{1L}}\right)^{n_1} + X\left(1 + \dfrac{[S]}{K_{2S}} + \dfrac{[L]}{K_{2L}}\right)^{n_2}} \tag{8.32}$$

where $V_h = n_h k_h[E_0]$, $h = 1,2$. If $X = 0$, equation (8.32) reduces to

$$v_0 = \frac{V_1[S]}{K_{1S}\{1 + ([L]/K_{1L})\} + [S]}$$

which is analogous to equation (3.1) describing the competitive inhibition of a Michaelis–Menten type enzyme. Equation (8.32) can be used to study not only the effect that compounds, which compete for the same binding sites as the substrate, have on the kinetics of a two-state Monod-type enzyme, but can also be used to describe the product inhibition of an allosteric enzyme and can be extended to cover the case of a reversible enzyme. For simplicity let $n_1 = n_2 = n$ and $k_1 = k_2$ so that $V_1 = V_2 = V$, then equation (8.32) can be written as

$$v_0 = \frac{V\left\{\dfrac{[S]}{K_{1S}}\left(1 + \dfrac{[S]}{K_{1S}} + \dfrac{[P]}{K_{1P}}\right)^{n-1} + X\dfrac{[S]}{K_{2S}}\left(1 + \dfrac{[S]}{K_{2S}} + \dfrac{[P]}{K_{2P}}\right)^{n-1}\right\}}{\left(1 + \dfrac{[S]}{K_{1S}} + \dfrac{[P]}{K_{1P}}\right)^{n} + X\left(1 + \dfrac{[S]}{K_{2S}} + \dfrac{[P]}{K_{2P}}\right)^{n}}$$

$$= -\frac{d[S]}{dt} = \frac{d[P]}{dt} \tag{8.33}$$

where [P] is the concentration of the product. For the reversible enzyme reaction, equation (8.33) can be written as

$$v = -\frac{d[S]}{dt} = \frac{d[P]}{dt} = \frac{\left\{\left(V_f\dfrac{[S]}{K_{1S}} - V_r\dfrac{[P]}{K_{1P}}\right)\left(1 + \dfrac{[S]}{K_{1S}} + \dfrac{[P]}{K_{1P}}\right)^{n-1} + X\left(V_f\dfrac{[S]}{K_{2S}} - V_r\dfrac{[P]}{K_{2P}}\right)\left(1 + \dfrac{[S]}{K_{2S}} + \dfrac{[P]}{K_{2P}}\right)^{n-1}\right\}}{\left(1 + \dfrac{[S]}{K_{1S}} + \dfrac{[P]}{K_{1P}}\right)^{n} + X\left(1 + \dfrac{[S]}{K_{2S}} + \dfrac{[P]}{K_{2P}}\right)^{n}} \tag{8.34}$$

where V_f and V_r are the forward and reverse maximum velocities. Equations (8.33) and (8.34) are useful as they enable the behaviour of allosteric enzymes to be described by rate equations involving product inhibition and hence they can be used for the computer simulation (see chapter 10) of metabolic pathways involving allosteric enzymes.

The last section deals with enzymes that cannot be treated by an equilibrium approach because the rates of isomerisation or polymerisa-ation are not rapid. The rate at which the isomerisation takes place must therefore be included in any kinetic equation (Hatfield, Ray & Umbarger, 1970). Frieden (1970) has called these hysteretic enzymes. They are not necessarily restricted to allosteric or regulatory enzymes and hence the following treatment can be applied to single- or multi-site enzymes. For an enzyme that undergoes a slow transition between two states during the course of the reaction, it is possible to produce a sigmoidal v against [S] curve. The enzyme may or may not be a true allosteric enzyme and hence sigmoidality should not be equated necessarily with allosterism. For example in the scheme shown below

$$\mathrm{F} \underset{k_{-1}}{\overset{k_1}{\rightleftharpoons}} \mathrm{E+S} \underset{k_{-2}}{\overset{k_2}{\rightleftharpoons}} \mathrm{ES} \xrightarrow{k_3} \mathrm{E+P} \tag{8.35}$$

the enzyme is assumed to exist in two forms, E and F, of which only the E form is active. The addition of the substrate causes a shift in the equilibrium between the E and F forms at a rate determined by the rate constants k_1 and k_{-1}. We can write the following differential equations:

$$\frac{\mathrm{d[F]}}{\mathrm{d}t} = -k_1[\mathrm{F}] + k_{-1}[\mathrm{E}], \tag{8.36}$$

$$\frac{\mathrm{d[E]}}{\mathrm{d}t} = k_1[\mathrm{F}] - k_{-1}[\mathrm{E}] - k_2[\mathrm{E}]\,[\mathrm{S_0}] + (k_{-2} + k_3)\,[\mathrm{ES}], \tag{8.37}$$

$$\frac{\mathrm{d[ES]}}{\mathrm{d}t} = k_2[\mathrm{E}]\,[\mathrm{S_0}] - (k_{-2} + k_3)\,[\mathrm{ES}]. \tag{8.38}$$

Let the quantity of F and E at $t = 0$ be $[\mathrm{F_0}]$ and $[\mathrm{E_0}]$ and at time t be $[\mathrm{F}_t]$ and $[\mathrm{E}_t]$ respectively, then

$$[\mathrm{E}_T] = [\mathrm{E_0}] + [\mathrm{F_0}] = [\mathrm{E}_t] + [\mathrm{F}_t] + [\mathrm{ES}]$$

where $[\mathrm{E}_T]$ is the total enzyme concentration. Equations (8.36)–(8.38) can be written in the Laplace–Carson operator form as

$$
\begin{aligned}
(\psi + k_1)[\mathrm{f}] \quad &- \quad k_{-1}[\mathrm{e}] \qquad\qquad\qquad\qquad\qquad 0 = \psi[\mathrm{F_0}] \\
-k_1[\mathrm{f}] + (\psi + k_{-1} + k_2[\mathrm{S_0}])[\mathrm{e}] &- \qquad (k_{-2} + k_3)[\mathrm{es}] = \psi[\mathrm{E_0}] \\
0 \qquad\qquad &-k_2[\mathrm{S_0}][\mathrm{e}] + (\psi + k_{-2} + k_3)[\mathrm{es}] = \quad 0
\end{aligned}
$$

where the lower case letters refer to the time-dependent species. The solutions for [e], [es] and [f] can be obtained by solving the three simultaneous equations. Using determinants,

$$
[\mathrm{es}] = \frac{\Delta_{\mathrm{es}}}{\Delta} = \frac{\begin{vmatrix} (\psi + k_1) & -k_{-1} & \psi[\mathrm{F_0}] \\ -k_1 & (\psi + k_{-1} + k_2[\mathrm{S_0}]) & \psi[\mathrm{E_0}] \\ 0 & -k_2[\mathrm{S_0}] & 0 \end{vmatrix}}{\begin{vmatrix} (\psi + k_1) & -k_{-1} & 0 \\ -k_1 & (\psi + k_{-1} + k_2[\mathrm{S_0}]) & -(k_{-2} + k_3) \\ 0 & -k_2[\mathrm{S_0}] & (\psi + k_{-2} + k_3) \end{vmatrix}}
$$

Hence

$$[\mathrm{es}] = \{k_1 k_2[\mathrm{S_0}][\mathrm{F_0}] + (\psi + k_1)k_2[\mathrm{E_0}][\mathrm{S_0}]\}/\Delta$$

where

$$
\begin{aligned}
\Delta = \{\psi^2 + \psi(k_2[\mathrm{S_0}] + k_1 + k_{-1} + k_{-2} + k_3) \\
+ (k_{-1} + k_1)(k_{-2} + k_3) + k_1 k_2[\mathrm{S_0}]\}
\end{aligned}
$$

$$
[\mathrm{es}] = \frac{k_1 k_2[\mathrm{S_0}][\mathrm{F_0}]}{(\psi + \lambda_1)(\psi + \lambda_2)} + \frac{(\psi + k_1)k_2[\mathrm{S_0}][\mathrm{E_0}]}{(\psi + \lambda_1)(\psi + \lambda_2)}
$$

where

$$
\begin{aligned}
\lambda_1 &= \{b + \sqrt{(b^2 - 4c)}\}/2 \\
\lambda_2 &= \{b - \sqrt{(b^2 - 4c)}\}/2 \\
b &= k_2[\mathrm{S_0}] + k_1 + k_{-1} + k_{-2} + k_3 \\
c &= k_1 k_2[\mathrm{S_0}] + (k_1 + k_{-1})(k_{-2} + k_3).
\end{aligned}
$$

The solution for [es] is therefore

$$
\begin{aligned}
[\mathrm{ES}] = k_1 k_2[\mathrm{S_0}][\mathrm{F_0}] \left\{ \frac{1}{\lambda_1 \lambda_2} - \frac{\mathrm{e}^{-\lambda_1 t}}{\lambda_1(\lambda_2 - \lambda_1)} - \frac{\mathrm{e}^{-\lambda_2 t}}{\lambda_2(\lambda_1 - \lambda_2)} \right\} \\
+ k_2[\mathrm{S_0}][\mathrm{E_0}] \left\{ \frac{k_1}{\lambda_1 \lambda_2} - \frac{(k_1 - \lambda_1)\mathrm{e}^{-\lambda_1 t}}{\lambda_1(\lambda_2 - \lambda_1)} - \frac{(k_1 - \lambda_2)\mathrm{e}^{-\lambda_2 t}}{\lambda_2(\lambda_1 - \lambda_2)} \right\}.
\end{aligned}
$$

The velocity of the reaction is given by

$$v = k_3[\mathrm{ES}]$$
$$= k_1 k_2 k_3[\mathrm{S_0}][\mathrm{E}_T]\left\{\frac{1}{\lambda_1\lambda_2} - \frac{\mathrm{e}^{-\lambda_1 t}}{\lambda_1(\lambda_2-\lambda_1)} - \frac{\mathrm{e}^{-\lambda_2 t}}{\lambda_2(\lambda_1-\lambda_2)}\right\}$$
$$+ k_2 k_3[\mathrm{S_0}][\mathrm{E_0}]\left\{\frac{\mathrm{e}^{-\lambda_1 t}}{(\lambda_2-\lambda_1)} + \frac{\mathrm{e}^{-\lambda_2 t}}{(\lambda_1-\lambda_2)}\right\}. \tag{8.39}$$

In the derivation of the Michaelis–Menten equation (equation (2.9)) using the steady-state approach, the zero time ($t = 0$) of the system is not the true zero time but is the point at which the exponentials in the complete solution (equation (7.31)) have become negligible, i.e. the point at which the steady state of enzyme complexes is reached. Using steady-state kinetic methods where it is not possible to observe changes that take place in less than ~30 s, we can define 'zero time' as the point at which the fast exponential, controlled in this scheme by λ_1 in equation (8.39), has decayed. Since $\lambda_1 \gg \lambda_2$ then $(\lambda_2 - \lambda_1) \sim -\lambda_1$ and $(\lambda_1 - \lambda_2) \sim \lambda_1$ and $\mathrm{e}^{-\lambda_1 t}$ becomes negligible after a very short time interval, then

$$v = \frac{k_1 k_2 k_3[\mathrm{S_0}][\mathrm{E}_T]}{\lambda_1\lambda_2}(1-\mathrm{e}^{-\lambda_2 t}) + \frac{k_2 k_3[\mathrm{S_0}][\mathrm{E_0}]\mathrm{e}^{-\lambda_2 t}}{\lambda_1}. \tag{8.40}$$

If $[\mathrm{S_0}]$ is maintained constant then as t increases the final velocity, v_f, will be

$$v_f = \frac{k_1 k_2 k_3[\mathrm{S_0}][\mathrm{E}_T]}{\lambda_1\lambda_2}.$$

From equation (8.40) the velocity, v_0, at $t = 0$ is

$$v_0 = \frac{k_2 k_3[\mathrm{S_0}][\mathrm{E_0}]}{\lambda_1}$$

and hence the initial velocity at $t = 0$ depends on the quantity of enzyme present in the active form, $[\mathrm{E_0}]$. Equation (8.40) can therefore be written as

$$v = v_f + (v_0 - v_f)\mathrm{e}^{-\lambda_2 t}. \tag{8.41}$$

Equation (8.41) assumes that the substrate concentration is maintained constant and that there is no change in the catalytic mechanism due to the build-up of products. Since the constant λ_2 contains the concentration of the substrate, the rate at which the isomerisation occurs will

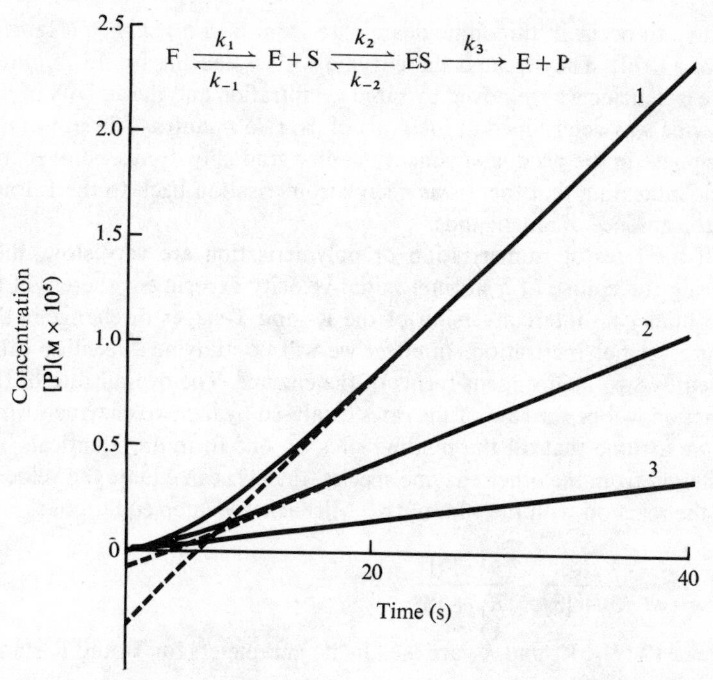

Fig. 8.20. A product against time plot for a hysteretic enzyme for the following set of parameters: $k_1 = 0.1\ s^{-1}$, $k_{-1} = 1\ s^{-1}$, $k_2 = 10^7\ M^{-1}\ s^{-1}$, $k_{-2} = 10^4\ s^{-1}$, $k_3 = 10^2\ s^{-1}$, $[E_T] = 1.1 \times 10^{-8}$ M ($[F_0] = 10^{-8}$ M, $[E_0] = 10^{-9}$ M). For 1, $[S_0] = 4 \times 10^{-2}$ M; 2, $[S_0] = 4 \times 10^{-3}$ M; and 3, $[S_0] = 1 \times 10^{-3}$ M. Note that the Michaelis constant is increased by the ratio k_{-1}/k_1. The intercept on the [P] axis tends to zero as [S] becomes small and to a maximum limiting value at high concentrations of [S].

depend on the initial concentration of substrate. This effect is shown in fig. 8.20. It is possible, therefore, to produce a sigmoidal v against [S] curve based on this simple model and other similar models (Rabin, 1967). A further modification to the model given in equation (8.35) would be to include the binding of an effector molecule to give the scheme below:

$$\mathrm{FI} \underset{k_{-1}[\mathrm{I}]}{\overset{k_1}{\rightleftharpoons}} \mathrm{F} \underset{k_{-2}}{\overset{k_2}{\rightleftharpoons}} \mathrm{E} \underset{k_{-3}}{\overset{k_3[\mathrm{S}]}{\rightleftharpoons}} \mathrm{ES} \xrightarrow{k_4} \mathrm{E} + \mathrm{P}.$$

If the isomerisation step, F ⇌ E, is still the rate-limiting step of the mechanism, the pre-equilibration of the enzyme with an effector molecule would increase the lag phase of the reaction since most of the enzyme would be in the F form. An example of this type of mechanism has been shown to occur in threonine deaminase from *Bacillus subtilis* (Hatfield *et al.*, 1970). They treated the enzyme with isoleucine for 15 minutes. The isoleucine was removed by rapid gel filtration and the activity of the enzyme was determined at intervals of over 75 minutes. The size of the lag phase in the product against time plot gradually decreased over this time indicating that there was a slow isomerisation back to the E form in the absence of any ligands.

If the rates of isomerisation or polymerisation are very slow, then during the course of a normal initial-velocity experiment there will be essentially no interconversion of the R- and T-stages or change in the degree of polymerisation. In effect we will be studying a reaction catalysed by two independent forms of the enzyme. The overall rate of the reaction will be the sum of the rates catalysed by the two enzyme forms. If we assume that all the binding sites on one form are identical, but different from the other enzyme species, then we can equate the velocity of the reaction with the sum of two Michaelis–Menten equations:

$$v = \frac{V_T[\mathrm{S}]}{K_T + [\mathrm{S}]} + \frac{V_R[\mathrm{S}]}{K_R + [\mathrm{S}]} \tag{8.42}$$

where V_T, K_T, V_R and K_R are the kinetic parameters for T- and R-states. Equation (8.42) can be rearranged to give

$$v = \frac{(V_T K_R + V_R K_T)[\mathrm{S}] + (V_T + V_R)\,[\mathrm{S}]^2}{K_T K_R + (K_T + K_R)\,[\mathrm{S}] + [\mathrm{S}]^2}$$

which is similar to the Adair equation for $n = 2$ (equation (8.29)) with K_T and K_R in this case being the Michaelis constants for the respective T- and R-states. If $1/v$ is plotted against $1/[\mathrm{S}]$, the curves are always concave downwards or linear for real non-negative values of the kinetic constants. The equation can describe negative co-operativity but never positive co-operativity. It is important to realise that the sum of two (or more) hyperbolas (equation (8.42)) can produce a reciprocal plot which is concave downwards. The experimental observation of such a plot therefore does not imply that an enzyme undergoes allosteric transitions. The real reason may be the presence of two isoenzymes in the enzyme preparation which catalyse the same substrate but with different kinetic constants. At high substrate concentrations the maximum velocity is the sum of the individual maximum velocities for each enzyme, $V = V_T + V_R$. In the past, plots which are concave downwards have been regarded as indicating activation by the substrate, since at

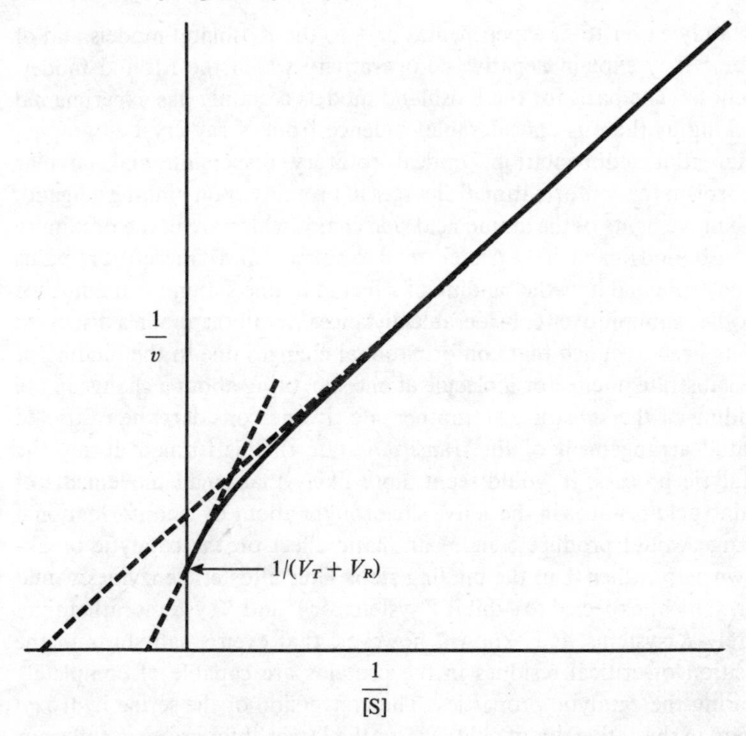

Fig. 8.21. Schematic plot of $1/v$ against $1/[S]$ for a system described by the sum of two Michaelis–Menten equations.

high substrate concentrations the reciprocal plot curves downwards to a higher maximum velocity than is indicated by the linear region (fig. 8.21).

In summary, it should be clear by now that the non-hyperbolic behaviour of certain regulatory enzymes and other proteins (haemoglobin) can be explained in a number of ways. It is also possible to describe sigmoidal systems and negative co-operative plots by mechanisms that are not restricted to allosteric enzymes. Negative and positive co-operativity that are indicated from reciprocal plots of $1/v$ against $1/[S]$ should therefore be regarded as not necessarily indicating that the enzyme undergoes allosteric transitions. The Monod model is by far the simplest conceptually but is also the most restrictive in the possible transitions that can occur. However, even though it is a more restricted model than the Adair or Koshland model, a surprisingly large number of enzymes can be shown to fit it. With the larger number of parameters, it is certainly easier to fit experimental data to the Koshland models and of course they explain negative co-operativity which the Monod models cannot.* The basis for the Koshland models certainly has experimental backing as there is considerable evidence from X-ray crystallography, differential sedimentation, optical rotatory dispersion and circular dichroism for conformational changes in proteins upon binding a ligand. The movements of the amino acid side chains which are in the proximity of the bound ligand in the protein tend to be small and it therefore remains to be explained how the binding of a ligand to one subunit can influence another subunit over considerable distances. In all the models discussed it has been assumed that conformational changes due to the binding of the substrate or effector molecule at one site, bring about a change in the binding of the substrate at another site. If one considers the restricted spatial arrangement of the transition state that is formed during the catalytic process, it would seem more likely that small movements of amino acid residues in the active site brought about by a conformational change would produce a more dramatic effect on the catalytic breakdown step rather than the binding step. Thus allosteric enzymes would normally be expected to exhibit V systems or V and K systems rather than purely K systems. It is known, however, that even small shifts in the position of critical residues in the proteins are capable of completely altering the catalytic properties. The conversion of the serine hydroxyl group in the active site of subtilisin to the larger thiol group is sufficient to decrease the enzyme activity. This was contrary to expectations as an increase in activity would be expected due to the increased nucleophilicity of the thiol group (Neet, Nanci & Koshland, 1968). Neither the Monod nor Koshland model can explain the dependence of the velocity or binding curve on the enzyme concentration. This can only be explained on the basis of a polymerising system. The polymerising system is also capable of a much wider control of the enzyme reaction than a simple isomerisation process (Monod model). As the polymerisation number (n) increases, the change in ligand concentration required to switch the

*A simple extension to the original Monod model will, however, explain both positive and negative co-operativity. For example, an enzyme molecule consisting of two subunits may possess two catalytic sites which are non-equivalent and independent, having association constants K_1 and K_2, and the whole molecule may exist in two conformations, R and T, described by an isomerisation constant $X = [T]/[R]$. If the T isomer is the non-binding form, then a large value of X will cause the v against [S] plot to be sigmoidal at low values of [S]. The difference between the two binding sites on the R form will cause the curve to exhibit negative co-operativity at high values of [S].

enzyme 'on' and 'off' becomes much less. A monomer–hexamer system is critically controlled over a very small range of ligand concentrations. The mechanisms of polymerisations induced by ligands is rather obscure at present.

In all the models discussed in this chapter, the non-hyperbolic behaviour of enzymes has been explained by the existence of the enzymes in two or more states which have different kinetic parameters. It is quite possible that an enzyme which behaves in a Michaelis–Menten fashion when studied *in vitro* could exhibit non-hyperbolic behaviour *in vivo* due to interactions with other components in the system. Thus an enzyme that formed a complex with another protein or membrane would satisfy the 'two-state' requirement if the kinetic constants were different in the bound and unbound forms (Nichol, Kuchel & Jeffrey, 1974). Thus interesting protein–protein interactions that may exist in the cell and alter the kinetic response of an enzyme, are lost on its subsequent isolation and purification for in-vitro studies.

9 Coupled enzyme systems

9.1. Steady-state behaviour

We have looked so far at the kinetic behaviour of enzymes in isolation and have derived equations based on various models that permit the evaluation of the kinetic parameters for any particular enzyme. The kinetic parameters, K_m and V, are not absolute constants but will vary depending upon the conditions under which the enzyme is studied. It is quite likely that results obtained for a particular enzyme in one buffer will be different from those obtained in another. Hence if we wish to study the enzyme *in vivo* then not only are the environmental conditions completely different from those in our isolated studies, but the enzyme is also part of an overall sequence or chain of coupled reactions. The kinetic behaviour of any particular enzyme in the sequence is determined not only by its own kinetic parameters in the reaction medium but also by the variation in the supply of substrate and removal of products. The latter factors may be controlled by a diffusion process or by other enzymes in the sequence. Considerable interest has therefore developed in the analysis of the kinetic behaviour of consecutive reactions catalysed either by enzymes in solution or immobilised on a supporting matrix (Goldman & Katchalski, 1971). The analysis of coupled systems has been approached in two ways. Garfinkel *et al.* (1970) have used computer simulation techniques involving numerical integration methods to study complex systems such as the glycolytic pathway. This type of analysis will be discussed in the next chapter on computer simulation methods. Other workers have attempted to produce analytical solutions for the set of non-linear differential equations by applying specific boundary conditions to the problem. (McClure, 1969; Barwell & Hess, 1970; Easterby, 1973; Kuchel *et al.*, 1974; Kuchel & Roberts, 1974). In these last examples, the solutions only apply to specific time domains during the course of the reaction and are relevant to the analysis of coupled enzyme assays usually involving two or three coupling enzymes. Unfortunately, with many enzyme systems, there is no convenient way of monitoring the change in concentration of any of the substrates, products or intermediates, since there is no convenient change in physical parameters of the system such as pH, optical density or nuclear magnetic resonance spectra. In order to study such enzymes, therefore, one has to couple some other reaction to the system which will give a product that is readily measurable. This can be achieved in two ways:

(1) by a purely chemical reaction in which the product of the enzymic process reacts with a chemical reagent to yield a new product that can be easily monitored;

(2) by coupling another enzyme (or enzymes) to the system that will convert the product of the first reaction into a new product.

These two methods can be carried out using either a batch or a continuous procedure. In the batch process, the first reaction is stopped at set times, for example by a pH change, and then the amount of product that has been formed is determined by one of the two methods above. A product versus time plot can be built up, thus enabling the initial velocity of the reaction (at zero time) to be measured from the data. This method is often used but is tedious and introduces many errors due to the various manipulations. In the continuous process a complete plot of the formation of the new product with time is obtained and is obviously the most ideal arrangement. The equations that we will now derive apply to a continuous coupled enzyme system. The simplest example of such a couple is shown in scheme (9.1) in which the first substrate, S_1, is converted by enzyme E_1 into a product, S_2, which is used by an enzyme E_2 as its substrate to give the new product S_p.

$$S_1 \xrightarrow{E_1} S_2 \xrightarrow{E_2} S_p \qquad (9.1)$$

A practical example of such a system is given in scheme (9.2).

$$\text{Glucose} \xrightarrow[\text{ATP} \to \text{ADP}]{\text{HK}} \text{Glucose-6-phosphate} \xrightarrow[\text{NADP} \to \text{NADPH}]{\text{G6PDH}} \text{6-phosphogluconate} \qquad (9.2)$$

(HK, hexokinase; G6PDH, glucose-6-phosphate dehydrogenase)

At any particular point in time there will be conservation of mass and so we can write

$$[S_1]_0 = [S_1]_t + [S_2]_t + [S_p]_t$$

where $[S_1]_0$ is the concentration of $[S_1]$ when $t = 0$ and $[S_1]_t$, $[S_2]_t$ and

$[S_p]_t$ are the concentrations at time t. In the conservation equation, we have assumed that $[E_1]$ and $[E_2]$ are both much lower than the substrates so that the concentration of substrates bound as intermediates $[E_1S_1]$ and $[E_2S_2]$ can be neglected. If we also assume that the enzymes obey Michaelis–Menten kinetics, then we can write

$$\frac{-d[S_1]}{dt} = \frac{V_1[S_1]}{K_1 + [S_1]} = v_0 \tag{9.3}$$

$$\frac{d[S_2]}{dt} = \frac{V_1[S_1]}{K_1 + [S_1]} - \frac{V_2[S_2]}{K_2 + [S_2]}, \tag{9.4}$$

$$\frac{d[S_p]}{dt} = \frac{V_2[S_2]}{K_2 + [S_2]}, \tag{9.5}$$

where V_1, K_1, V_2 and K_2 are the kinetic parameters for the first and second enzymes respectively and v_0 is the velocity of the first reaction. These equations assume that (*a*) there is no change in the catalytic mechanism by the build-up of products which are not used in the sequential system, (*b*) there are no interactions between enzyme species, and (*c*) the various E_iS_i intermediates, though not necessarily each intermediate substrate, are in a steady state. We are interested in obtaining a relation between the observed velocity, $(d[S_p]/dt)$, and the velocity of the first reaction. Suppose that the rate of the second reaction is much faster than the first reaction, which can be achieved by having a high concentration of the coupling enzyme, then the concentration of the intermediate, $[S_2]$, will be very low. Under these conditions, where $[S_2] \ll K_2$, and assuming that the reactions are irreversible we can write

$$\psi[S_p] = \frac{V_2[S_2]}{K_2} = \frac{[S_2]}{\tau_2} \tag{9.6}$$

where V_2/K_2 (or $1/\tau_2$) is the first order rate constant for the reaction and ψ is the Laplace–Carson operator. From equation (9.4),

$$\psi[S_2] = v_0 - \frac{[S_2]}{\tau_2}$$

where $v_0(= -d[S_1]/dt)$ is taken to be constant during the early (zero order) portion of the first reaction. Hence

$$[S_2] = \frac{v_0}{\{\psi + (1/\tau_2)\}} \tag{9.7}$$

for which the solution is

$$[S_2] = v_0\,\tau_2(1 - e^{-t/\tau_2}). \tag{9.8}$$

From equations (9.6) and (9.7),

$$[S_p] = \frac{v_0}{\tau_2\,\psi\{\psi + (1/\tau_2)\}},$$

$$[S_p] = v_0\left(\frac{1}{\psi} - \frac{1}{\{\psi + (1/\tau_2)\}}\right),$$

and hence the solution for $[S_p]$ is

$$[S_p] = v_0\{t - \tau_2(1 - e^{-t/\tau_2})\}. \tag{9.9}$$

If v_0 continues to be constant and $[S_2]$ remains low so that the con-

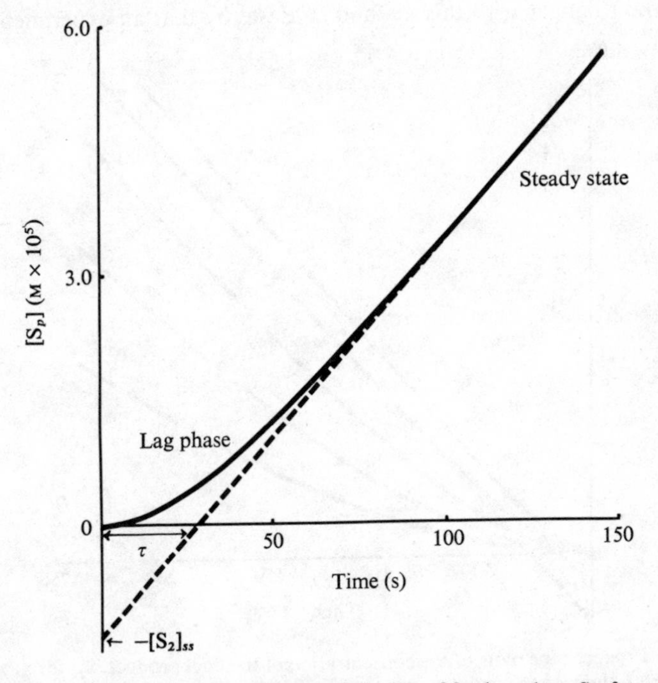

Fig. 9.1. A product/time plot of the concentration of final product, S_p, for a coupled enzyme reaction illustrating the lag phase and the eventual steady-state region of the reaction. The parameters used were $[S_0] = 10^{-3}$ M, $[E_1]_0 = 10^{-8}$ M, $[E_2]_0 = 3.2 \times 10^{-7}$ M, $K_1 = 1.01 \times 10^{-3}$ M, $k_{cat_1} = 10^2$ s^{-1}, $V_1 = 10^{-6}$ M s^{-1}, $K_2 = 1.01 \times 10^{-3}$ M, $k_{cat_2} = 10^2$ s^{-1}, $V_2 = 3.2 \times 10^{-5}$ M s^{-1}. Note that even with $V_2 = 32 \times V_1$, over 5% depletion of substrate has occurred before the steady-state region of the reaction is reached.

dition $K_2 \gg [S_2]$ remains valid, then as t increases equation (9.9) reduces to

$$[S_p] = v_0(t - \tau_2) \tag{9.10}$$

and the steady-state concentration of the intermediate $[S_2]$, $[S_2]_{SS}$, will be

$$[S_2]_{SS} = v_0 \tau_2. \tag{9.11}$$

Equation (9.10) is the equation of a straight line with an abscissal intercept of τ_2 and an ordinal intercept of $-v_0\tau_2$. A plot of the complete equation (9.9) is shown in fig. 9.1. Thus, from a series of experiments with various values of $[S_1]_0$, we can plot v_0 against $[S_1]_0$ or in reciprocal form $1/v_0$ against $1/[S_1]_0$ and hence obtain the parameters V_1 and K_1 for the first enzyme. Without independent studies of the second enzyme we can also obtain the ratio K_2/V_2, but not their individual values. There is an inherent danger with this method of Easterby that an experimenter

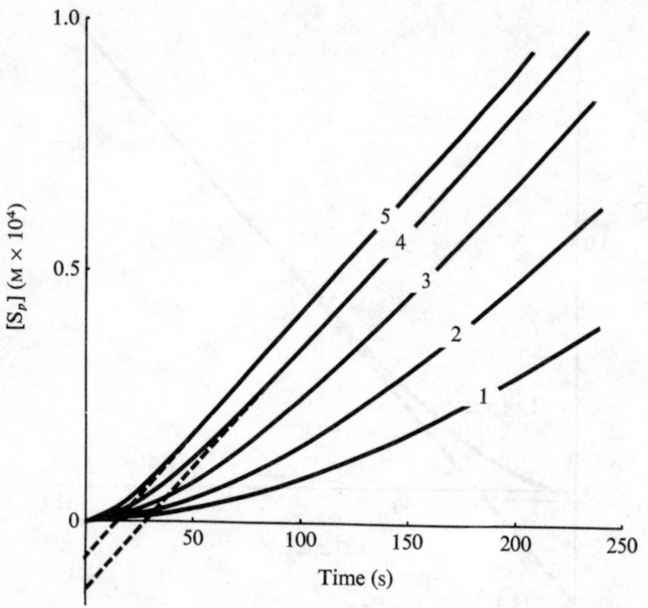

Fig. 9.2 Product/time plots of the concentration of the final product, S_p, for a range of concentrations of the coupling enzyme $[E_2]$. The values of the kinetic constants, concentration of substrate and first enzyme $[E_1]_0$, are the same as in fig. 9.1. For 1, $[E_2]_0 = 4.0 \times 10^{-8}$ M $V_2 = 4.0 \times 10^{-6}$ M s^{-1}; 2, $[E_2]_0 = 8 \times 10^{-8}$ M $V_2 = 8.0 \times 10^{-6}$ M s^{-1}; 3, $[E_2]_0 = 1.6 \times 10^{-7}$ M $V_2 = 1.6 \times 10^{-5}$ M s^{-1}; 4, $[E_2]_0 = 3.2 \times 10^{-7}$ M $V_2 = 3.2 \times 10^{-5}$ M s^{-1}; and 5, $[E_2]_0 = 6.4 \times 10^{-7}$ M $V_2 = 6.4 \times 10^{-5}$ M s^{-1}. It is apparent that only curves 4 and 5 have reached the steady-state velocity, v_0, with less than 5% depletion of substrate.

encountering an example of the above fig. 9.1 may erroneously equate the slope and intercept with v_0 and K_2/V_2 whereas they only approximate to these values if $V_2/K_2 \gg V_1/K_1$. Since the parameters V_1 and K_1 are unknown, there is no practical guide on how to choose $[E_2]$. For example, if the catalytic rate constant for the first enzyme, k_1, is much greater than that for the second enzyme, k_2, then a considerable molar excess of E_2 over E_1 would be required to ensure that $V_2 \gg V_1 (k_2[E_2] \gg k_1[E_1])$. One method to determine the amount of $[E_2]$ required to satisfy the above condition is to carry out a series of experiments at increasing $[E_2]$ until the different curves with decreasing transient times (K_2/V_2) converge to a constant v_0 (fig. 9.2). From the transient time, τ_2, and v_0, it is possible to determine the steady-state concentration of the intermediate $[S_2]$ (equation (9.11)); thus if the parameters of the second enzyme are known (which in this method is not essential) it is possible to determine whether the condition $[S_2] \ll K_2$ is fulfilled.

It may be necessary to use more than one coupling enzyme before a suitable product can be monitored. Such an example is given below in which the reaction scheme (9.2) is coupled through its other product ADP using pyruvate kinase and lactate dehydrogenase. This latter coupling system has been frequently used for systems generating ADP because of the high sensitivity of the change NADH → NAD^+.

$$\text{Glucose} + \text{ATP} \xrightarrow{\text{HK}} \text{glucose-6-P} + \text{ADP}.$$

$$\text{PEP} + \text{ADP} \xrightarrow{\text{PK}} \text{pyruvate} + \text{ATP}.$$

$$\text{Pyruvate} + \text{NADH} + \text{H}^+ \xrightarrow{\text{LDH}} \text{lactate} + \text{NAD}^+.$$

(HK, hexokinase; PK, pyruvate kinase; LDH, lactate dehydrogenase.) In general, we can write a series of n equations for the sequence

$$S_1 \xrightarrow{E_1} S_2 \xrightarrow{E_2} S_3 \longrightarrow \cdots \longrightarrow S_n \xrightarrow{E_n} S_p$$

and if $V_n \gg V_{n-1} \gg \cdots \gg V_2 \gg v_0$ then we can write

$$\frac{-d[S_1]}{dt} = \frac{V_1[S_1]}{K_1 + [S_1]} = v_0$$

$$\frac{d[S_2]}{dt} = \frac{V_1[S_1]}{K_1 + [S_1]} - \frac{V_2[S_2]}{K_2} = v_0 - \frac{[S_2]}{\tau_2}$$

$$\vdots \qquad \vdots \qquad \vdots \qquad \vdots$$

$$\frac{d[S_n]}{dt} = \frac{V_{n-1}[S_{n-1}]}{K_{n-1}} - \frac{V_n[S_n]}{K_n} = \frac{[S_{n-1}]}{\tau_{n-1}} - \frac{[S_n]}{\tau_n}$$

$$\frac{d[S_p]}{dt} = \frac{V_n[S_n]}{K_n} = \frac{[S_n]}{\tau_n}.$$

The solution for $[S_2]$, given by equation (9.8), can be substituted in the corresponding equation for $d[S_3]/dt$ and hence

$$[S_3] = v_0 \tau_3 \left\{ 1 + \frac{\tau_2 e^{-t/\tau_2}}{(\tau_3 - \tau_2)} + \frac{\tau_3 e^{-t/\tau_3}}{(\tau_2 - \tau_3)} \right\}.$$

By successive substitutions of $[S_j]$ in the next differential equation for $d[S_{j+1}]/dt$, we can obtain a general expression for the variation of $[S_p]$ with time for a sequence of n enzymes. The concentration of each intermediate $[S_i]$ is given by

$$[S_i] = v_0 \tau_i \left\{ 1 - \sum_{j=2}^{i} \tau_j^{i-2} e^{-t/\tau_j} \prod_{\substack{k=2 \\ k \neq j}}^{i} \frac{1}{(\tau_j - \tau_k)} \right\} \quad i = 2, n$$

and the concentration of the final product $[S_p]$ is given by

$$[S_p] = v_0 \left\{ t + \sum_{i=2}^{n} \tau_i^{n-1} e^{-t/\tau_i} \prod_{\substack{j=2 \\ j \neq i}}^{n} \frac{1}{(\tau_i - \tau_j)} - \sum_{i=2}^{n} \tau_i \right\}. \quad (9.12)$$

Thus for a system with three enzymes ($n = 3$) and two coupling enzymes, the variation of $[S_p]$ with time is given by

$$[S_p] = v_0 \left\{ t + \frac{\tau_2^2 e^{-t/\tau_2}}{(\tau_2 - \tau_3)} + \frac{\tau_3^2 e^{-t/\tau_3}}{(\tau_3 - \tau_2)} - \tau_2 - \tau_3 \right\}.$$

In all cases as t becomes large, a 'steady state' is reached when

$$[S_p] = v_0 \left(t - \sum_{i=2}^{n} \tau_i \right),$$

which describes a line intersecting the t axis at $\sum_{i=2}^{n} \tau_i$. The transient time for any sequence is therefore the sum of the transients for the individual coupling enzymes. There will therefore be an increase in the lag time as the number of coupling enzymes increases. The term 'steady state' refers to the linear portion of fig. 9.1 and not to the steady state that is maintained between a particular enzyme, substrate and enzyme–substrate complex. In deriving the general expression for a linear sequence of reactions, the concentrations of the coupling enzymes must be arranged so that $V_n/K_n \gg V_{n-1}/K_{n-1} \gg \cdots \gg V_2/K_2 \gg v_0$. If this is not so, there is the possibility that the concentration of one of the intermediates may rise to a value approaching its K_m. This will invalidate the assumption made in deriving equation (9.12) for $[S_p]$, namely $[S_i] \ll K_i$, and hence the apparent velocity v_0 will contain a contribution due to the partially rate-limiting step in the coupling enzymes. In order to obtain a complete plot of v_0 against $[S_1]_0$ for the first enzyme, a fairly wide range of $[S_1]_0$ concentrations, at least $0.1K_1$–$5.0K_1$, must be studied. At the low substrate concentrations, the linear portion of the $[S_p]$ against t plot may be of such short duration that it is impossible to obtain accurate estimates of the steady-state velocity v_0. Easterby's method makes no use of the early curved portion of the $[S_p]$ against t plot and obviously it would be very useful if this portion of the graph could be used to extract additional information about the first enzyme. Such an analysis has been presented by Kuchel *et al.* (1974).

Their approach to solving equations (9.3)–(9.5) is an example of integration by series to obtain a solution to a differential equation or set of differential equations that cannot be integrated by normal methods. The integral of the derivative of a function, $f'(t)$, between particular limits simply describes the behaviour of the function between the same limits. This enables us to predict the value of the function at some value $t + \Delta t$ knowing the value at t. We can achieve a similar result using the Maclaurin's series. This states that if $f(t)$ can be expanded as a *convergent* power series for a given range of values of t then,

$$f(t) = f(0) + f'(0)\,t + f''(0)\frac{t^2}{2!} + f'''(0)\frac{t^3}{3!} + \cdots + f^{\langle n\rangle}(0)\frac{t^n}{n!}$$

where $f'(0), f''(0), \cdots, f^{\langle n\rangle}(0)$ are the first, second and nth derivatives of the function giving t the value zero. The time domain over which the series will represent the true solution depends on the number of terms that are included in the series. We can therefore express each reactant by its respective Maclaurin polynomial; thus for $[S_1]$:

$$[S_1]_t = [S_1]_0 + [S_1]'_0\,t + [S_1]''_0\frac{t^2}{2!} + [S_1]'''_0\frac{t^3}{3!} + \cdots + [S_1]_0^{\langle n\rangle}\frac{t^n}{n!}.$$

At any time, t, the following conservation equation is valid:

$$[S_1]_0 = [S_1]_t + [S_2]_t + [S_p]_t. \qquad (9.13)$$

For $[S_1]$, we may write

$$[S_1]'_0 = -\frac{V_1[S_1]_0}{(K_1 + [S_1]_0)},$$

$$[S_1]''_0 = \frac{V_1^2 K_1[S_1]_0}{(K_1 + [S_1]_0)^3}.$$

For $[S_2]$,

$$[S_2]' = \frac{V_1[S_1]}{(K_1 + [S_1])} - \frac{V_2[S_2]}{(K_2 + [S_2])}$$

$$[S_2]'_0 = \frac{V_1[S_1]_0}{(K_1 + [S_1]_0)} \qquad (\text{since } [S_2]_0 = 0)$$

and $$[S_2]''_0 = -\frac{V_1 V_2[S_1]_0}{K_2(K_1 + [S_1])} + \frac{V_1^2 K_1[S_1]_0}{(K_1 + [S_1]_0)^3}.$$

For $[S_p]$,

$$[S_p]' = \frac{V_2[S_2]}{(K_2 + [S_2])}$$

hence $[S_p]' = 0 \qquad (\text{since } [S_2]_0 = 0)$

and $$[S_p]''_0 = \frac{V_1 V_2[S_1]_0}{K_2(K_1 + [S_1]_0)}.$$

We can therefore express the concentration of each species in terms of its Maclaurin polynomial up to terms in t^2:

$$[S_1]_t = [S_1]_0 - \frac{V_1[S_1]_0 t}{(K_1 + [S_1]_0)} + \frac{V_1^2 K_1[S_1]_0}{(K_1 + [S_1]_0)^3}\frac{t^2}{2}$$

$$[S_2]_t = 0 + \frac{V_1[S_1]_0 t}{(K_1 + [S_1]_0)} - \left\{\frac{V_1 V_2[S_1]_0}{K_2(K_1 + [S_1]_0)} + \frac{V_1^2 K_1[S_1]_0}{(K_1 + [S_1]_0)^3}\right\}\frac{t^2}{2}$$

$$[S_p]_t = 0 + 0 + \frac{V_1 V_2[S_1]_0}{K_2(K_1 + [S_1]_0)}\frac{t^2}{2}$$

As a check, it can be seen that the sum of the terms on each side of the above equations equals $[S_1]_0$ as required by the conservation equation (9.13). Hence, the concentration of the product, S_p, is described by the equation:

$$[S_p] = \frac{V_1 V_2[S_1]_0}{K_2(K_1 + [S_1]_0)}\frac{t^2}{2}. \tag{9.14}$$

In general, when there are n coupled reactions:

$$[S_p] = \frac{V_1 V_2 \cdots V_n[S]\, t^n}{K_2 K_3 \cdots K_n(K_1 + [S_1]_0)\, n!} \qquad (p = n + 1). \tag{9.15}$$

These equations will remain true at times small enough to permit the polynomial for the final product to be truncated at the first non-zero term of the series.

The practical use of equation (9.14) may be illustrated immediately in relation to a coupled enzymic assay for which the experimenter has obtained values of $[S_p]$ at various time intervals. It follows from equation (9.14) that:

$$\lim_{t \to 0} \frac{d[S_p]}{d(t^2)} = \frac{V_1 V_2 [S_1]_0}{2K_2(K_1 + [S_1]_0)}. \tag{9.16}$$

The initial slope of a plot of $[S_p]$ against t^2 is therefore

$$\text{slope} = K \cdot \frac{V_1 [S_1]_0}{(K_1 + [S_1]_0)} \tag{9.17}$$

where $K = V_2/2K_2$. This slope when plotted against $[S_1]_0$ is just one point of a rectangular hyperbola. The slopes of a number of $[S_p]$ against t^2 plots for varying initial concentrations of $[S_1]_0$ can therefore be plotted in double reciprocal form to yield a straight line with an abscissal intercept of $-1/K_1$ and an ordinal intercept $2K_2/V_1 V_2$. The second enzyme is of course chosen because its product may be readily assayed. The values of V_2 and K_2 can therefore be determined by an independent study of the reaction of S_2 with the second enzyme under the *same* conditions, e.g. pH, ionic strength, buffer type, that are used in the coupled assay. It follows that the kinetic parameters, V_1 and K_1, for the first enzyme can be determined. This method of analysis can be extended to any number of coupled enzymes using equation (9.15); thus for a linear sequence of three enzymes the slope of a plot of $[S_p]$ against t^3 would be determined. A plot of $[S_p]$ against t and t^2 for two coupled reactions is shown in fig. 9.3. Obviously the t^2 plot starts to deviate quite markedly from linearity when the linear region of the $[S_p]$ against t plot is reached, but it is the initial slope at $t = 0$ of the t^2 plot that is important. This method of analysis is more suitable for systems where the catalytic rates of the two enzymes are approximately the same. There is no need, therefore, to determine an optimum concentration for the second enzyme. If the concentrations of the two enzymes are similar (assuming similar turnover rates per mole) then a curve with a long transient time will be obtained and this treatment can be applied satisfactorily. If the concentration of the second enzyme is much higher than the first enzyme, then there will be a short transient time and the data can be analysed by the Easterby treatment. Kuchel, Nichol & Jeffrey (1974) showed that

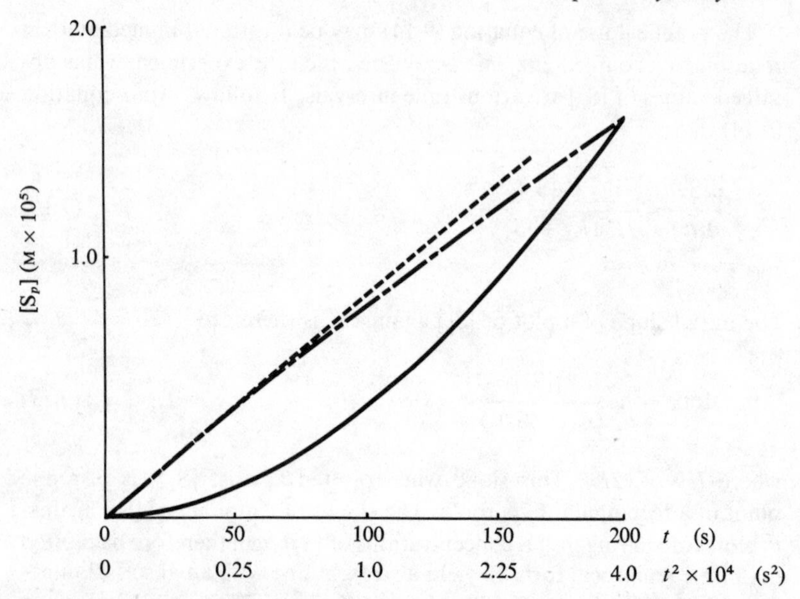

Fig. 9.3. A product/time plot (solid line) of the concentration of the final product, S_p, for the same set of parameters as used in fig. 9.2 with the exception that $[E_2]_0 = 2 \times 10^{-8}$ M (i.e. $[E_2]_0 = 2 \times [E_1]_0$). The two dashed lines illustrate the plot of $[S_p]$ versus t^2 (----) and the limiting tangent at $t = 0$ of the t^2 plot (-- --).

although the inclusion of additional terms in the Maclaurin polynomial, e.g. t^3, improved the fit of the $[S_p]$ against t plot, the limit as $t \to 0$ of $d[S_p]/d(t^2)$ was given adequately by equation (9.16). They further showed that the behaviour of the coupled system always reflected the behaviour of the first enzyme. Thus a Michaelis–Menten enzyme monitored by a control enzyme gave a hyperbolic plot of the slope given by equation (9.16) against $[S_1]_0$ (equation (9.17)). Similarly a control enzyme monitored by a Michaelis–Menten enzyme gave an equation corresponding to equation (9.16) of the control type (see equation (8.6) and fig. 8.3*a*). It can also be shown (Nichol *et al.*, 1974) that if two Michaelis–Menten enzymes interact when catalysing consecutive reactions, then the overall response of the system can be similar to that of a system involving at least one 'sigmoidal' enzyme. Hence a control-type response can be generated from two enzymes that seemingly lack any control effect.

It is interesting to note that expansion of the exponential term in the

Easterby treatment (equation (9.9)) to terms in t^2 also gives equation (9.14):

$$[S_p] = v_0(t + \tau_2\,e^{-t/\tau_2} - \tau_2)$$

$$= v_0\left\{t + \tau_2\left(1 - \frac{t}{\tau_2} + \frac{t^2}{\tau_2^2}\right) - \tau_2\right\}$$

$$= \frac{v_0\,t^2}{2\tau_2}$$

$$= \frac{V_1[S_1]_0\,V_2}{(K_1 + [S_1]_0)\,K_2}\,\frac{t^2}{2}.$$

These two methods of analysing the experimental data obtained from steady-state coupled assays are therefore complementary.

9.2. Pre-steady-state and transient behaviour of coupled enzyme systems

The previous treatments apply only to the steady-state region of the coupled systems. The only parameters that can be determined from such experiments are V and K_m. Only by an analysis of the transient phase of the coupled reactions can individual rate constants be obtained. In chapter 7 we considered the behaviour of an enzyme following the three-step mechanism in the pre-steady-state region under two conditions $[S_0] \gg [E_0]$ and $[E_0] \gg [S_0]$. Under certain circumstances, we can also solve the differential equations that describe the similar case of a coupled system of two enzymes, each of which operates by a two-step mechanism:

$$E + S \underset{k_{-1}}{\overset{k_1}{\rightleftharpoons}} ES \xrightarrow{k_2} E + P$$

$$F + P \underset{k_{-3}}{\overset{k_3}{\rightleftharpoons}} FP \xrightarrow{k_4} F + R$$

In the transient phase of an enzymic reaction, the amount of product formed is of the same order of magnitude as the enzyme concentration. If the concentration of the coupling enzyme is sufficiently high, then during this transient phase (and perhaps longer) the amount of F bound as FP will be much smaller than the concentration of free F, [F]. Under these conditions, $[F] \approx [F_0]$. The differential equations that describe the above system are given below, in which the differential (d/dt) has been replaced by the Laplace–Carson operator ψ. The time-dependent transformed species are shown in lower case letters. The additional assumption is made that $[S_0] \gg [E_0]$ and so $[S] \approx [S_0]$.

$$\psi[\text{es}] = k_1[\text{e}][S_0] - k_E[\text{es}] \quad (\text{where } k_E = k_{-1} + k_2). \tag{9.18}$$

$$\psi[\text{p}] = k_2[\text{es}] - k_3[F_0][\text{p}] + k_{-3}[\text{fp}]. \tag{9.19}$$

$$\psi[\text{fp}] = k_3[F_0][\text{p}] - k_F[\text{fp}] \quad (\text{where } k_F = k_{-3} + k_4). \tag{9.20}$$

$$\psi[\text{r}] = k_4[\text{fp}]. \tag{9.21}$$

$$[E_0] = [\text{e}] + [\text{es}]. \tag{9.22}$$

We can derive transforms for the various species in terms of the initial concentrations, rate constants and the operator ψ. Thus

$$[\text{es}] = \frac{k_1[E_0][S_0]}{(\psi + \lambda_1)},$$

where

$$\lambda_1 = k_1[S_0] + k_E. \tag{9.23}$$

$$[\text{p}] = \frac{k_1 k_2[E_0][S_0](\psi + k_F)}{(\psi + \lambda_1)(\psi + \lambda_2)(\psi + \lambda_3)},$$

where

$$\lambda_2 = \{A + \sqrt{(A^2 - 4B)}\}/2$$

$$\lambda_3 = \{A - \sqrt{(A^2 - 4B)}\}/2$$

$$A = k_3[F_0] + k_F$$

$$B = k_3 k_4[F_0].$$

$$[\text{fp}] = \frac{k_1 k_2 k_3[E_0][S_0][F_0]}{(\psi + \lambda_1)(\psi + \lambda_2)(\psi + \lambda_3)}.$$

$$[\text{r}] = \frac{k_1 k_2 k_3 k_4[E_0][S_0][F_0]}{\psi(\psi + \lambda_1)(\psi + \lambda_2)(\psi + \lambda_3)}.$$

The original or solution for [r] is therefore

$$[R] = \prod_{i=1}^{4} \frac{k_i[E_0][S_0][F_0]}{\lambda_1 \lambda_2 \lambda_3} t - \sum_{i=1}^{3} C_i(1 - e^{-\lambda_i t}), \tag{9.24}$$

where

$$C_i = \frac{\prod_{n=1}^{4} k_n[E_0][F_0][S_0]}{\lambda_i^2 \prod_{j=1}^{3} (\lambda_j - \lambda_i) \quad j \neq i}.$$

Substitution for the λ_i in the first term of equation (9.24) gives

$$[\mathrm{R}] = v_0 t - \sum_{i=1}^{3} C_i (1 - \mathrm{e}^{-\lambda_i t}), \tag{9.25}$$

where

$$v_0 = \frac{k_2[\mathrm{E_0}]\,[\mathrm{S_0}]}{(K_m + [\mathrm{S_0}])} \quad \text{and} \quad K_m = (k_{-1} + k_2)/k_1.$$

If the concentration of the second enzyme is high enough then it is probable that the concentration of the first product, [P], will remain lower than the K_m of the second enzyme even during the steady-state region of the reaction. Under these conditions the amount of F bound as FP will remain much smaller than the concentration of free enzyme and equation (9.25) will remain valid. As t increases, the exponential terms become vanishingly small (λ_1, λ_2 and λ_3 are all positive) and equation (9.25) reduces to

$$[\mathrm{R}] = v_0 t - \sum_{i=1}^{3} C_i.$$

As t increases, therefore, a plot of [R] against t becomes linear with a slope of v_0 equal to the velocity of the first reaction. The linear region when extrapolated back cuts the t axis at

$$\tau = \sum_{i=1}^{3} C_i/v_0$$

and intersects the [R] axis at

$$[\mathrm{R}] = -\sum_{i=1}^{3} C_i = -v_0 \tau = \beta.$$

A plot of equation (9.25) is given in fig. 9.4 and compared to a numerically integrated solution of the differential equations (see chapter 10). Since the second enzyme is used because its final product can be conveniently monitored, an independent study of the reaction of F and P should give the two constants λ_2 and λ_3. The remaining constant λ_1 can be determined by fitting equation (9.25) by computer for various estimates of λ_1 until an optimum fit is achieved. From equation (9.25),

$$[\mathrm{R}] - v_0 t - \beta = \sum_{i=1}^{3} C_i \mathrm{e}^{-\lambda_i t}, \tag{9.26}$$

and so an alternative approach to determining the constants is to plot

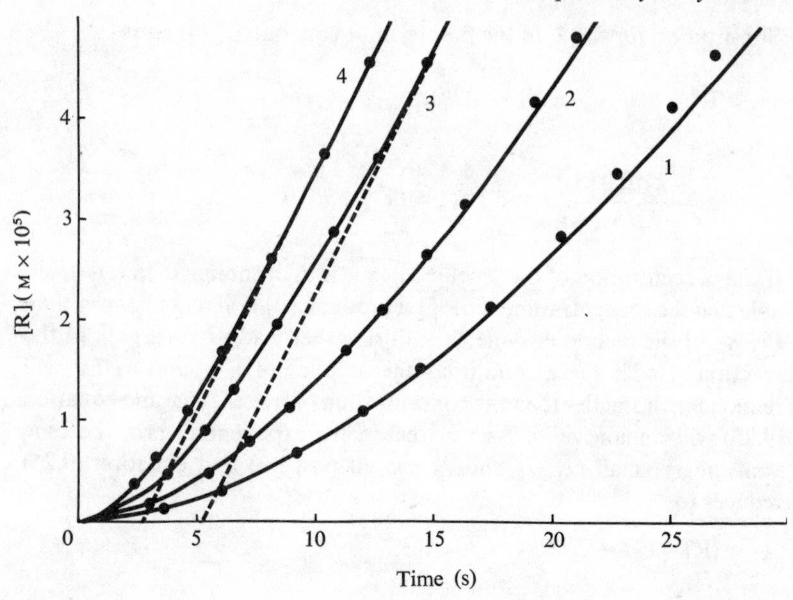

Fig. 9.4. A comparison of the analytical solution (●) given by equation (9.25) with the solution (——) obtained by the numerical integration of the basic set of differential equations that describe the coupled system. For all curves: $k_1 = 10^7$ M^{-1} s^{-1}; $k_{-1} = 10^4$ s^{-1}; $k_2 = 10^3$ s^{-1}; $k_3 = 2 \times 10^7$ M^{-1} s^{-1}; $k_{-3} = 2 \times 10^4$ s^{-1}; $k_4 = 2 \times 10^3$ s^{-1}; $[S_0] = 10^{-3}$ M and $[E_0] = 10^{-8}$ M. The various concentrations of the second enzyme, $[F_0]$, are: 1, 2×10^{-8} M; 2, 4×10^{-8} M; 3, 1×10^{-7} M; and 4, 2×10^{-7} M. The slopes of the dashed lines in curves 3 and 4 give the velocity (v_0) of the respective reactions after the attainment of a steady-state of [P] and the intercept on the t-axis yields τ. (From Kuchel & Roberts (1974) *Biochim. Biophys. Acta* **364**, 181–92.)

$\ln([R] - v_0 t - \beta)$ against t. Over long times, equation (9.26) reduces to a single exponential and hence the values of λ_i and C_i corresponding to the slowest exponential decay can be determined from the slope and ordinal intercept of the log plot. This can be repeated until all three λ_i are determined. The various rate constants for the first enzyme can be determined from the steady-state information V and K_m and from the variation of λ_1 with initial substrate concentration (equation (9.23)). With the assumptions made in the derivation of equation (9.21), the behaviour of [R] with time is very similar to the Easterby steady-state treatment, but since the transient phase is now expressed in terms of the various exponentials, additional information can be extracted from the analysis of the pre-steady-state and steady-state regions. This analysis can be applied in an exactly similar manner to two three-step mechanisms:

$$E + S \underset{k_{-1}}{\overset{k_1}{\rightleftharpoons}} ES \xrightarrow{k_2} EP \xrightarrow{k_3} E + P$$
$$\downarrow P_1$$

$$F + P \underset{k_{-4}}{\overset{k_4}{\rightleftharpoons}} FP \xrightarrow{k_5} FR \xrightarrow{k_6} F + R.$$
$$\downarrow R_1$$

With the three-step mechanisms, there is the additional possibility of coupling the second enzyme through the first product, P_1, and also monitoring the reaction via R_1 or R. In any event there will be more exponential terms to analyse. Thus for the particular scheme given above,

$$[R] = \frac{\prod_{i=1}^{6} k_i[E_0][S_0][F_0]\, t}{\prod_{j=1}^{5} \lambda_j} - \sum_{i=1}^{5} C_i(1 - e^{-\lambda_i t}), \tag{9.27}$$

where

$$C_i = \frac{\prod_{n=1}^{6} k_n[E_0][S_0][F_0]}{\lambda_i^2 \prod_{j=1}^{5} (\lambda_j - \lambda_i) \quad j \neq i}$$

$$\lambda_1 = \{A + \sqrt{(A^2 - 4B)}\}/2$$

$$\lambda_2 = \{A - \sqrt{(A^2 - 4B)}\}/2$$

$$A = k_1[S_0] + k_{-1} + k_2 + k_3$$

$$B = k_1(k_2 + k_3)([S_0] + K_m)$$

$$K_m = k_3(k_{-1} + k_2)/\{k_1(k_2 + k_3)\}$$

$$\lambda_3 = \{C + \sqrt{(C^2 - 4D)}\}/2$$

$$\lambda_4 = \{C - \sqrt{(C^2 - 4D)}\}/2$$

$$C = k_4[F_0] + k_{-4} + k_5$$

$$D = k_4 k_5[F_0]$$

$$\lambda_5 = k_6.$$

In order to analyse equation (9.27) it is obvious that an independent study of the second reaction under the same conditions as the coupled reaction is essential. Elimination of the three exponential terms associated with the second reaction (λ_3, λ_4 and λ_5) considerably eases the determination of the additional information regarding the first enzyme. The behaviour is similar to equation (9.25) with a steady state, equal to the velocity of the first reaction, reached as t becomes large.

The coupled system can also be studied under the conditions of $[E_0], [F_0] \gg [S_0]$. Under these conditions, in which the concentrations of the enzymes are quite high, it is possible that enzyme–enzyme interactions may become important. Thus if the two enzymes can be studied independently by both steady-state and fast reaction methods, it is possible to predict their behaviour as a coupled system. Deviations from the predicted behaviour would indicate a change in mechanism such as an enzyme–enzyme interaction. Equations (9.18)–(9.22) can be solved by the Laplace–Carson procedure. The conservation equation (9.22) for $[E_0]$ is replaced by one for $[S_0]$:

$$[S_0] = [s] + [es] + [p] + [fp] + [r].$$

The concentrations of [E] and [F] are now regarded as constants ($[E] = [E_0]$, $[F] = [F_0]$) and $[S_0]$ becomes [s] in equation (9.18). The transform for [r] is then found to be

$$[r] = \frac{k_1 k_2 k_3 k_4 [E_0][S_0][F_0]}{(\psi + \lambda_1)(\psi + \lambda_2)(\psi + \lambda_3)(\psi + \lambda_4)}, \tag{9.28}$$

where

$$\lambda_1 = \{A + \surd(A^2 - 4B)\}/2$$
$$\lambda_2 = \{A - \surd(A^2 - 4B)\}/2$$
$$A = k_1[E_0] + k_{-1} + k_2$$
$$B = k_1 k_2 [E_0]$$
$$\lambda_3 = \{C + \surd(C^2 - 4D)\}/2$$
$$\lambda_4 = \{C - \surd(C^2 - 4D)\}/2$$
$$C = k_3[F_0] + k_{-3} + k_4$$
$$D = k_3 k_4 [F_0].$$

The solution for [r] from equation (9.28) is therefore

$$[R] = \prod_{i=1}^{4} \frac{k_i}{\lambda_i} [E_0][F_0][S_0] - \sum_{i=1}^{4} C_i e^{-\lambda_i t}. \tag{9.29}$$

Substitution for $\lambda_1\lambda_2$ and $\lambda_3\lambda_4$ in the first term of equation (9.29) gives

$$[\mathrm{R}] = [\mathrm{S_0}] - \sum_{i=1}^{4} C_i \mathrm{e}^{-\lambda_i t},$$

$$C_i = \frac{\prod_{n=1}^{4} k_n [\mathrm{E_0}][\mathrm{S_0}][\mathrm{F_0}]}{\lambda_i \prod_{j=1}^{4} (\lambda_j - \lambda_i)_{j \neq i}}.$$

Fig. 9.5 shows that there is no steady-state under these experimental conditions; the final product rises from zero in a complicated exponential fashion to a final concentration, $[\mathrm{S_0}]$. A similar expression can be derived for two coupled three-step mechanisms.

From the foregoing treatment of both steady-state and pre-steady-

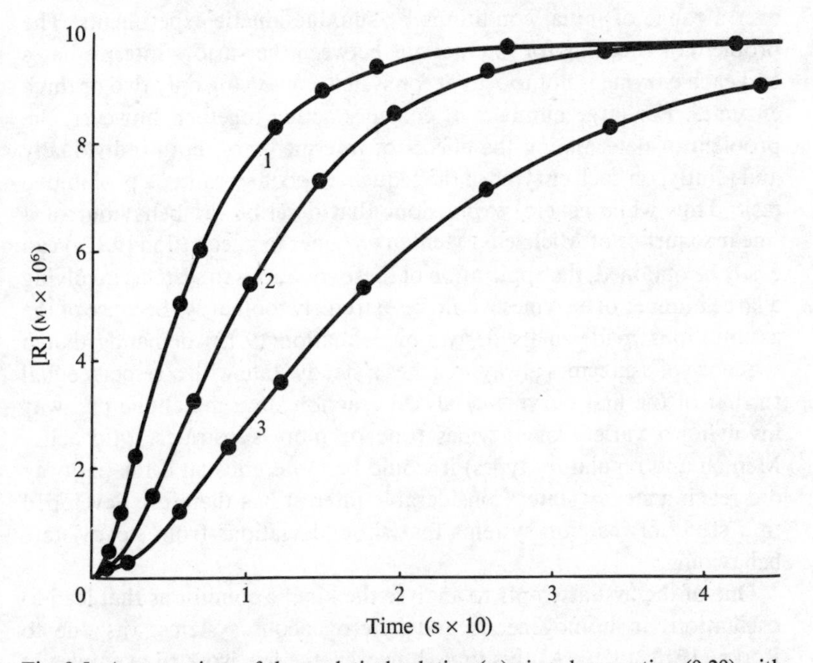

Fig. 9.5. A comparison of the analytical solution (●) given by equation (9.29) with the solution (——) obtained by numerical integration of the basic set of differential equations that describe the coupled system. The same set of parameters are used as in fig. 9.4 with the exception that $[\mathrm{S_0}] = 10^{-5}$ M and $[\mathrm{F_0}] = 10^{-5}$ M. The various concentrations of the first enzyme, $[\mathrm{E_0}]$, are: 1, 7.5×10^{-6} M; 2, 2×10^{-5} M; and 3, 8×10^{-5} M. (From Kuchel & Roberts (1974) *Biochim. Biophys. Acta* **364**, 181–92.)

state regions of coupled enzymic reactions and given favourable experimental conditions, it is theoretically possible to determine most if not all of the individual rate constants for the first enzyme-catalysed reaction. In all cases the second reaction should be studied independently and consideration given to the possible inhibitory nature of factors relating to the first enzyme, e.g. co-factors, metal ions, second product.

9.3. Oscillatory behaviour

In the discussion on the steady-state behaviour of coupled enzyme reactions, we made a rather arbitrary assumption that there was no change in the mechanism during the course of the reaction. It should be realised by now that a lot of apparently insoluble problems associated with enzyme kinetics can be solved analytically by making reasonable assumptions relating to the particular problem under investigation. The resulting solution should of course always be checked for its validity over a range of initial conditions by suitable kinetic experiments. The problem of checking for interactions between the various intermediates and each enzyme is not too great for systems involving only two or three enzymes. For large numbers of enzymes acting together, however, the problem of determining the effects of intermediates, both individually and jointly, on each enzyme in the sequence becomes rather a prohibitive task. Thus while general expressions that describe the behaviour of a linear sequence of Michaelis–Menten enzymes (e.g. equation (9.12)) can easily be obtained, the application of these equations to systems involving a large number of enzymes would be extremely foolhardy. Because of the assumptions made in its derivation, equation (9.12) demands that a sequence of n enzymes always reaches a steady state with a velocity equal to that of the first enzyme-catalysed reaction. In a metabolic pathway involving a variety of enzymes (one or more substrates, Michaelis–Menten and regulatory types) it would be more unusual if the pathway did reach a steady state. Considerable interest has therefore developed in a study of reaction systems that show deviations from steady-state behaviour.

One of the first attempts to analyse the kinetic conditions that lead to oscillations in homogeneous and heterogeneous systems was due to Lotka (1910, 1920). At this time, however, the few isolated examples of oscillatory processes were viewed sceptically and it was generally believed that all physicochemical systems lead invariably to a steady state. This was thought to follow immediately from the second law of thermodynamics. The possibility of sustained oscillations was therefore ruled out. In the 1940s Schrödinger (1944) and Prigogine (1947) introduced the concept of biological systems as open systems and this enabled rhythmic behaviour to be postulated theoretically. The investigation of biological periodicities such as the circadian clock (Bünning, 1964; Brown, Hastings & Palmer, 1970) and the rhythmic behaviour of the central nervous system shows that, at least on the large scale, oscillations are in fact quite a common occurrence. The discovery of oscillatory behaviour at the cellular level in the glycolytic pathway in yeast cells (Ghosh & Chance, 1964) in the peroxidase-catalysed reaction (Yamazaki, Yokota & Nakajima, 1965) and in the biosynthesis of certain proteins (Knorre, 1968), added considerable impetus to the analysis of such systems. Chance and co-workers (1964*a*,*b*) also showed that the oscillatory behaviour of glycolytic pathway was still present in cell-free extracts. The rhythmic activity, therefore, was not due to interactions with any of the physical structures in the cell and was purely a manifestation of certain interactions between the different enzymes and intermediates in the pathway.

Lotka (1910) formulated the following scheme:

$$\begin{aligned} &\longrightarrow \mathrm{A} \\ \mathrm{A} + \mathrm{B} &\xrightarrow{k_1} 2\mathrm{B} \\ \mathrm{B} &\xrightarrow{k_2} \end{aligned}$$

in which the substance A is introduced at a constant rate and is converted by an autocatalytic reaction into B. The latter product is lost by a first order process. Lotka showed that the steady-state concentration of A is k_2/k_1 and this concentration may be approached via damped oscillations. In a second paper published in 1920, he introduced the concept of cross-coupling between reactions and showed that under certain conditions continuous oscillations could occur. This mechanism can be written as

$$\begin{aligned} [\mathrm{G}] + \mathrm{A} &\xrightarrow{k_1} 2\mathrm{A} \\ \mathrm{A} + \mathrm{B} &\xrightarrow{k_2} 2\mathrm{B} \\ \mathrm{B} &\xrightarrow{k_3} \end{aligned} \qquad (9.30)$$

The mechanism was proposed as a model of host–parasite, prey–predator interactions and thus simulated ecological oscillations. In this model, animal A has an abundant supply of grass, [G], and hence multiplies. Animal B, in turn, eats animal A and multiplies. The rate of growth of B therefore depends on the quantity of A and the number of encounters between A and B. Animal B eventually dies. These equations can be expressed as differentials by

$$\frac{d[A]}{dt} = k_1'[A] - k_2[A][B]$$

$$\frac{d[B]}{dt} = k_2[A][B] - k_3[B]$$

where $k_1' = k_1[G]$. Unfortunately, these equations cannot be solved by any of the simple methods that we have used so far. Bak (1963) has obtained an expression for the frequency of the oscillations using a method devised by Kryloff & Bogoliuboff (1947) involving Fourier expansions. The solutions for A and B contain sine and cosine terms. These equations can, however, be solved by digital or analogue computer techniques and one example is shown in fig. 9.6. Many models have been published since Lotka's original paper but they are all essentially only modifications of the original scheme. Nearly all models are impossible

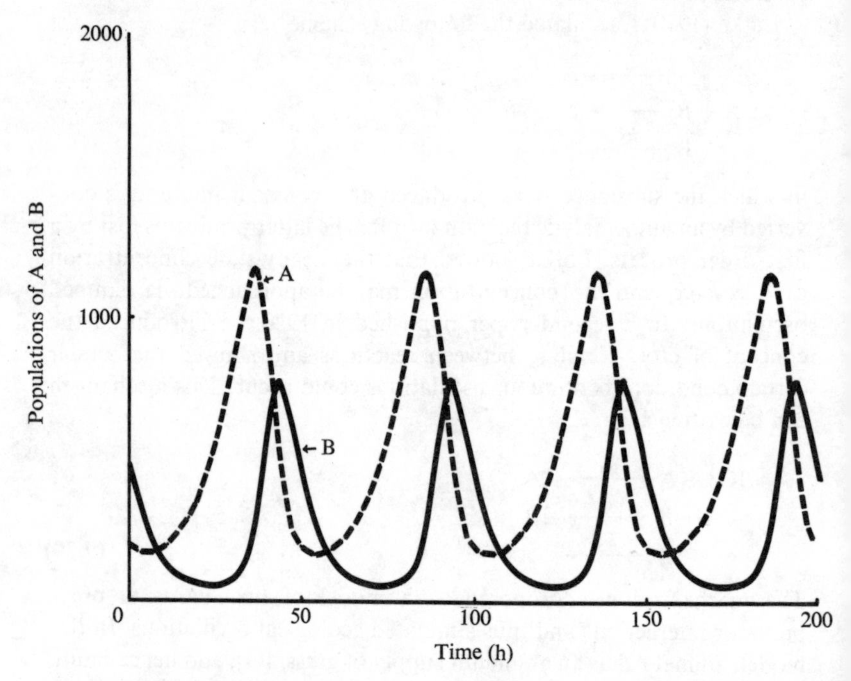

Fig. 9.6. Typical behaviour of the Lotka model (equation (9.30)). $k_1 = 0.05\ h^{-1}$; $k_2 = 2 \times 10^{-4}\ h^{-1}$; $k_3 = 0.1\ h^{-1}$. For explanation see text.

to solve analytically but they can be scanned for periodic solutions by the help of computers. In such ways many workers have found periodic solutions in a variety of hypothetical but kinetically plausible reactions (Higgins, 1967; Lindblad & Degn, 1967; Morales & McKay, 1967; Spangler & Snell, 1967; Sel'kov, 1968). Since a linear sequence of n reactions, in which there are no interactions, cannot oscillate, all models contain feedback or feedforward loops. Four general types of loops that can feed information forwards or backwards can exist. These are indicated below:

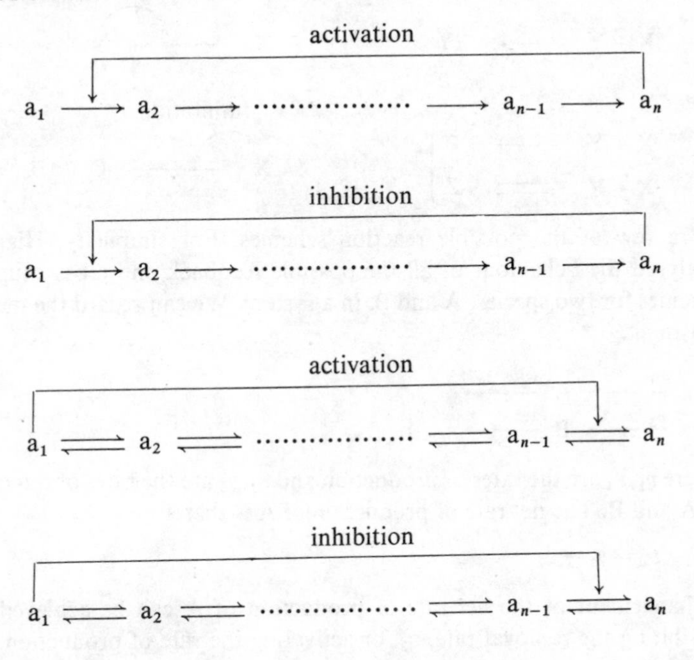

The last two schemes are not effective unless the reaction steps are reversible. Lotka's model (1920) is an example of the first scheme:

activation activation

$$[G] \longrightarrow A \longrightarrow B$$

Higgins (1967) has analysed many possible reaction schemes in terms of activation or inhibition feedback or feedforward loops. Thus,

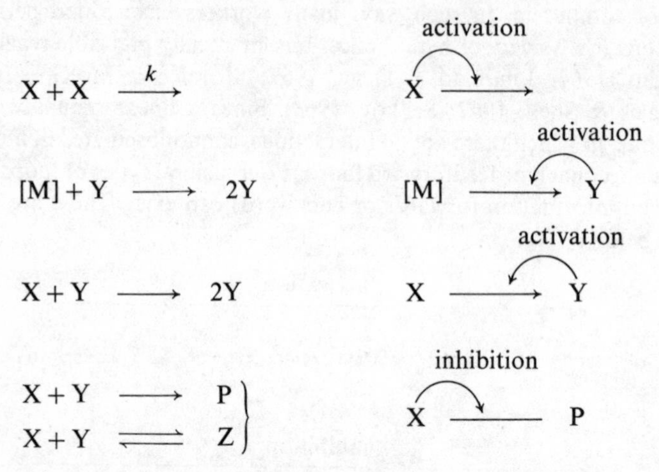

are a few of the possible reaction schemes. For simplicity, Higgins analysed the behaviour of all the possible feedback and cross-coupled schemes for two species, A and B, in a system. We can regard the mechanism as

$$\xrightarrow{v_1} \mathrm{A} \xrightarrow{v_2}$$
$$\xrightarrow{v_3} \mathrm{B} \xrightarrow{v_4}$$

where v_1, v_3 are the rates of production and v_2, v_4 are the rates of removal of A and B. The net rate of production of A is therefore

$$v_a = v_1 - v_2.$$

Enhancement of the *net* rate of production of A can be achieved by inhibiting the removal rate, v_2, or activating the rate of production, v_1. The mechanism can be written as a net flux diagram in which the arrows →⌒ indicate possible inhibition or activation of the net rates of production of A and B.

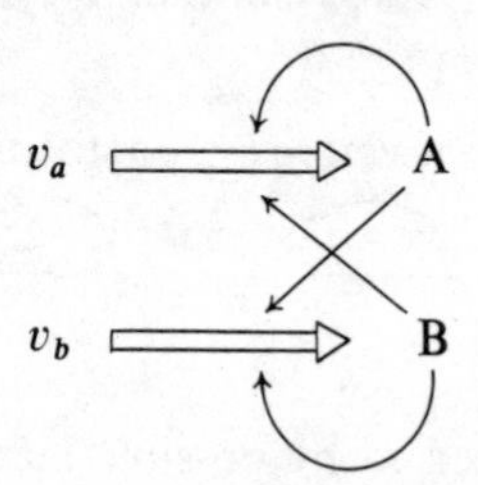

Thus the possible schemes for a sequential scheme in which A precedes B are

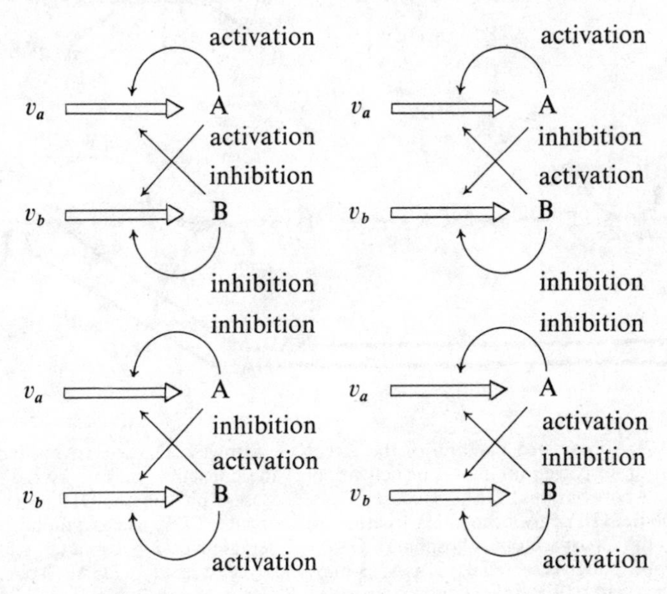

Higgins (1964) has summarised the conditions under which oscillations may occur for a two species system:

(1) One of the chemicals must activate its own production, i.e. increasing its concentration will tend to increase its net rate of production.

(2) The other chemical must tend to inhibit its own net rate of production. This is normally true for most chemical reactions since increasing the concentration of the chemical will increase its net rate of removal.

(3) There must be cross-coupling of opposite character as shown in the diagrams above. If an increase in A increases the net rate of production of B, then increasing B must decrease the net rate of production of A.

Higgins (1964, 1967), Chance (Ghosh & Chance, 1964; Chance *et al.*, 1964*a*,*b*) have analysed the oscillations in the glycolytic pathway by analogue and digital computer simulation and have suggested that the oscillations are due to the product activation of the allosteric enzyme, phosphofructokinase. This enzyme is also activated by AMP, ADP and

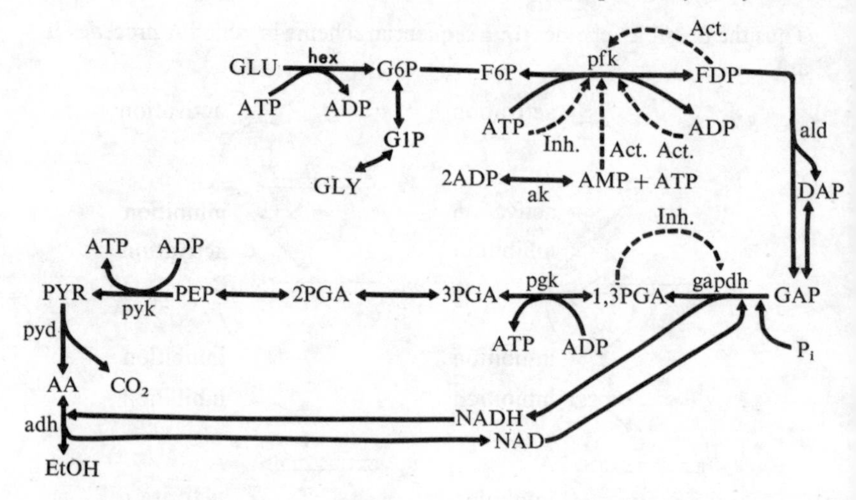

Fig. 9.7. Abbreviated diagram of the glycolytic pathway. Dashed arrows indicate known or suspected inhibition or activation of the particula enzyme by the metabolite. Abbreviations: GLU, glucose; G6P, glucose-6-phosphate; G1P, glucose-1-phosphate; GLY, glycogen; F6P, fructose-6-phosphate; FDP, fructose diphosphate; DAP, dihydroxyacetone phosphate; GAP, glyceraldehyde-3-phosphate; 1,3PGA, 1,3-diphosphoglyceric acid; 3PGA, 3-phosphoglyceric acid; 2PGA, 2-phosphoglyceric acid; PEP, phosphoenolpyruvate; PYR, pyruvate; AA, acetaldehyde; EtOH, ethanol; AMP, ADP, ATP, adenosine mono-, di-, triphosphate; NAD, NADH, nicotinamide adenine dinucleotide; P_i, phosphate. Enzymes: hex, hexokinase; pfk, phosphofructokinase; ak, adenylate kinase; ald, aldolase; gapdh, glyceraldehyde-3-phosphate dehydrogenase; pgk, phosphoglycerate kinase; pyk, pyruvate kinase; pyd, pyruvate decarboxylase; adh, alcohol dehydrogenase. (Adapted from Higgins (1967).)

inhibited by ATP. A number of experiments have shown that the observed oscillations have a double periodicity (Pye & Chance, 1966; Pye, 1969). Dynnik & Sel'kov (1973) have suggested that there is another oscillator in the lower half of the glycolytic pathway. This may be due to the allosteric enzyme glyceraldehyde-3-phosphate dehydrogenase which is inhibited by its product 1,3-diphosphoglycerate. It has also been shown to exist in an inactive form in the absence of NAD (Kirschner, 1971). An abbreviated diagram of the glycolytic pathway is shown in fig. 9.7. The first stage of the pathway can be summarised by the equation

$$\text{GLU} \longrightarrow \text{F6P} \xrightarrow[E_2]{\curvearrowleft \text{activation}} \text{FDP} \xrightarrow[E_3]{} \text{GAP}.$$

This system has been studied by computer simulation using the following equations:

$$
\begin{aligned}
\mathrm{GLU} &\longrightarrow \mathrm{F6P} \\
\mathrm{F6P} + \mathrm{E_2} &\rightleftharpoons \mathrm{E_2 \cdot F6P} \\
\mathrm{E_2 \cdot F6P} &\longrightarrow \mathrm{E_2 + FDP} \\
\mathrm{FDP} + \mathrm{E_2^*} &\rightleftharpoons \mathrm{E_2} \\
\mathrm{FDP} + \mathrm{E_3} &\rightleftharpoons \mathrm{E_3 \cdot FDP} \\
\mathrm{E_3 \cdot FDP} &\longrightarrow \mathrm{E_3 + GAP}
\end{aligned}
$$

The critical equation in the system is the reaction of FDP with an inactive form of the enzyme, E_2^*, to give the active species E_2. If this equation were not present and the enzyme E_2 only existed in an active form, then the equations would describe a linear sequence of catalysed reactions with no interactions. Analogue computer studies have shown that the above model exhibits sustained oscillations provided there is a constant supply of glucose (GLU).

The role of metabolic oscillations at the cellular level is unclear at the moment. It may simply be an accident of nature without any special significance. For instance, glycolysis, which is oscillatory in yeast, is not oscillatory in *Dictyostelium discoideum*, a slime mould. It is possible that any sequence containing a number of allosteric enzymes will inevitably oscillate under certain conditions because of the co-operative properties of the enzymes. It is certainly intriguing to argue that oscillations at the cellular level constitute an important part of the higher level circadian oscillations (Sel'kov, 1971), although the rapidity of the glycolytic oscillations (a few seconds) makes it difficult to connect it with the longer period circadian clocks. Further work is certainly needed on this interesting phenomenon.

10 Computer simulation of biochemical systems

10.1. Introduction

This chapter describes methods for the computer simulation of biochemical systems where the aim is to investigate the behaviour of the system and not necessarily to fit experimental data. The biochemistry of systems of physiological significance (e.g. multi-enzyme systems, catalysis by intact cells) is complicated and involves a number of reactions occurring simultaneously. Kinetic models of such systems inevitably lead to large numbers of non-linear differential equations with their associated problems of analysis. Before one can use any simulation techniques, it is first necessary to define the system under investigation. The model is formulated from all the knowledge currently available, i.e. the chemical reactions that are known and the probable complexes that are formed, values of rate constants, possible interactions and pathways between reactants, products and effector molecules. The determination of all these interrelated parameters is of course a major experimental exercise and the impossibility of measuring some of them is one of the reasons for using computer simulation. Although a model may be built up by the summation of the behaviour of its constituent parts, the physiological behaviour may be completely different. This may be due to the unexpected interactions between constituent parts of the model or because the conditions of the complete model are radically different from those under which the individual parts have been studied. Simulation experiments of glycolysis (Garfinkel, Frenkel & Garfinkel, 1968; Achs & Garfinkel, 1968) seem to indicate that only one out of a dozen enzymes in the pathway behaves in a manner indicated by results published in the literature. It has also been shown that the conditions prevailing in the cell invalidate some of the traditional simplifications of enzyme kinetics, particularly that the substrate is present in amounts much larger than the enzyme (Hess, Boiteux & Kruger, 1969). Model building, therefore, consists of assembling ideas, approximations and experimental results into an overall package that can be suitably studied on a computer. The model will inevitably grow by the addition of new ideas and new data but will not necessarily fit the known facts any better. A successful model does not constitute proof of the original ideas; this can only be achieved from experimental evidence. The failure of a model to describe accurately the observed facts can, however, be regarded as a disproof. It must be stressed that model building cannot be done in isolation from experiments and that the model is only as good as the data that is available. Whilst theoretical models can be simulated, eventually some experimental evidence for their validity must be sought.

The model under study can be reduced to a number of differential equations and conservation of mass equations. The solution of the differential equations can be obtained analytically by classical methods, numerical integration or by electronic analogue computation. Classical solutions of simple models are possible if the model is composed of ordinary linear and/or partial differential equations and for certain classes of non-linear differential equations. Frequently, this technique can be applied to limiting cases of complex models if approximations are acceptable. An example of such an approximation is found in the earlier chapters where the concentration of substrate is assumed to be constant over the time range of interest when $[S_0] \gg [E_0]$. Analytical solutions of non-linear models are rare and hence variable substitutions are often made to linearise the model as required. Eventually, as the models become more complicated, only the use of computers allows the solution to be obtained.

10.2. Computer simulation using analogue computers

Most people associate the word computer with digital computers that are found in most large firms, universities, banks etc. and are used essentially for the processing of data. Another and sometimes more useful computer for simulation work, however, is the analogue computer. Indeed, the first biochemical kinetic simulation was carried out with an analogue computer more than thirty years ago, before digital computers existed (Chance, 1943). Analogue computation is undervalued in most computer centres because they are staffed exclusively by digital computer experts. This is unfortunate, as the analogue computer is usually quite a small device involving only a moderate capital outlay and hence can frequently be used exclusively by one department or even one research worker in a laboratory. Such exclusive use could not be tolerated with even the smallest digital computer as the capital outlay is usually considerably larger. For the biochemist who may not be very adept at solving a number of interlinked differential equations, the analogue computer provides a rapid means of producing an experimental solution.

An analogue computer, as its name implies, makes use of the similarity between the laws of nature. For example, consider the similarity between mechanical, electrical and thermal systems:

$$F = \frac{m}{g} \cdot \frac{dv}{dt} \quad \text{Force acting on mass}$$

$$i = C \cdot \frac{de}{dt} \quad \text{Current flow through a capacitor}$$

$$Q = W \cdot C \cdot \frac{dT}{dt} \quad \text{Heat flow in a solid}$$

The similar mathematical form of these equations allows, with suitable scaling factors, the heat flow in a solid to be investigated using an electrical circuit or mechanical system as an analogue of the case under study. Early work by Lord Kelvin (Thomson, 1876*a*, *b*) showed that mechanical devices could be designed that would function as integrators and summers and that these devices could be interconnected to solve continuously a second order differential equation. By far the most useful and versatile, however, is the electrical system in which the physical parameters are represented by continuously variable voltages. The electronic analogue computer is an assembly of electronic elements which are able to add, subtract, multiply, divide and integrate, the last being the most important function. These computer units can be arranged in such a way by the operator that they describe the mathematical equations of the system of interest. The solution to the particular problem is read out as a variation in voltage with respect to time. For simple analogue computers there are four basic computing elements. These are:

(1) coefficient multiplier or potentiometer;
(2) summer and/or inverter;
(3) integrator;
(4) multiplier.

The first three are linear components and are sufficient to solve linear differential equations with constant coefficients. The multiplier is a non-linear device (i.e. it can operate on at least two variables) which allows the analogue computer to simulate the non-linear systems that frequently occur. In Table 10.1 is a list of the various computing elements and a brief discussion of their functions. All the connections to the various

Table 10.1. Summary of the most common analogue computer units and their symbols

Computing element	Electrical symbol	Computing symbol	Function
Potentiometer or Pot.	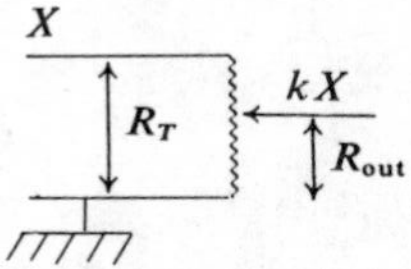 X; R_T; kX; R_{out}; $k = \frac{R_{out}}{R_T}$	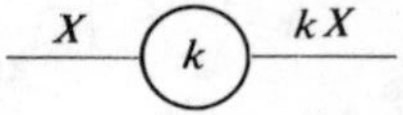 X → (k) → kX	Multiplication of a variable by a constant, k. k lies in the range, $0 \leqslant k \leqslant 1$.
Inverter	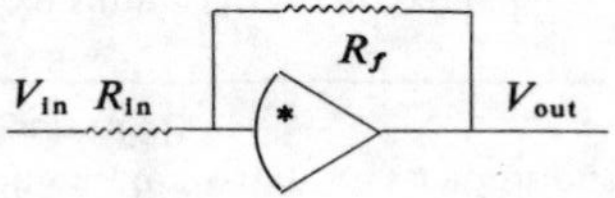 V_{in} R_{in}; R_f; * ; V_{out} * high gain amplifier Gain $= V_{out}/V_{in} = -R_f/R_{in}$ (Proof in appendix III) For an inverter, $R_f = R_{in}$; gain = 1	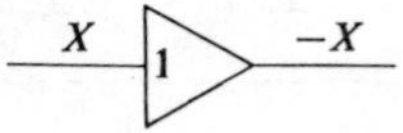 X → [1] → $-X$	The variable X is inverted to $-X$
Summer	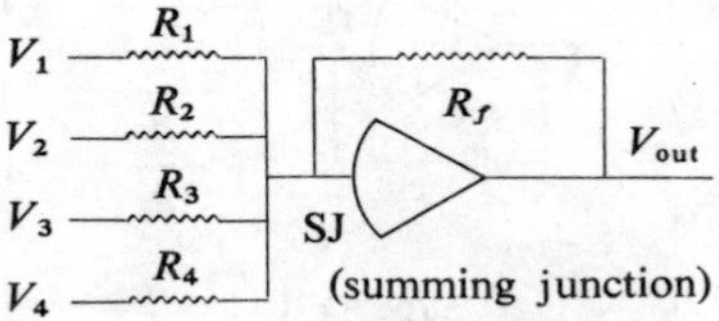 V_1 R_1; V_2 R_2; V_3 R_3; V_4 R_4; R_f; V_{out}; SJ (summing junction) $V_{out} = -R_f\left[\frac{V_1}{R_1} + \frac{V_2}{R_2} + \frac{V_3}{R_3} + \frac{V_4}{R_4}\right]$ Usually $R_1 = R_2 = R_f$ and $R_3 = R_4 = R_f/10$. Hence $V_{out} = -[V_1 + V_2 + 10V_3 + 10V_4]$	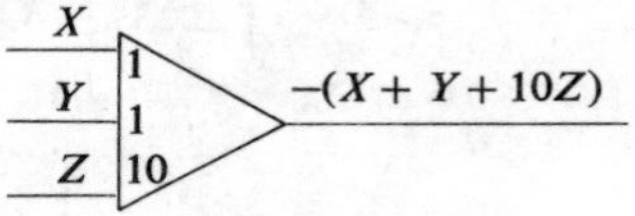 X (1), Y (1), Z (10) → $-(X + Y + 10Z)$	Sums a number of variables and inverts the sign

Computing element	Electrical symbol	Computing symbol	Function
Integrator	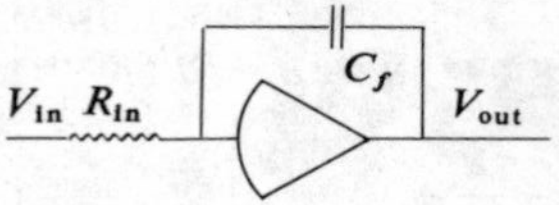 V_{in} R_{in} C_f V_{out} $V_{out} = \frac{-1}{R_{in} C_f} \int V_{in} \cdot dt$ 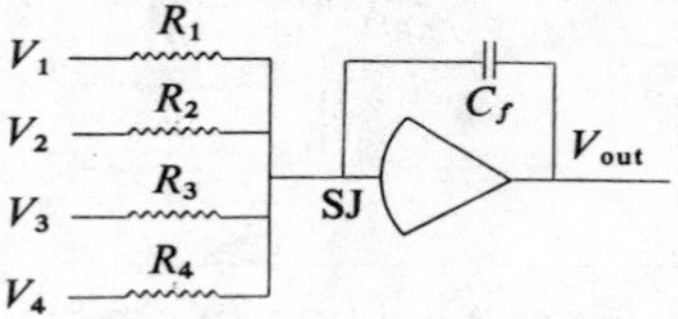 V_1 R_1, V_2 R_2, V_3 R_3, V_4 R_4, SJ, C_f, V_{out} Additional gain can be obtained by lowering the feedback capacitor C_f. Thus if C_f is reduced by 10, the integrator will have an additional gain of 10.	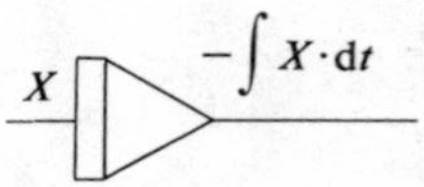 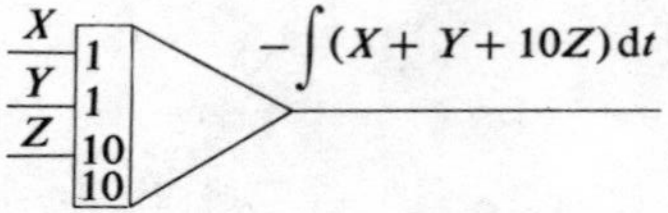 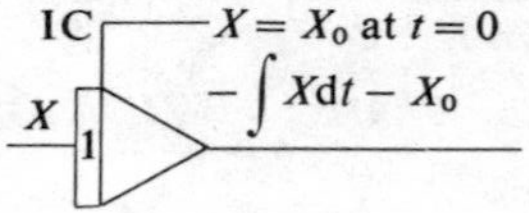 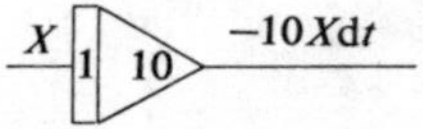 X → $-\int X \cdot dt$ X 1, Y 1, Z 10 10 → $-\int (X + Y + 10Z)\,dt$ IC — $X = X_0$ at $t = 0$; X 1 → $-\int X dt - X_0$ X 1 10 → $-10X dt$	Integrates the variable $X = X(t)$ with respect to time. Usually the integrator has multiple inputs with the same input resistors as the summer, i.e. $R_1 = R_2$, $R_3 = R_4 = R_1/10$. The integrator can produce the integral of the sum of a number of variables. The initial conditions can be inserted into the IC terminal. This extra gain is indicated by the 10 (in this case) in the apex of the computing symbol.
Multiplier	No simple electrical symbol	(*a*) 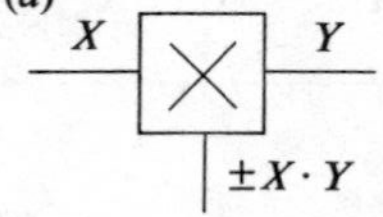 X, Y → $\pm X \cdot Y$ (*b*) X, $-X$, Y 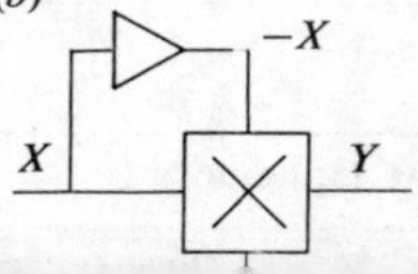	The multiplier allows the multiplication of two variables. The input to the multiplier varies with the type of computer. Thus only X and Y are required in (*a*) but X, $-X$ and Y are required in (*b*). The output can be +

computing elements are brought out to a central control panel called a patch panel. Connections from one unit to another are made by wires or jumper plugs. The process of connecting up a number of computer units in such a way that they describe an analogue circuit, such as those in the following section, is called patching.

10.3. Programming the analogue computer

10.3.1. Examples of first order equations.

(*a*) $$\frac{d[X]}{dt} = -0.5[X] \quad \text{given } [X] = 0.5 \text{ at } t = 0 \tag{10.1}$$

This equation describes a number of different physical or chemical phenomena, e.g. radioactive decay, first order chemical reaction. The two variables in the equation are [X] and d[X]/dt. Clearly we require an integrator to solve the equation. If we feed d[X]/dt into an integrator, the output is –[X].

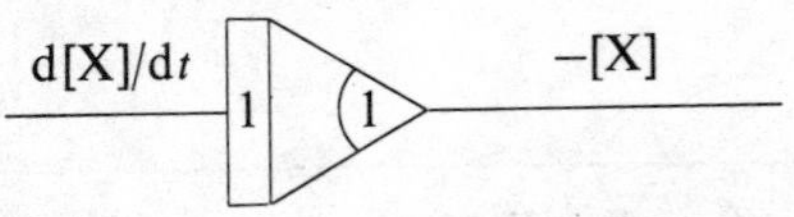

Where can we obtain the input variable d[X]/dt? From equation (10.1), we know that it is –0.5[X]. Since –[X] is available at the output of the integrator we can use a potentiometer to obtain –0.5[X],

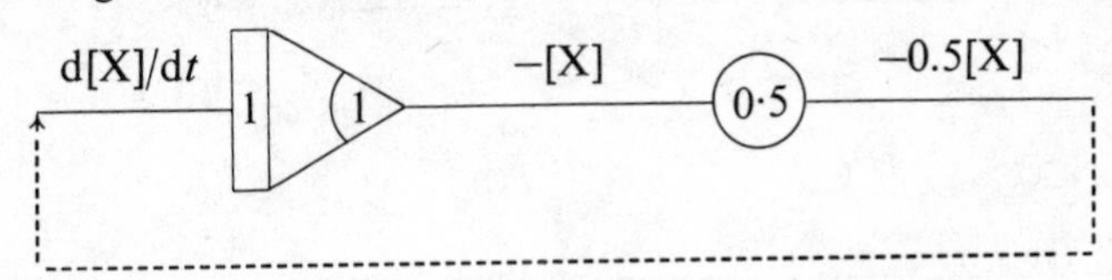

and since this is required at the input, we can connect the output from the potentiometer to the input of the integrator. The initial conditions for the problem are [X] = 0.5 at t = 0. Hence we can enter [X] = 0.5 into the IC (initial condition) terminal since +1 and –1 are always available on the computer and the complete circuit becomes

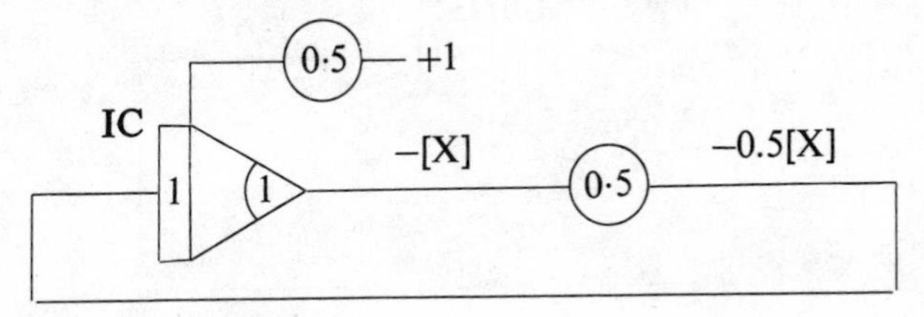

(*b*) $$\frac{4\mathrm{d}[\mathrm{X}]}{\mathrm{d}t}+5[\mathrm{X}]=6 \qquad (10.2)$$

Firstly we must arrange equation (10.2) so that the highest derivative (in this case d[X]/d*t*) is on the left-hand side and is not multiplied by any term:

$$\frac{\mathrm{d}[\mathrm{X}]}{\mathrm{d}t}=\tfrac{6}{4}-\tfrac{5}{4}[\mathrm{X}].$$

There are three circuits that can be arranged to satisfy this equation. The first two are essentially the same but the order of the computing elements is different.

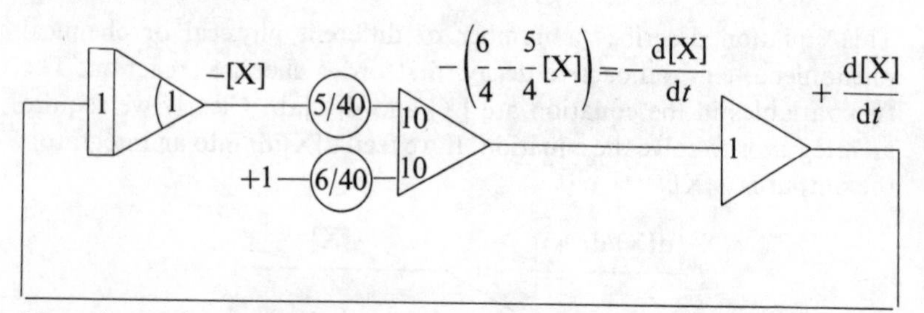

Note that since $\frac{5}{4}$ and $\frac{6}{4}$ are both greater than 1, we have to set the potentiometers to $\frac{5}{40}$ and $\frac{6}{40}$ and use gains of 10 on the summer.

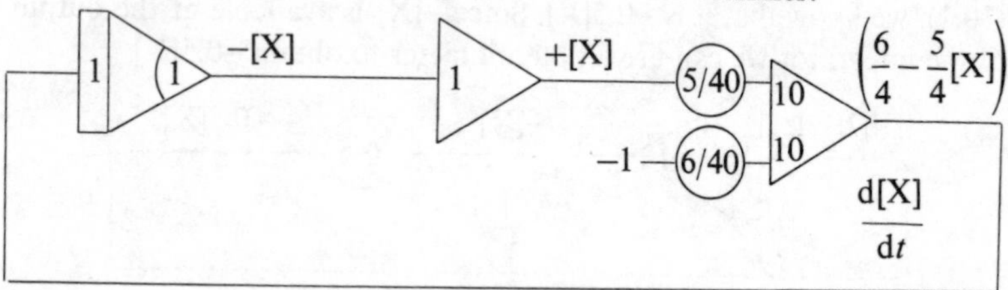

Since in most cases the variation of the highest derivative (d[X]/d*t* in this example) in the problem is not of interest, then it is possible to use the dual purpose of the integrator, i.e. as a summer and integrator. The circuit is then much simpler and only uses one integrator.

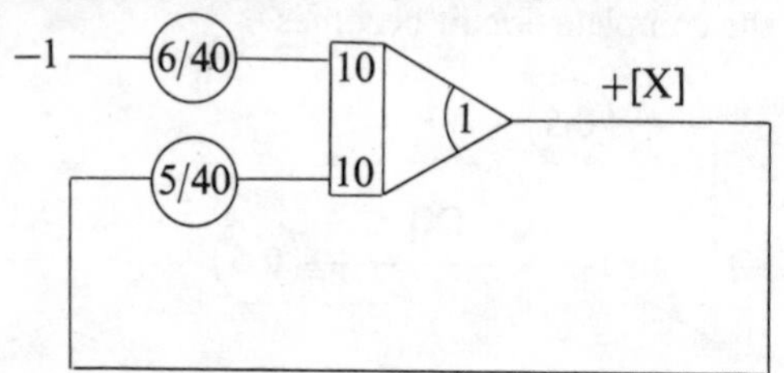

The left-hand side is summed to give $-6/40 + 5[X]/40$ which is $-0.1\mathrm{d}[X]/\mathrm{d}t$, but the ×10 inputs are used and so the effective input is $-\mathrm{d}[X]/\mathrm{d}t$.

10.3.2. Second order equations.

$$\frac{2\mathrm{d}^2[X]}{\mathrm{d}t^2} + \frac{3\mathrm{d}[X]}{\mathrm{d}t} + 4[X] = 1 \tag{10.3}$$

Firstly rearrange equation (10.3) to obtain $\mathrm{d}^2[X]/\mathrm{d}t^2$ on the left-hand side:

$$\frac{\mathrm{d}^2[X]}{\mathrm{d}t^2} = \frac{1}{2} - 2[X] - \frac{3}{2}\frac{\mathrm{d}[X]}{\mathrm{d}t}. \tag{10.4}$$

If we assume that $\mathrm{d}^2[X]/\mathrm{d}t^2$ is available and we integrate, we obtain $-\mathrm{d}[X]/\mathrm{d}t$; further integration yields $+[X]$.

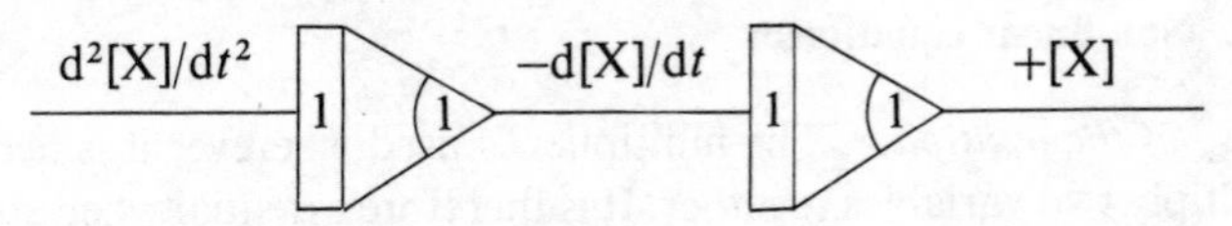

The necessary input, $\mathrm{d}^2[X]/\mathrm{d}t^2$, is provided from equation (10.4).

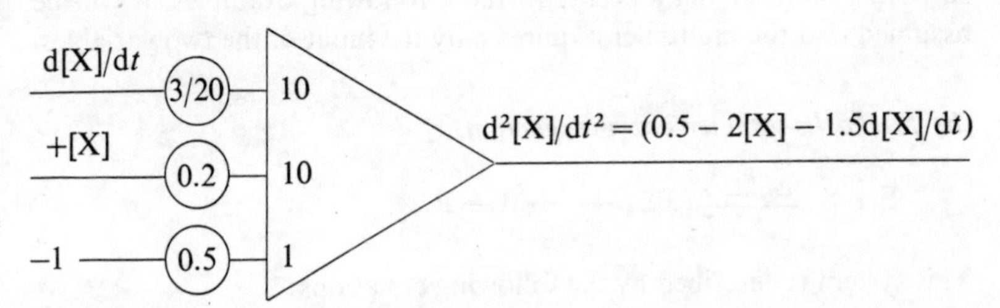

It is a simple matter to add an additional inverter to obtain $\mathrm{d}[X]/\mathrm{d}t$ and to link the last two circuits.

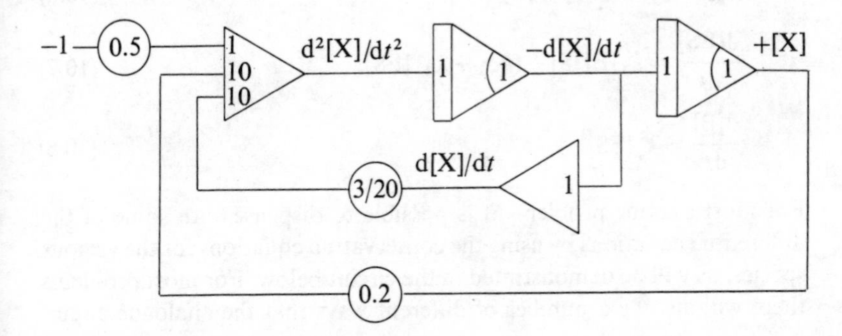

Since there are two integrators, two initial conditions must be specified (the initial conditions of [X] and d[X]/dt). We can simplify the previous circuit by using the summing facility of the integrator, e.g.

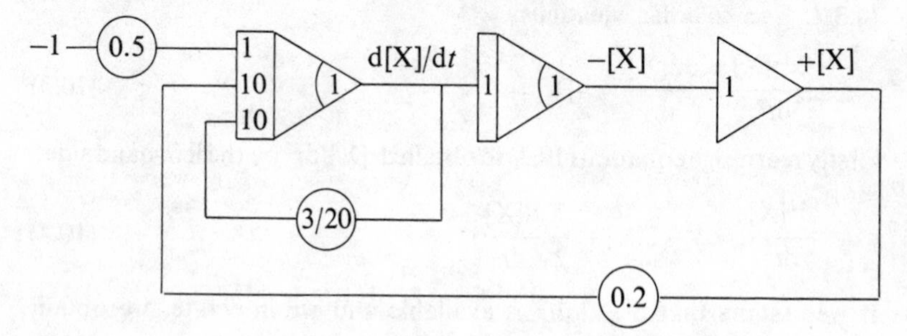

10.3.3. Non-linear equations.

(*a*) *Use of the multiplier*. The multiplier is used wherever it is necessary to multiply two variables together. It is therefore essential when studying enzyme systems where, for example, products of concentration terms such as [E][S] or [E][I] may occur. In these following examples, it will be assumed that the multiplier requires only the input of the two variables.

(*b*) *Michaelis–Menten enzyme reaction.*

$$\mathrm{E} + \mathrm{S} \underset{k_{-1}}{\overset{k_1}{\rightleftharpoons}} \mathrm{ES} \xrightarrow{k_2} \mathrm{E} + \mathrm{P}$$

This system is described by the following equations:

$$[\mathrm{E_0}] = [\mathrm{E}] + [\mathrm{ES}], \tag{10.5}$$

$$[\mathrm{S_0}] = [\mathrm{S}] + [\mathrm{ES}] + [\mathrm{P}], \tag{10.6}$$

$$\frac{\mathrm{d[ES]}}{\mathrm{d}t} = k_1[\mathrm{E}][\mathrm{S}] - (k_{-1} + k_2)[\mathrm{ES}], \tag{10.7}$$

$$\frac{\mathrm{d[P]}}{\mathrm{d}t} = k_2[\mathrm{ES}]. \tag{10.8}$$

For most enzyme problems it is possible to dispense with some of the differential equations by using the conservation equations for the various species, as will be demonstrated in the circuit below. For most problems there will also be a number of different ways that the analogue circuit can be arranged. The most convenient circuit depends on the variables that are under investigation. The circuit is arranged so that these variables are all positive in sign and therefore when they are monitored, on say an X–Y plotter, the Y-axis or concentration axis will be in the usual direction, upwards. The analogue computer diagram for this system is shown on p. 264.

The summer (1) generates $-[E]$ from $-[E_0]+[ES]$ (equation (10.5)); summer (3) generates $+[S]$ from $-[S_0]+[ES]+[P]$ (equation (10.6)); [E] and [S] are multiplied by (2) to give $-[E][S]$ which is used by the integrator (4) to generate [ES]. This is used by (1) and (3) and is integrated by (5) to give $-[P]$, which is inverted by (6) to give [P]. The various species [E], [ES], [S] and [P] can be monitored by taking outputs (Mon.) from their respective amplifiers. Note that no steady-state assumptions have been made regarding the concentration of [S], which are necessary to produce an analytical solution of the differential equations. The analogue computer solution generated by the above circuit will describe, with suitable scaling factors, the behaviour of the enzyme system from $t=0$, where $[S]=[S_0]$ and $[E]=[E_0]$, to the point where all the substrate has been used.

(*c*) *Host–parasite problem.* This is an example of the oscillating system that was mentioned in chapter 9 (see fig. 9.6). The differential equations are

$$\frac{d[H]}{dt} = k_1[H] - k_2[H][P],$$

$$\frac{d[P]}{dt} = k_2[H][P] - k_3[P].$$

At $t=0$, $[H]=[H_0]$ and $[P]=[P_0]$. The analogue circuit is simply

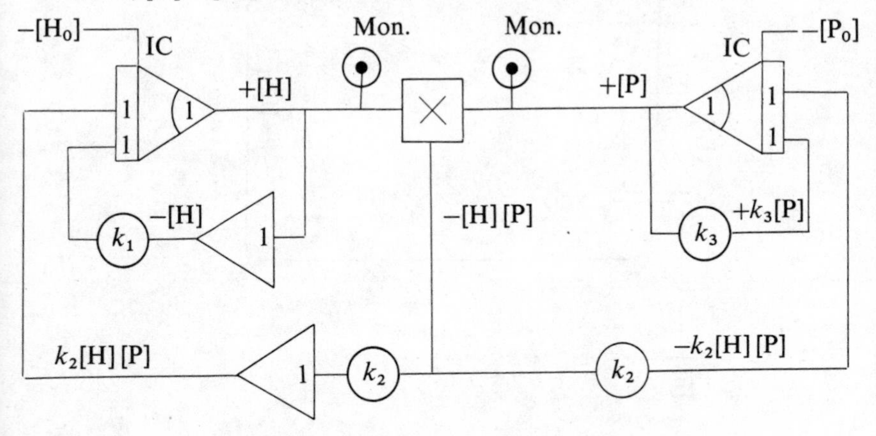

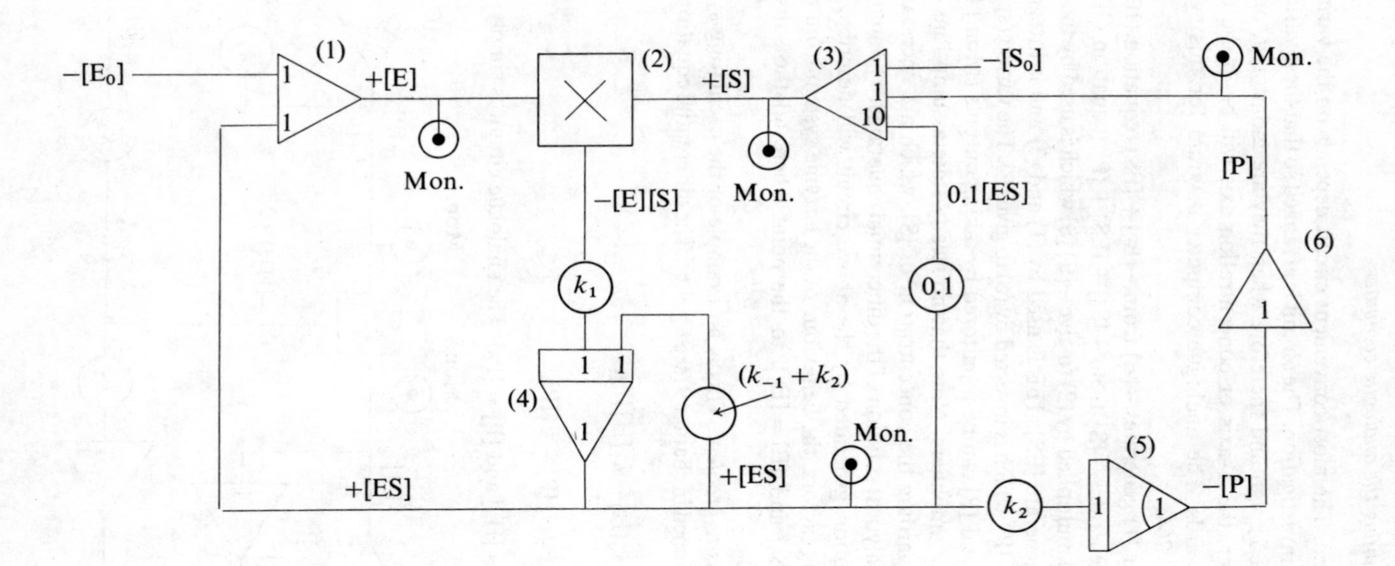

−[E₀]
(1)
+[E]
Mon.
(2)
−[E][S]
+[S]
Mon.
(3)
−[S₀]
10
0.1[ES]
Mon.
[P]
(6)
k₁
0.1
(k₋₁ + k₂)
(4)
Mon.
(5)
+[ES]
+[ES]
k₂
−[P]

(*d*) *Relaxation experiment*. The analogue computer can also be used to simulate relaxation kinetic experiments, produced by, for example, a temperature jump. For the simplest equilibrium experiment

$$\mathrm{A} \underset{k_{-1}}{\overset{k_1}{\rightleftharpoons}} \mathrm{B}$$

$$\frac{\mathrm{d[A]}}{\mathrm{d}t} = -k_1[\mathrm{A}] + k_{-1}[\mathrm{B}]$$

$$[\mathrm{A_0}] = [\mathrm{A}] + [\mathrm{B}].$$

In a temperature jump relaxation experiment, the equilibrium constant for the reaction is changed by a fast temperature rise. This change initiates a relaxation reaction during which the components assume their new equilibrium concentrations. We can simulate this change by allowing the system to reach equilibrium and then altering the rate constants k_1 and/or k_{-1}, and initiating a time base at the same instant. The analogue diagram is

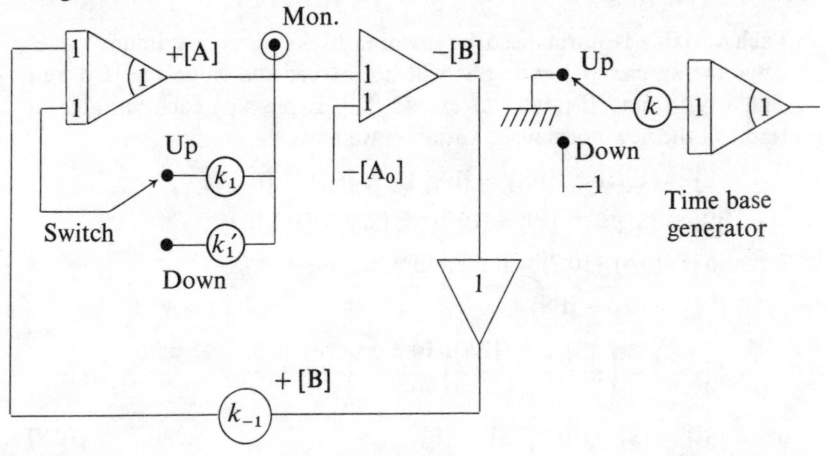

In this case we have assumed that k_{-1} is unaltered by the temperature jump and that k_1 changes to k_1'. Two potentiometers are arranged to give k_1 and k_1' and are connected to a change-over switch. Two contacts on the switch are also connected to an integrator which is used as a time base generator. The system is allowed to reach equilibrium with the switch in the up position. The concentrations of A and B will be determined by the equilibrium constant $K = k_1/k_{-1}$. The switch is rapidly placed in the down position; this starts the time base which drives the

X–Y plotter (at a rate determined by k) and the system changes to the new equilibrium position determined by the ratio k'_1/k_{-1}. For this simple example, a single exponential curve will be obtained, given by equation (7.36).

10.4. Amplitude and time scaling

Although all the previous circuit diagrams satisfy the differential equations, it will usually be necessary to scale the problem before it can be run satisfactorily on the computer. The analogue computer is not able to handle very large or very small numbers and hence it would be impossible to use typical enzyme concentrations of 10^{-6}–10^{-9} M. An example of scaling is given below; further examples can be found in any text book on analogue computing (Charlesworth & Fletcher, 1967). We will scale the Michaelis–Menten problem for the following set of values:

$$[E_0] = 10^{-6}\ \text{M};\ [S_0] = 10^{-4}\ \text{M};\ k_1 = 10^6\ \text{M}^{-1}\ \text{s}^{-1};\ k_{-1} = 10^3\ \text{s}^{-1};$$
$$k_2 = 10^2\ \text{s}^{-1}.$$

Each variable is normalised by dividing by its likely maximum value. Thus the species [E] and [ES] will not exceed the value for $[E_0]$ and similarly [S] and [P] will not exceed $[S_0]$. Expressing each variable in terms of the new normalised variables we have,

$$[E]' = [E]/10^{-6},\ [ES]' = [ES]/10^{-6},\ [E_0]' = [E_0]/10^{-6},$$
$$[S]' = [S]/10^{-4},\ [P]' = [P]/10^{-4},\ [S_0]' = [S_0]/10^{-4}.$$

Equations (10.5)–(10.7) will therefore become

$$[E_0]' = [E]' + [ES]'$$

$$\left(\frac{[S_0]}{10^{-4}}\right) = \left(\frac{[S]}{10^{-4}}\right) + \left(\frac{[ES]}{10^{-6}}\right)\frac{10^{-6}}{10^{-4}} + \left(\frac{[P]}{10^{-4}}\right)$$

or $$[S_0]' = [S]' + 10^{-2}[ES]' + [P]' \tag{10.9}$$

$$\frac{d}{dt}\left(\frac{[ES]}{10^{-6}}\right) = k_1\left(\frac{[E]}{10^{-6}}\right)\left(\frac{[S]}{10^{-4}}\right)10^{-4} - (k_{-1}+k_2)\left(\frac{[ES]}{10^{-6}}\right)$$

or $$\frac{d[ES]'}{dt} = (10^{-4}k_1)[E]'[S]' - (k_{-1}+k_2)[ES]'$$
$$= k'_1[E]'[S]' - (k_{-1}+k_2)[ES]' \tag{10.10}$$

where

$$k'_1 = (10^{-4}k_1) = 10^2, \qquad k_{-1} = 10^3, k_2 = 10^2.$$

Similarly

$$\frac{d}{dt}\left(\frac{[P]}{10^{-4}}\right) = k_2\left(\frac{[ES]}{10^{-6}}\right)\frac{10^{-6}}{10^{-4}} \tag{10.11}$$

$$\frac{d[P]'}{dt} = k_2'[ES]' \tag{10.12}$$

where $k_2' = 10^{-2}k_2 = 1$. Whatever scaling factors are used, the final equations must always agree arithmetically with the original unscaled equations; hence the factors 10^{-2}, 10^{-4} and 10^{-2} in equations (10.9), (10.10) and (10.12). Setting $[E_0] = 1$ and $[S_0] = 1$ on the computer will now be equivalent to using a reaction mixture that is 10^{-6} M and 10^{-4} M with respect to $[E_0]$ and $[S_0]$. The circuit diagram that will be set up on the computer is shown on p. 268.

This form of scaling is called amplitude scaling. The scaled problem can now be run on the computer and computation will occur in real time, and the rate at which the substrate is consumed or product is formed will be the same as in an actual experiment. We could therefore plot the formation of [P] and the loss of [S] on an X–Y plotter but we could not follow the early stage of the reaction in the pre-steady state as this would occur at a rate that would be too fast for the X–Y plotter to follow. Therefore, if we wish to simulate the pre-steady-state portion of the reaction the problem has to be slowed down by time scaling.

Let $\tau = \beta t$, then $1/dt = \beta/d\tau$ where β is the time scaling factor, τ the computer time and t the real time. Equation (10.10) can thus be written as

$$\frac{d[ES]'}{d\tau} = \frac{100}{\beta}[E]'[S]' - \frac{1000}{\beta}[ES]' - \frac{100}{\beta}[ES]'.$$

Equation (10.12) can be treated in a similar manner. The value of β is chosen such that the problem is slow enough to follow on the recording instrument being used. If $\beta = 10^3$, then 1 second of computer time will be equivalent to 1 millisecond of real time. It will probably be necessary to alter the amplitude scaling parameters since, for example, the amount of product formed during the pre-steady-state region is quite small and therefore would not be detected using the scaling factors for the previous example. Time scaling for simple problems is therefore just a matter of altering the gain of all the integrators by a common factor, β. The time scaling of more complex problems, e.g. a number of coupled enzyme reactions, is not much more difficult than this simple problem.

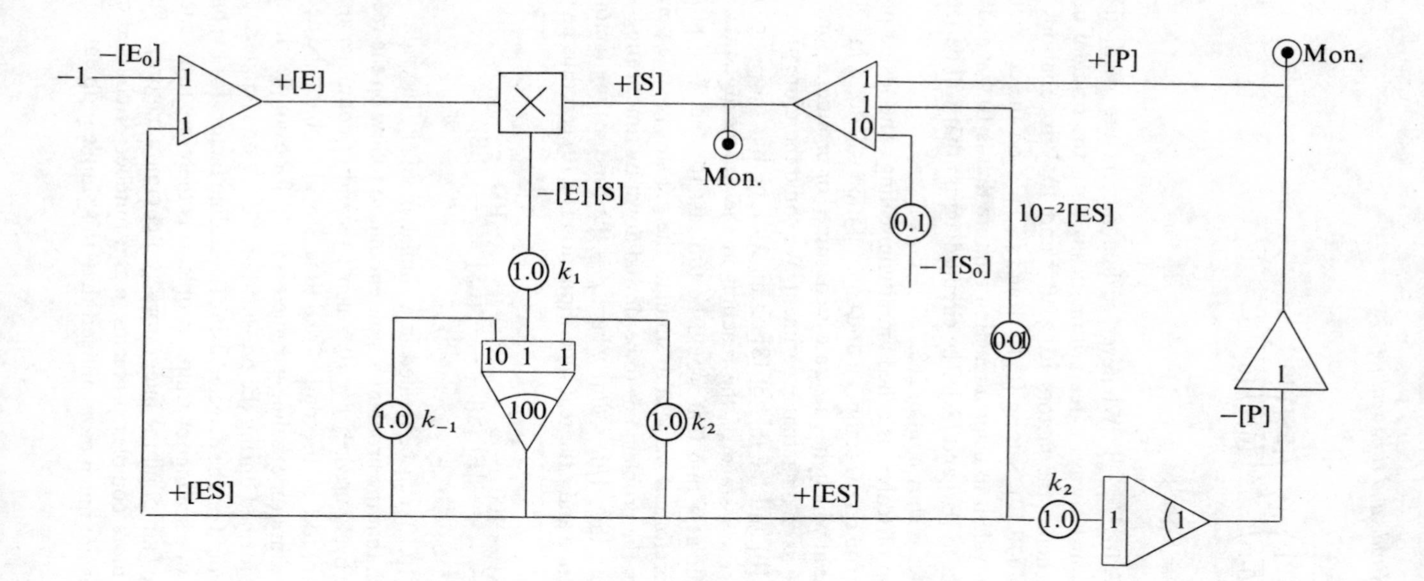
−[E_0]
−1
1
1
+[E]
+[S]
−[E] [S]
1.0
k_1
10 1 1
100
1.0
k_{-1}
1.0
k_2
Mon.
1
1
10
0.1
−1[S_0]
10^{-2}[ES]
0·01
+[P]
Mon.
1
−[P]
+[ES]
+[ES]
k_2
1.0
1
1

10.5. Computer simulation using digital computers

Although an analogue computer has the advantages of simplicity, lower cost and ease of operation, it is usually limited in size and thus many biochemical systems cannot be simulated without some simplification of the problem. The capacity and speed of modern digital computers is quite sufficient for the study of even the most complicated biochemical systems. One of the main drawbacks of a digital computer, however, is that programs must be written for each particular type of problem and these may take many months of work before operating satisfactorily. Although a number of general simulation programs have been published (Garfinkel, 1968; Kibby, 1969; Curtis & Chance, 1974), it is often easier to write one's own program (see footnote p. 276) than to modify one to suit a particular computer.

Suppose we wish to simulate the simple Michaelis–Menten system:

$$E + S \underset{k_{-1}}{\overset{k_1}{\rightleftharpoons}} ES \tag{10.13}$$

$$ES \underset{k_{-2}}{\overset{k_2}{\rightleftharpoons}} E + P. \tag{10.14}$$

We could write the differential equations for the problem directly into a simulation program, but this would mean changing the program for each new problem. It is preferable, therefore, to read the data in the form of equations together with the values of the associated rate constants and initial reactant concentrations. There are nine basic types of equations that are likely to be involved:

(1) $\quad A \underset{k_{-1}}{\overset{k_1}{\rightleftharpoons}} B$

(2) $\quad A + B \underset{k_{-1}}{\overset{k_1}{\rightleftharpoons}} C$

(3) $\quad A \underset{k_{-1}}{\overset{k_1}{\rightleftharpoons}} B + C$

(4) $\quad A + B \underset{k_{-1}}{\overset{k_1}{\rightleftharpoons}} C + D$

(5) $\quad A + B + C \underset{k_{-1}}{\overset{k_1}{\rightleftharpoons}} D$

(6) $$A \underset{k_{-1}}{\overset{k_1}{\rightleftharpoons}} B+C+D$$

(7) $$A+B+C \underset{k_{-1}}{\overset{k_1}{\rightleftharpoons}} D+E$$

(8) $$A+B \underset{k_{-1}}{\overset{k_1}{\rightleftharpoons}} C+D+E$$

(9) $$A+B+C \underset{k_{-1}}{\overset{k_1}{\rightleftharpoons}} D+E+F$$

Reactions that are more complicated probably occur in distinct steps and can therefore be simulated using two or more simpler equations. The relevant equations for a particular problem are converted into flux equations. For example, the flux equations for reactions of types (1) and (2) would be respectively:

$$F(1) = k_1[A] - k_{-1}[B]$$

and

$$F(2) = k_1[A][B] - k_{-1}[C].$$

In the Michaelis–Menten scheme there would be two flux equations:

$$F(1) = k_1[E][S] - k_{-1}[ES] \quad (10.15)$$

and

$$F(2) = k_2[ES] - k_{-2}[E][P]. \quad (10.16)$$

The rate at which any individual species changes in concentration is obtained by summing the fluxes of all the reactions in which that species occurs. Thus for [E], the rate of change of $[E] = d[E]/dt$ is given by the sum of F(1) + F(2). Since the initial conditions at $t = 0$ are known, it is quite easy to determine the initial fluxes for each reaction and hence to calculate the initial rates of change of each species. The next step in the simulation program is to increase the time by a small amount Δt and to determine the new concentrations of each species and consequently the new fluxes of each reaction and the new rates of change of each species. A numerical integration procedure uses the rate of change of concentration of each species, determined by summing the fluxes of all the equations in which that species occurs, as the derivative, $d[x]/dt$, with respect to time for that species.

10.6. Numerical integration methods

Firstly, what is meant by a numerical solution of a differential equation? Suppose we have a function $y = f(x)$, then for each value of x there will be a value of y (provided y is always single-valued and continuous). If, however, we are presented not with the function y but with its derivative y' and certain initial conditions, then the numerical solution is one that within allowable errors agrees with the 'true' solution. The errors that arise are due to the mathematical limitation of the particular numerical integration procedure that is used and to the limitation in the size of numbers that the computer can handle (round-off errors). Since numerical integration methods are used because particular sets of differential equations have no analytical solution, then these methods should be automatic; if errors build up in the solution then the integration should be restarted with a smaller step size.

In broad outline, a numerical solution is obtained as follows. In the simple two-dimensional case, we know initially only one point in the x–y plane, namely the initial conditions x_0, y_0. The differential equation describes the slope of the function at any point x, y and we can determine the initial slope at the point x_0, y_0. The slope at this point is determined

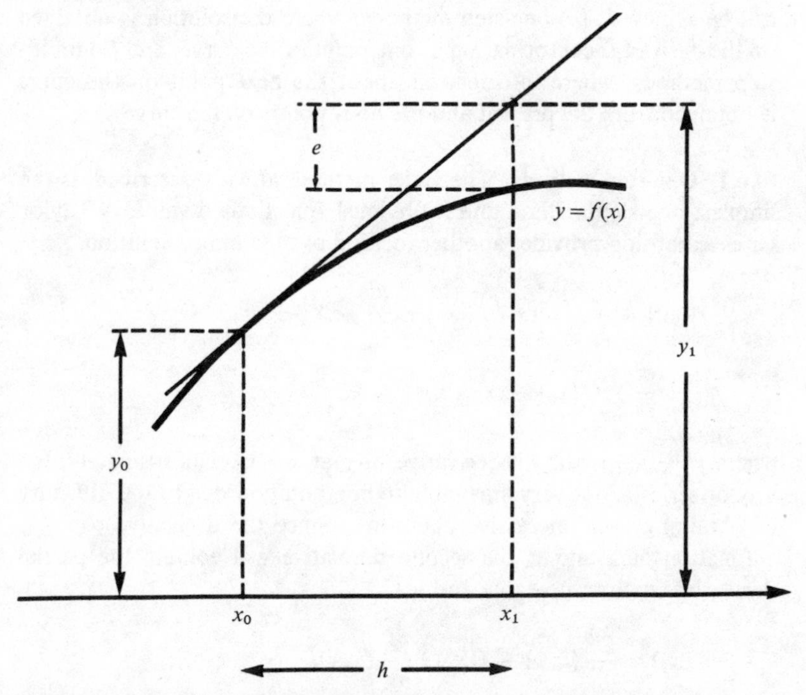

Fig. 10.1. Geometrical representation of Euler's method. (Adapted from McCracken & Dorn (1964) *Numerical Methods and Fortran Programming*. Wiley.)

and the x co-ordinate is increased by a small amount h, to a new point x_1 where $x_1 = x_0 + h$. The new value of y, y_1, is calculated at this point x_1. The new co-ordinates x_1, y_1 are then used to determine the new slope and the process is repeated. The true curve is therefore approximated by a series of short straight lines which, hopefully, is a true reflection of the real curve. This simple approach is called the Euler method and is shown geometrically in fig. 10.1.

Obviously, there are pitfalls in this method. If the step size, h, is too large, then the sequence of 'short' straight lines will deviate considerably from the 'true' solution as suggested in fig. 10.1 in which the error, e, is growing. This deviation is called the instability of the method. The problem would be much more stable and the error smaller if the step size h were reduced, but this would of course increase considerably the amount of computing time required. Obviously, the stability of the problem would be improved if the curvature of the 'true' solution could be taken into account rather than simply approximating it by a sequence of short straight lines. There are two basic methods by which this can be achieved: (*a*) one-step methods, where the solution is obtained on the basis of the information at one point of the curve; and (*b*) multistep methods, where information about the next point on the curve is obtained from the present and previous points on the curve.

10.6.1. One-step methods. The Euler method, already described, is the simplest one-step method and it obviously has limited value. A Taylor series expansion provides another method of obtaining a solution.

$$y(x_0 + h) = y(x_0) + hy'(x_0) + \frac{h^2}{2!} y''(x_0) + \cdots$$

$$+ \frac{h^j}{j!} y^{\langle j \rangle}(x_0) + \cdots$$

where $y^{\langle j \rangle}(x_0)$ is the jth derivative of y at $x = x_0$. Unfortunately, the Taylor series is not very amenable to computation due to the difficulty of obtaining each successive derivative. Since the derivative $y'(x_0)$ is a function of x and y, the second derivative will contain the partial derivatives with respect to x and y:

$$y''(x_0) = \frac{\delta f}{\delta x}(x, y) + f(x, y) \frac{\delta f}{\delta y}(x, y),$$

and succeeding derivatives are even more complicated. The Taylor series is used, however, as a means of judging the extent to which other methods agree with the Taylor series expansion. Thus the Euler method agrees as far as terms in h and is therefore called a first order method. It is in fact an example of a first order Runge–Kutta method. Runge–Kutta methods have three properties in common:

(*a*) They are one-step methods; i.e. to find y_{m+1} it is only necessary to have the information from the previous point, y_m.

(*b*) They agree with the Taylor series expansion to terms in h^n where n is the order of the method (the Runge–Kutta methods that are commonly used on the larger and faster computers are fourth or sixth order).

(*c*) They do not require the evaluation of any higher derivatives, as in the Taylor series.

An improvement of the Euler method can be achieved by taking the average of the slopes at x_m, y_m and x_{m+1}, y_{m+1}. This is shown geometrically in fig. 10.2. First, the slope, s_m, at x_m, y_m is computed. The value of x_m

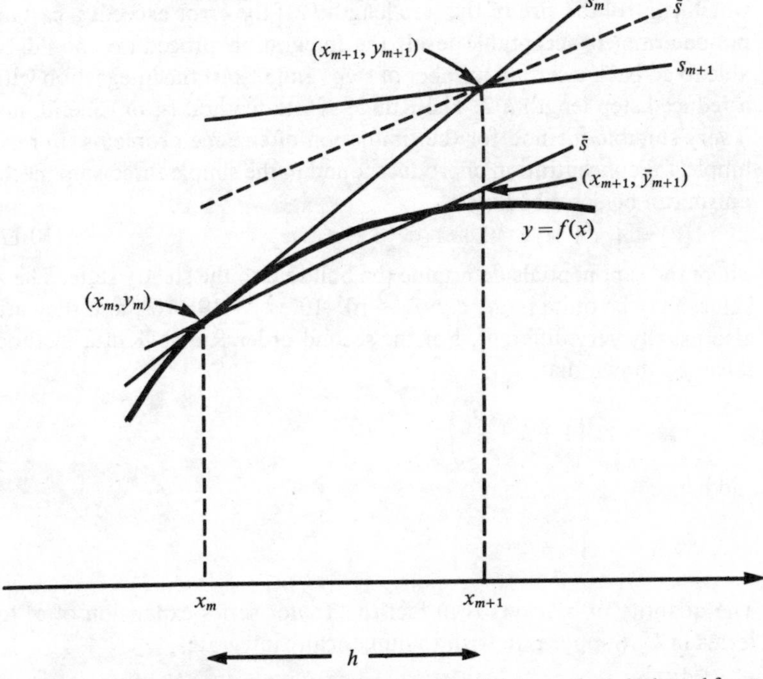

Fig. 10.2. Geometrical representation of the improved Euler method. (Adapted from McCracken & Dorn (1964) *Numerical Methods and Fortran Programming*. Wiley.)

is increased by the step length h to x_{m+1} and the new value y_{m+1} is determined. The new slope, s_{m+1}, is determined at x_{m+1}, y_{m+1}. The average of the two slopes is then determined and a line with this mean slope $\bar{s}$ is then drawn through the point x_m, y_m to give the new value $\bar{y}_{m+1}$. The error, $\bar{e}$, between this point and the true solution is smaller than the error, e, for the simple Euler method. This improved Euler method is an example of a second order Runge–Kutta method. Third and fourth order Runge–Kutta methods can be developed in a similar manner. The fourth order Runge–Kutta method is very widely used and is frequently referred to as simply 'The Runge–Kutta Method' without any indication of the order. It can be seen that the error of the method is determined by the terms in the Taylor series that are not included. Thus for the simple Euler method the error at each step is kh^2 where k is some constant. In general, the error is kh^{n+1} where n is the order of the method. The determination of this truncation error for the Runge–Kutta methods is not a simple matter and this is one of the drawbacks of these methods. The ability to determine the error at each step is essential as this is one of the main criteria which control the size of the step length h. If the error exceeds a certain pre-determined acceptable level, the integration procedure should be able to go back a certain number of steps and restart the integration with a reduced step length $h/2$. The Runge–Kutta method is, in general, not a very suitable method for the simulation of enzyme problems. For example, the concentration of product found in the simple three-step mechanism can be described by

$$[\mathrm{P}] = A_1 t + A_2 \mathrm{e}^{-\lambda_1 t} + A_3 \mathrm{e}^{-\lambda_2 t} + A_4, \tag{10.17}$$

where the exponentials determine the build up to the steady state. The λ values may be quite large, e.g. $\lambda_1 \sim 10^4$–10^6, $\lambda_2 \sim 10^1$–10^3, and they are also usually very different. For the second order Runge–Kutta method it can be shown that

$$y_{m+1} = y_m \left(1 + h + \frac{h^2}{2}\right)$$

and hence

$$y_{m+1} = y_0 \left(1 + h + \frac{h^2}{2}\right)^{m+1}.$$

The quantity in brackets is in fact the Taylor series expansion of e^h to terms in h^2. Suppose we wish to numerically integrate,

$$\frac{\mathrm{d}y}{\mathrm{d}x} = -10x \quad \text{given } y = 1 \text{ at } x = 0. \tag{10.18}$$

The true solution is of course $y = e^{-10x}$. From equation (10.18), it can be seen that

$$y_{m+1} = (1 - 10h + 50h^2)^{m+1}. \qquad (10.19)$$

The term in brackets in equation (10.19) is greater than 1 for $h > 0.2$ and therefore for large values of m (i.e. after a large number of steps), y becomes very large. We know, however, that the true solution $y = e^{-10x}$ becomes very small. The numerical solution is therefore unstable unless small steps are taken. The situation is even more critical with enzyme systems in which terms such as $e^{-10^5 t}$ and $e^{-10^2 t}$ occur during the pre-steady-state region. The differential equations that describe enzyme systems are regarded as 'stiff' because the Eigen values (λ_1 and λ_2 in equation (10.17)) are very different in numerical magnitude. The large values of λ_1 and λ_2 force the numerical integration method to take exceedingly small steps during the pre-steady-state region of the reaction, in order to achieve accuracy and stability. Runge–Kutta methods are therefore not very suitable for numerically integrating differential equations relating to enzyme problems or other problems that are described by 'stiff' equations.

10.6.2. Multi-step methods. In the Runge–Kutta methods, the next point y_{m+1}, x_{m+1} is obtained from the previous point y_m, x_m but not earlier points, e.g. y_{m-1}, x_{m-1}. In the second and higher order methods, the Runge–Kutta method requires the evaluation of the derivative at additional points but does not utilise any of the previously computed points. Obviously after the integration has proceeded for some time, there is a considerable amount of useful information available about the shape of the curve from the previous points, and for solutions that are rapidly changing this information is very valuable. Methods that use this information are called predictor–corrector procedures. As the name implies, a value of y_{m+1} is first predicted and is then corrected and further refined by iteration. The iteration process can be repeated many times but this is usually not required and suitable tests of the convergence of the solution determine the number of iterations that are needed. One of the problems of the predictor–corrector methods is getting the integration started since the methods rely on information from prior points. This problem is easily overcome by starting the integration with a Runge–Kutta method and then changing over to the predictor–corrector method after a specified number of steps. A geometric representation of a second order predictor is shown in fig. 10.3. The predictor formula first determines the slope (s_m) at y_m, x_m. A line with this slope is then drawn

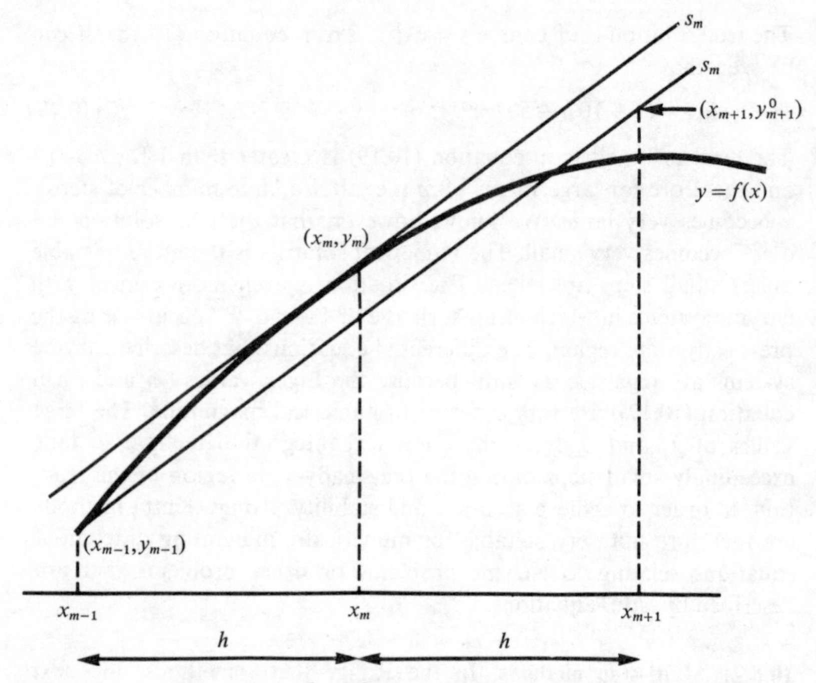

Fig. 10.3. Geometrical representation of the second order predictor. (Adapted from McCracken & Dorn (1964) *Numerical Methods and Fortran Programming*. Wiley.)

through the point y_{m-1}, x_{m-1} until it intersects the line $x = x_{m+1}$. This is then the predicted value of y, y^0_{m+1}. The second order corrector formula is shown geometrically in fig. 10.4. Since we know the approximate new value of y, y^0_{m+1}, we can calculate the slope (s_2) at this point x_{m+1}, y^0_{m+1}. The line s_1 is the slope at the point x_m, y_m. The average of the two slopes is $\bar{s}$ and this average is then used from the point x_m, y_m to find the corrected value $y^{(1)}_{m+1}$. If desired this corrected value of $y^{(1)}_{m+1}$ can be reiterated to obtain a new corrected value $y^{(2)}_{m+1}$. The iterations are stopped when the difference between two successive values of y_{m+1} is less than a previously set error, e. A variable order predictor–corrector method (Gear, 1969) is particularly suitable for the solution of stiff equations and is the basis of a program used by the author.*

* This FORTRAN program was developed by Professor D. T. Elmore of Queen's University, Belfast and the author. The program accepts input in the form of chemical equations and uses this information to formulate automatically the required differential rate equations. The program incorporates the variable order predictor–corrector subroutine DIFSUB, developed by Gear (1969), which is specifically designed for the numerical integration of 'stiff' differential equations.

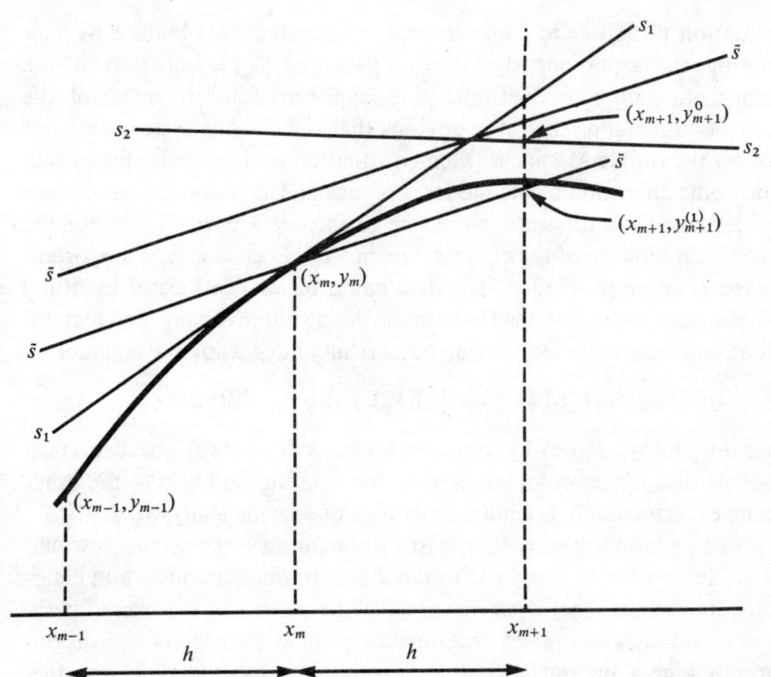

Fig. 10.4. Geometrical representation of the second order corrector. (Adapted from McCracken & Dorn (1964) *Numerical Methods and Fortran Programming*. Wiley.)

10.7. Summary

The aim of this chapter has been to examine some of the methods that are available for the analysis of specific enzyme models using computer simulation techniques. The purpose of numerical integration techniques is simply to provide an additional method for studying the behaviour of enzyme systems. The usual technique for analysing the behaviour of an enzyme or sequence of enzymes involves the determination of the *initial velocity* of the reaction at various concentrations of substrate(s). These data are then fitted by some statistical method (see appendix II) to one of the various rate equations. The reason for determining only the initial velocity of the reaction is that it overcomes the problems of changes in the proposed enzyme mechanism as the reaction proceeds. If the behaviour of an enzyme can be described by the Michaelis–Menten equation $v = V[\mathrm{S}]/(K_m + [\mathrm{S}])$, then it should be possible to fit the data from a complete product/time plot (i.e. from $[\mathrm{S}] = [\mathrm{S}_0]$ to $[\mathrm{S}] = 0$) either to the integrated version of the Michaelis–Menten equation

(equation (2.14)) or to a numerically integrated curve obtained by integrating the equations $d[S]/dt = -V[S]/(K_m + [S]) = -d[P]/dt$. If the data from such a product/time plot cannot be fitted by either of the latter two methods then it is obvious that the enzyme system does not follow the simple Michaelis–Menten equation at times other than $t = 0$, and some alteration of the model is required. The most obvious change to the model, in this case, would be to include a term to describe the product inhibition of the enzyme, which would become more important as the reaction proceeded. The data could be analysed either by fitting to the integrated Michaelis–Menten equation involving product inhibition (equation (3.35)), or by numerically integrating the equations

$$d[S]/dt = -V[S]/\{K_m(1 + [P]/K_p) + [S]\} = -d[P]/dt.$$

It is this ability to study by computer the behaviour of various theoretical models that makes computer simulation a useful addition to the other more experimentally orientated methods of enzyme analysis.

For a metabolic pathway, consisting of a number of enzymes, it would be quite feasible to study each individual enzyme separately and determine its various kinetic parameters. The behaviour of the overall pathway could also be studied experimentally. A knowledge of the various enzyme kinetic parameters and their mechanisms would allow a mathematical model to be constructed comprising the differential equations associated with each enzyme. This model could then be numerically integrated to yield theoretical concentration levels of the metabolites as a function of time, thus allowing a direct comparison with the experimentally determined values. Any large discrepancies would indicate the need for a change in the theoretical model. It should be realised that a study of an enzyme in isolation does not necessarily provide sufficient information to describe the behaviour of the enzyme in the presence of other enzymes and their metabolites. There is certainly experimental evidence to show that at the higher cellular concentration levels, enzymes may polymerise or associate with other enzymes, proteins or membranes and that these associations can dramatically affect the kinetic behaviour of the individual enzymes (Garfinkel *et al.*, 1968).

It is to be hoped that the use of computer simulation of metabolic pathways, such as the glycolytic pathway (Garfinkel *et al.*, 1968), will eventually play a diagnostic role in medical science, and that this role will increase in importance as more information is available on human enzymes or those from closely related species.

Appendix I Theory and use of the Laplace–Carson operator method

I.1. Theory

It is an essential feature of the operator method that we do not attempt to find the unknown function that satisfies the differential; instead we attempt to find a transform of the unknown function. This transform will involve the operator ψ, where ψ is the derivative (d/dt) with respect to time. (ψ is used rather than the more commonly used symbols, s or P, due to the possible confusion with S = substrate and P = product.) Knowing the transform, we can find the original or solution to our problem. The Laplace–Carson operator method is an extension of the ordinary Laplace transform. Providing $f(t)$ is 'well behaved', its Laplace transform, $\mathscr{L}\{f(t)\} = F(\psi)$, can be defined by the following integral:

$$\mathscr{L}\{f(t)\} = F(\psi) = \int_0^\infty f(t) \cdot e^{-\psi t} \, dt. \tag{I.1}$$

Under the integral sign, there are two factors $e^{-\psi t}$ and $f(t)$. Integration between the limits, 0 and c (Lim $c \to \infty$), produces a function that no longer contains the variable t, i.e. $F(\psi)$. Using equation (I.1), it is possible to tabulate the transformed function, $F(\psi)$, for a large number of expressions, $f(t)$, and these tables are listed in mathematical textbooks. The inverse procedure is also possible; given the transform, $F(\psi)$, the original function, $f(t)$, can also be determined from tables.

The Laplace–Carson procedure is very similar. It is in fact the Laplace integral multiplied by the operator ψ:

$$\mathscr{L}C\{f(t)\} = F(\psi) = \psi \int_0^\infty f(t) \cdot e^{-\psi t} \, dt. \tag{I.2}$$

Let us consider the transformations of three simple functions.

(a) The function, $f(t)$, equals a constant, then

$$F(\psi) = \psi \int_0^\infty k \cdot e^{-\psi t} \, dt \tag{I.3}$$

$$F(\psi) = \psi k \left| \frac{e^{-\psi t}}{-\psi} \right|_{0}^{c} \quad \text{Lim } c \to \infty$$

$$F(\psi) = k. \tag{I.4}$$

The transform of a constant is therefore the constant itself.

(*b*) The function, $f(t)$, equals t.

$$F(\psi) = \psi \int_0^{\infty} t \cdot e^{-\psi t} \, dt. \tag{I.5}$$

Using integration by parts $[\int uv dx = u \int v dx - \int (du/dx \int v dx) dx]$

$$F(\psi) = \psi \left[t \int_0^{\infty} e^{-\psi t} \, dt \right] - \psi \left[\int_0^{\infty} \left(\int_0^{\infty} e^{-\psi t} \, dt \right) dt \right]$$

$$F(\psi) = \left[t \frac{(e^{-\psi t})}{-\psi} \right]_0^c - \psi \left[\int_0^{\infty} \frac{(e^{-\psi t})}{-\psi} \, dt \right] \quad \text{Lim } c \to \infty$$

$$F(\psi) = 0 + \int_0^{\infty} e^{-\psi t} \, dt$$

$$F(\psi) = - \left[\frac{e^{-\psi t}}{-\psi} \right]_0^c \quad \text{Lim } c \to \infty$$

$$F(\psi) = \frac{1}{\psi}. \tag{I.6}$$

(*c*) The transform of the exponential function, $f(t) = e^{-at}$, is expressed by

$$F(\psi) = \psi \int_0^{\infty} e^{-at} \cdot e^{-\psi t} \, dt \tag{I.7}$$

$$F(\psi) = \psi \int_0^{\infty} e^{-(\psi + a)t} \, dt$$

$$F(\psi) = -\psi \left[\frac{e^{-(\psi + a)t}}{(\psi + a)} \right]_0^c \quad \text{Lim } c \to \infty$$

$$F(\psi) = \frac{\psi}{\psi + a}. \tag{I.8}$$

The aim, therefore, is to produce a transform of the original differential equation and then determine from tables the original function $f(t)$; thus if the following transform were obtained

$$x = \frac{\psi}{\psi + a}$$

then the solution (or original), from equation (I.8), must be

$$x = \mathrm{e}^{-at}.$$

Finally, the transform of a sum of functions is the sum of individual transforms and of course the inverse is also true; the original of a sum of transforms is the sum of the originals for the individual transforms.

I.2. Use of the Laplace–Carson procedure

We can consider two types of problems; those in which the value of the function and its derivatives are zero at $t = 0$ and those for which this is not the case.

Example 1.

$$\frac{\mathrm{d}x}{\mathrm{d}t} + ax = 1. \tag{I.9}$$

The variables are x and t; a is a coefficient constant. Set $\mathrm{d}/\mathrm{d}t = \psi$ and treat the operator, ψ, as a constant. The equation is now an ordinary algebraic expression with one unknown:

$$\psi x + ax = 1$$

or

$$x = \frac{1}{(\psi + a)}. \tag{I.10}$$

The term $1/(\psi + a)$ is the transform. The solution of equation (I.10) is determined from the table of transforms (Table I.1, equation (3)). Hence,

$$x = \frac{1}{a}(1 - \mathrm{e}^{-at}). \tag{I.11}$$

Example 2.

$$\frac{\mathrm{d}^2 x}{\mathrm{d}t^2} + (a + b)\frac{\mathrm{d}x}{dt} + abx = 1. \tag{I.12}$$

Since the operator is independent of x, then

$$\frac{\mathrm{d}^2 x}{\mathrm{d}t^2} = \frac{\mathrm{d}}{\mathrm{d}t}\left(\frac{\mathrm{d}x}{\mathrm{d}t}\right) = \frac{\mathrm{d}}{\mathrm{d}t}(\psi x) = \psi^2 x.$$

Hence, equation (I.12) can be written as

$$\psi^2 x + (a+b)\,\psi x + abx = 1$$

or $\quad x(\psi^2 + (a+b)\,\psi + ab) = 1$

and hence

$$x = \frac{1}{(\psi + a)(\psi + b)} \tag{I.13}$$

where a and b are the roots of the quadratic in ψ. Substitution for the original of the transform (Table I.1, equation (6)) gives

$$x = \frac{1}{ab} - \frac{e^{-at}}{a(b-a)} - \frac{e^{-bt}}{b(a-b)}.$$

If the initial conditions are such that the function, $f(t)$, and its derivatives are non-zero at $t=0$ then the application of the method is slightly different. In this case

$$\frac{dx}{dt} = \psi x - \psi x_0,$$

where x_0 is the value of the function at $t=0$. Then

$$\frac{d^2 x}{dt^2} = \frac{d}{dt}\left(\frac{dx}{dt}\right)$$

$$\frac{d^2 x}{dt^2} = \frac{d}{dt}(\psi x) - \frac{d}{dt}(\psi x_0)$$

$$\frac{d^2 x}{dt^2} = \psi^2 x - (\psi^2 x_0 + \psi x_1)$$

$$\ldots$$

$$\frac{d^n x}{dt^n} = \psi^n x - (\psi^n x_0 + \psi^{n-1} x_1 + \psi^{n-2} x_2 + \cdots + \psi x_{n-1})$$

where

$$x_0 = f(t) \text{ at } t=0$$

$$x_1 = f'(t) \text{ at } t=0$$

$$x_2 = f''(t) \text{ at } t=0$$

$$\ldots$$

$$x_{n-1} = f^{\langle n-1\rangle}(t) \text{ at } t=0.$$

TABLE I.1 *Relationship between various Laplace–Carson transforms and their original functions*

	Transform	Original
(1)	$\dfrac{1}{\psi}$	t
(2)	$\dfrac{1}{\psi^n}$	$\dfrac{t^n}{n!}$
(3)	$\dfrac{1}{(\psi \pm a)}$	$\dfrac{1}{a}(1 - e^{\mp at})$
(4)	$\dfrac{\psi}{(\psi \pm a)}$	$e^{\mp at}$
(5)	$\dfrac{(\psi + b)}{(\psi + a)}$	$\dfrac{b}{a} - \dfrac{(b-a)e^{-at}}{a}$
(6)	$\dfrac{1}{(\psi + a_1)(\psi + a_2)(\psi + a_3)\cdots(\psi + a_n)}$	$\dfrac{1}{a_1 a_2 a_3 \cdots a_n} - \dfrac{e^{-a_1 t}}{a_1(a_2 - a_1)(a_3 - a_1)\cdots(a_n - a_1)} - \dfrac{e^{-a_2 t}}{a_2(a_1 - a_2)(a_3 - a_2)\cdots(a_n - a_2)} \cdots - \dfrac{e^{-a_n t}}{a_n(a_1 - a_n)(a_2 - a_n)\cdots(a_{n-1} - a_n)}$
(7)	$\dfrac{\psi}{(\psi + a_1)(\psi + a_2)\cdots(\psi + a_n)}$	$\dfrac{e^{-a_1 t}}{(a_2 - a_1)(a_3 - a_1)\cdots(a_n - a_1)} + \dfrac{e^{-a_2 t}}{(a_1 - a_2)(a_3 - a_2)\cdots(a_n - a_2)} \cdots + \dfrac{e^{-a_n t}}{(a_1 - a_n)(a_2 - a_n)\cdots(a_{n-1} - a_n)}$
(8)	$\dfrac{(\psi + b)}{(\psi + a_1)(\psi + a_2)\cdots(\psi + a_n)}$	$\dfrac{b}{a_1 a_2 \cdots a_n} - \dfrac{(b - a_1)e^{-a_1 t}}{a_1(a_2 - a_1)(a_3 - a_1)\cdots(a_n - a_1)} - \dfrac{(b - a_2)e^{-a_2 t}}{a_2(a_1 - a_2)(a_3 - a_2)\cdots(a_n - a_2)} \cdots - \dfrac{(b - a_n)e^{-a_n t}}{a_n(a_1 - a_n)(a_2 - a_n)\cdots(a_{n-1} - a_n)}$
(9)	$\dfrac{\psi(\psi + b)}{(\psi + a_1)(\psi + a_2)\cdots(\psi + a_n)}$	$\dfrac{(b - a_1)e^{-a_1 t}}{(a_2 - a_1)(a_3 - a_1)\cdots(a_n - a_1)} + \dfrac{(b - a_2)e^{-a_2 t}}{(a_1 - a_2)(a_3 - a_2)\cdots(a_n - a_2)} \cdots + \dfrac{(b - a_n)e^{-a_n t}}{(a_1 - a_n)(a_2 - a_n)\cdots(a_{n-1} - a_n)}$

In example 1, let x have the value x_0 at $t = 0$, then replacing the derivative according to the rules set out above, we obtain

$$\psi x - \psi x_0 + ax = 1$$

$$(\psi + a)x = 1 + \psi x_0$$

or
$$x = \frac{1}{(\psi + a)} + \frac{\psi x_0}{(\psi + a)}. \tag{I.14}$$

Each individual transform in equation (I.14) is replaced by its original, thus

$$\frac{1}{(\psi + a)} = \frac{1}{a}(1 - e^{-at}) \quad \text{(Table I.1, equation (3))}$$

$$\frac{\psi}{(\psi + a)} = e^{-at} \quad \text{(Table I.1, equation (4))}$$

hence

$$x = \frac{1}{a}(1 - e^{-at}) + x_0 e^{-at}. \tag{I.15}$$

Appendix II The computation of enzyme parameters by curve fitting procedures

II.1. Introduction

For an enzyme-catalysed reaction obeying Michaelis–Menten kinetics, the steady-state velocity, v, at a given substrate concentration, [S], is given by

$$v = \frac{V[\mathrm{S}]}{K_m + [\mathrm{S}]}, \tag{II.1}$$

where the parameters that characterise this equation are V and K_m. One of the aims of enzyme kinetics is to determine these parameters at different values of pH and temperature or in the presence of inhibitor(s), metal ions etc. It is therefore essential that these parameters are as free as possible from human bias and that estimates of their reliability are also available. This effectively means that kinetic plots should not be fitted by eye; rather the data should be fitted to suitable statistical equations.

Experimentally, it is usual to determine the initial velocity, v, for a series of differing initial substrate concentrations $[\mathrm{S}_0]$. Since, in equation (II.1), the relationship between the independent variable, [S], and the dependent variable, v, is curvilinear, it is usual to plot the data in one of the following linear transformations:

$$\frac{1}{v} = \frac{K_m}{V}\frac{1}{[\mathrm{S}]} + \frac{1}{V} \tag{II.2}$$

$$\frac{[\mathrm{S}]}{v} = \frac{K_m}{V} + \frac{[\mathrm{S}]}{V} \tag{II.3}$$

$$\frac{v}{[\mathrm{S}]} = \frac{V}{K_m} - \frac{v}{K_m} \tag{II.4}$$

$$v = V - \frac{vK_m}{[\mathrm{S}]}. \tag{II.5}$$

Besides the obvious purpose of determining the kinetic parameters,

K_m and V, a plot of the experimental data in one of the linear transformations is more convenient for detecting deviations from Michaelis–Menten behaviour. Of these transformations, the most popular is the Lineweaver–Burk or double reciprocal plot (equation (II.2)). Since equations (II.2)–(II.5) are all variations of the hyperbolic function (Equation (II.1)), it might be imagined that any one of them could be used to estimate V and K_m with equal accuracy. This would certainly be true if both v and [S] were free from error, but although the concentration of the substrate, [S], can usually be controlled precisely by the operator by the use of high quality volumetric apparatus, the velocity, v, is subject to much larger errors. Under these circumstances, (with [S] assumed to be free from error), the four linear transformations no longer provide equally accurate estimates of V and K_m. Equations (II.4) and (II.5) are inconvenient for statistical purposes, as variables $v/[S]$ and v are both subject to error.

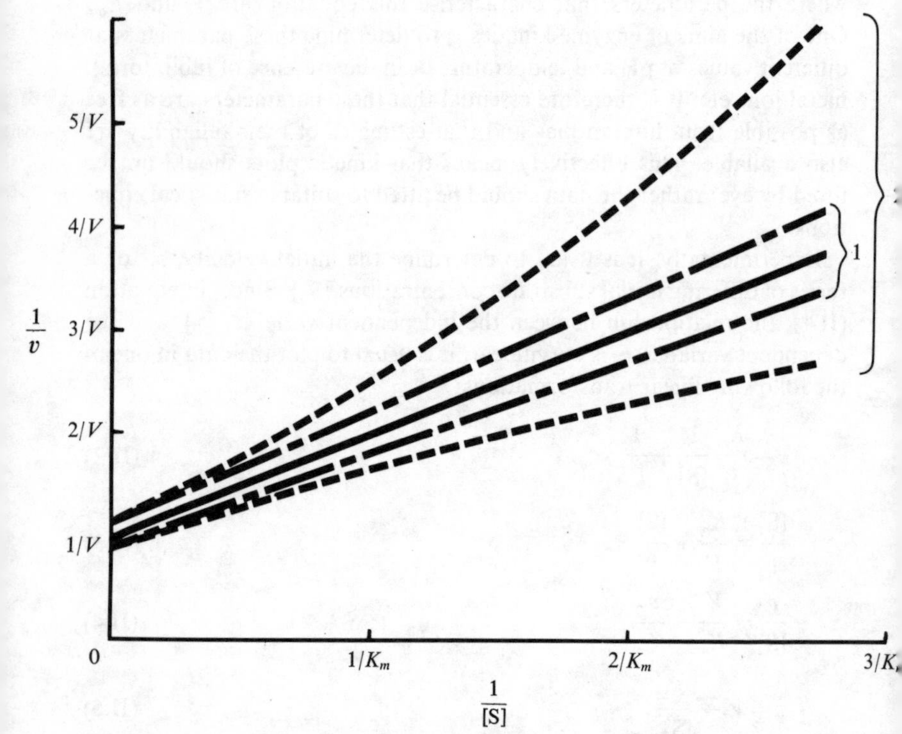

Fig. II.1. Error envelope for a Lineweaver–Burk plot. 1, a constant error in v and 2, an error proportional to v.

While the double reciprocal plot is the most frequently used, it is also the most unreliable when fitting data by eye, since reciprocation distorts the error span and places undue emphasis on the data points with smallest values of velocity, v, which also have the largest percentage error. This distortion of the error span is shown more clearly in fig. II.1 in terms of an error envelope for two cases; 1, a constant error in v and 2, an error proportional to v. It is obvious from fig. II.1 that in order to obtain reliable estimates of the two parameters, K_m and V, some account must be taken of the relative importance of the points at high and low values of $1/[S]$, i.e. the line must be properly *weighted*. The possible methods of computing these parameters using linear and non-linear regression methods are given in the next section.

II.2. Statistical analysis: least squares methods

The least squares methods can be divided into three groups:

(*a*) unweighted linear regression analysis;
(*b*) weighted linear regression analysis; and
(*c*) non-linear regression analysis.

Although this appendix is concerned essentially with the analysis of enzyme kinetic data, the method of least squares analysis is a general method with much wider applications and in that context it is useful to discuss the procedures in general terms and then relate them more specifically to the analysis of enzyme kinetic data. Before we can discuss the analysis of data, it is necessary to elucidate some of the terms that occur in the following sections.

The *normal distribution* was first derived by Demoivre in 1733 when dealing with problems associated with the tossing of coins. The distribution was also determined independently by Laplace and Gauss and is sometimes referred to as a *Gaussian distribution*. The equation is of the form

$$y = A \cdot e^{-h^2(x-m)^2},$$

where A, h and m are constants. The area under the curve is given by

$$\int_{-\infty}^{\infty} A \cdot e^{-h^2(x-m)^2}\,dx = \frac{A\sqrt{\pi}}{h}.$$

If $A = h\sqrt{\pi}$, the area under the curve is unity and hence the equation becomes

$$y = \frac{h}{\sqrt{\pi}} \cdot e^{-h^2(x-m)^2}. \tag{II.6}$$

The *probability* that an observation lies between x and $x+\delta x$ is

$$\frac{h}{\sqrt{\pi}}\cdot \mathrm{e}^{-h^2(x-m)^2}\,\delta x.$$

The first derivative (dy/dx) of equation (II.6) is zero when $x=m$; the curve therefore has a maximum at this point and m is in fact the *mean value* (usually written as $\bar{x}$) of the distribution in x. The second derivative ($\mathrm{d}^2y/\mathrm{d}x^2$) is zero when $x=m\pm 1/h\sqrt{2}$, i.e. there are two inflexion points. The constant, h, is sometimes called the *precision constant* and if $h^2=1/2\sigma^2$, then the inflexion points occur at $x=m\pm\sigma$, where σ is called the *standard deviation*. The complete equation can therefore be written as

$$y=\frac{1}{\sigma\sqrt{2\pi}}\cdot \mathrm{e}^{-(x-\bar{x})^2/2\sigma^2} \tag{II.7}$$

and the shape of the normal error curve for $\sigma=1$ is shown in fig. II.2. The area under the curve between the points $\pm 1\sigma$ is 0.68 and between $\pm 2\sigma$ is 0.95. This means that, on average, two out of three observations

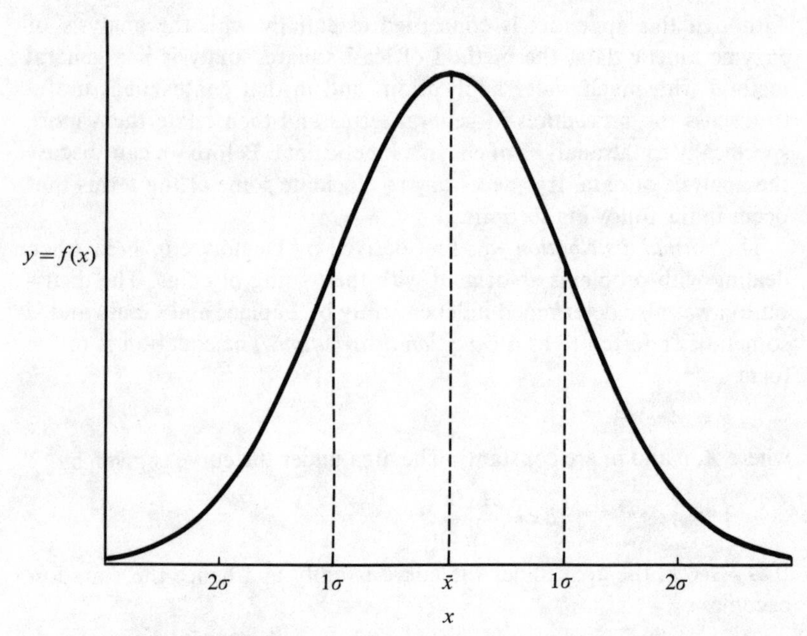

Fig. II.2. A normal or Gaussian distribution curve. Approximately two out of three observations lie within $\pm 1\sigma$ of the mean $x(\bar{x})$.

will lie between $\pm 1\sigma$ from the mean, whereas only one observation in twenty will lie outside $\pm 2\sigma$.

The *deviation*, d_i, of any point x_i from the mean, $\bar{x}$, is $d_i = x_i - \bar{x}$ and since the curve is symmetrical about the mean, the sum of the deviations is zero:

$$d_1 + d_2 + d_3 + \cdots + d_n = x_1 + x_2 + x_3 + \cdots + x_n - n\bar{x} = 0\,.$$

The *mean deviation* is defined as the mean of the absolute values of the deviations:

$$|d_1| + |d_2| + |d_3| + \cdots + |d_n| = \frac{1}{n} \cdot \sum_{i=1}^{n} |d_i|.$$

The *standard deviation*, σ, is defined in terms of the squares of the deviations from the mean:

$$\sigma^2 = (d_1^2 + d_2^2 + \cdots + d_n^2)/n$$

$$\sigma = \sqrt{\left(\frac{(x - \bar{x})^2}{n}\right)}.$$

σ^2 is known as the *variance* of the data and σ is sometimes referred to as the *root-mean-square deviation*. These definitions are for a normal distribution where n is large (Lim $n \to \infty$). If we select at random a finite sample of n observations then it is obvious that in general the mean of this smaller sample will not be the same as the mean of the whole population, but will approach it as the size of the sample increases. For a finite sample it is usual to define a *sample standard deviation* (s), given by

$$s = \sqrt{\left(\frac{\sum (x - \bar{x})^2}{n - 1}\right)}$$

where the term $n - 1$ is called the *number of degrees of freedom* (for two variables (x_i, y_i) this would be $n - 2$, for three variables (x_i, y_i, z_i) $n - 3$ and so on). The standard deviation of the *sample mean* is given by

$$s_{\bar{x}} = \frac{s}{\sqrt{n}}\cdot$$

II.2.1. Unweighted linear least squares analysis. Suppose that in an experiment, a range of values of a variable y has been determined at a series of values of the independent variable x. Assume that each value of y_i, determined at a particular value of x_i, is a member of a set normally distributed about a mean, $\alpha + \beta x_i$, i.e. y is linearly related to x with a constant variance σ^2, and that the value of x_i is known precisely. This

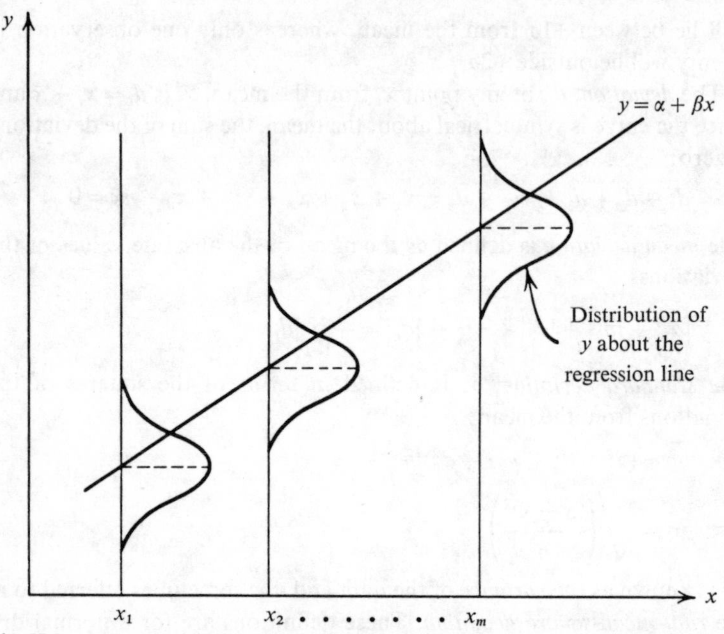

Fig. II.3. A linear regression line indicating the distribution of y at each value of x (x is assumed to be error-free). (From Carnahan, Luther & Wilkes (1969) *Applied Numerical Methods*. Wiley.)

situation is indicated in fig. II.3, in which $y = \alpha + \beta x$ is the *regression line*. Using a limited set of observations (x_i, y_i; $i = 1, m$), we wish to obtain estimates a, b and s^2 of the parameters α, β and σ^2. The probability that a value y lies within the range y and $y + \delta y$ is given by

$$\Pr(y) = \frac{1}{\sigma\sqrt{2\pi}} \mathrm{e}^{-(y-\alpha-\beta x)^2/2\sigma^2} = k\mathrm{e}^{-(y-\alpha-\beta x)^2/2\sigma^2}. \qquad \text{(II.8)}$$

The probability, P, of all values of y occurring simultaneously is

$$P = \Pr(1)\cdot\Pr(2)\cdots\Pr(m) = k^m \exp\left(-\sum_{i=1}^{m}(y_i - \alpha - \beta x_i)^2/2\sigma^2\right). \qquad \text{(II.9)}$$

The regression line is regarded as having the best fit to the data when the probability is a *maximum*, i.e. when the values of α and β are chosen to minimise the sum of squares:

$$s = \sum_{i=1}^{m}(y_i - \alpha - \beta x_i)^2. \qquad \text{(II.10)}$$

Since α and β define the calculated value of y_i, (y'_i), at each value of the independent variable x_i, the term $y_i - \alpha - \beta x_i$ is in fact the deviation,

d_i, of the point (x_i, y_i) from the line and hence the method of least squares is often referred to as an attempt to minimise the *sum of the squares of deviation*. From equation (II.10), $s/\delta\alpha = -2\Sigma(y - \alpha - \beta x)$ and $\delta s/\delta\beta = -2\Sigma x(y - \alpha - \beta x)$. (The subscript i is dropped since the summations are assumed to be over all m observations.) For s to be a minimum, $\delta s/\delta\alpha = \delta s/\delta\beta = 0$ and replacing α and β by their estimates a and b then

$$\sum(y - a - bx) = 0, \qquad \sum x(y - a - bx) = 0.$$

These equations can be rearranged to give the simultaneous *normal equations*

$$ma + b\sum x = \sum y \tag{II.11}$$

and $$a\sum x + b\sum x^2 = \sum xy, \tag{II.12}$$

whence

$$b = \frac{m\sum xy - \sum x \sum y}{m\sum x^2 - (\sum x)^2} \tag{II.13}$$

or $$b = \frac{\sum(x - \bar{x})(y - \bar{y})}{\sum(x - \bar{x})^2} = \frac{\operatorname{cov}(x, y)}{\operatorname{var}(x)},$$

where $\operatorname{cov}(x,y)$ is the *covariance* of x and y, i.e. a measure of the interdependence of x and y, and $\bar{x}$, $\bar{y}$ are the mean values of the set of observations x_i, y_i, $i = 1, m$, i.e. $\bar{x} = \Sigma x_i/m$, $\bar{y} = \Sigma y_i/m$. Equation (II.13) gives the slope of the line that has the best fit to the data; the intercept a can be determined from this equation and the knowledge that the line passes through the mean value point $(\bar{x}, \bar{y})$. Hence

$$a = \bar{y} - b\bar{x}. \tag{II.14}$$

The estimated value, s^2, of the variance, σ^2, of the regression line is given by

$$s^2 = \frac{\sum(y - a - bx)^2}{m - 2}, \tag{II.15}$$

where $m - 2$ is the number of degrees of freedom. The estimated values of the variance and the standard deviation (SD) of the intercept, a, and slope, b, are obtained from the normal equations (II.11) and (II.12).

$$\frac{s_b^2}{m} = \frac{s_a^2}{\sum x^2} = \frac{s^2}{\Delta}$$

where Δ is the determinant

$$\Delta = \begin{vmatrix} m & \sum x \\ \sum x & \sum x^2 \end{vmatrix} = m\sum x^2 - (\Sigma x)^2.$$

Hence, for the slope,

$$s_b^2 = \frac{s^2}{\sum x^2 - \{(\sum x)^2/m\}} \quad \text{and} \quad \text{SD}_b = \sqrt{(s_b^2)}; \tag{II.16}$$

for the intercept

$$s_a^2 = \frac{s^2 \sum x^2}{m\sum x^2 - (\sum x)^2} \quad \text{and} \quad \text{SD}_a = \sqrt{(s_a^2)}. \tag{II.17}$$

For the simple linear least squares regression it is relatively easy to tabulate x_i, y_i, $x_i y_i$, x_i^2 and y_i^2, compute the various sums and hence determine the slope and intercept of the line that best fits the data.

For enzyme kinetic data, any of the linear transformations mentioned in section II.1 could be fitted using the unweighted least squares procedure; thus equation (II.2) is related to the above equations by the transformations $y = 1/v$, $x = 1/[S]$, $b = K_m/V$ and $a = 1/V$, but because of the distortion of the error span due to reciprocation and the bias this places on points at high values of $1/[S]$, the simple unweighted fit is not recommended. This problem of the distortion of the error span may be found with any non-linear function that is plotted as a transformed linear function, e.g. $y = ke^{ax}$ which is plotted as $\ln(y)$ against x.

II.2.2. Weighted least squares linear regression analysis. The problem of error distortion may be overcome by suitably weighting the linear regression line. The undue bias of the points at high values of $1/[S]$, in the case of the double reciprocal plot, is overcome by weighting the points at the other end of the regression line. The weighted regression procedure is equivalent to fitting an unweighted regression to an *enlarged* set of data, each point (x_i, y_i) in the set, being repeated w_i times where w_i is the *weighting* factor. The optimum regression line is now the one that minimises the *weighted* sum of squares of the deviations:

$$s = \sum_{i=1}^{m} w_i(y_i - \alpha - \beta x_i)^2.$$

The *normal equations*, in terms of the estimates a and b, are now

$$a\sum w + b\sum wx = \sum wy \tag{II.18}$$

$$a\sum wx + b\sum wx^2 = \sum wxy \tag{II.19}$$

which can be solved for a and b to give

$$b = \frac{\sum w\sum wxy - \sum wy\sum wx}{\sum w\sum wx^2 - (\sum wx)^2} \tag{II.20}$$

$$a = \bar{y} - b\bar{x}, \tag{II.21}$$

where $\bar{x}$, $\bar{y}$ are now the *weighted means* $\bar{x} = \Sigma wx/\Sigma w$, $\bar{y} = \Sigma wy/\Sigma w$. Whilst arbitrary weighting factors may be chosen to favour particular points, the most efficient statistical weight and the least subjective, is the reciprocal of the variance, $w_i = 1/\sigma^2_{y_i}$, where w_i is the weighting factor for the point (x_i, y_i) and $\sigma^2_{y_i}$ is the estimated or determined variance of y_i. In the case of plots that have been linearised by a transformation of a non-linear function, the weights must be those of the *transformed variable*. Two examples should clarify this point.

In the case of the Lineweaver–Burk plot, the new variable y is the reciprocal of the velocity, thus $y = 1/v$ and $\delta y = -\delta v/v^2$. Equating δy and δv with the standard deviations σ_y and σ_v, it is apparent that the variance of the transformed variable σ^2_y equals σ^2_v/v^4. The correct weighting factor for the double reciprocal plot is therefore (Elmore, Kingston & Shields, 1963)

$$w_i = \frac{1}{\sigma^2_{y_i}} = \frac{v^4}{\sigma^2_{v_i}} \tag{II.22}$$

where σ_{v_i} is the standard deviation of the original velocity v_i determined at a substrate concentration $[S_i]$. It is obvious from the weighting factor (equation (II.22)), that for a constant variance σ^2_v, the double reciprocal plot is increasingly weighted as the velocity increases (i.e. at low $1/[S]$), thus offsetting the bias in the unweighted linear regression at low values of v (i.e. at high $1/[S]$). The second example of a transformed variable is an exponential function $y = k \cdot e^{ax}$, which is plotted in linear form as $\ln(y) = \ln(k) + ax$. The new variable in this case is $z = \ln(y)$ and hence $\delta z = \delta y/y$, i.e. $\sigma_z = \sigma_y/y$ and the correct weighting factor for this plot in terms of the variance of the original observation, y, is $w_z = y^2/\sigma^2_y$.

The variance and standard deviation of the intercept and slope of a weighted least squares regression line are given by

$$\frac{s_b^2}{\sum w} = \frac{s_a^2}{\sum wx^2} = \frac{s^2}{\Delta}$$

where

$$s^2 = \frac{\sum w(y-a-bx)^2}{m-2}$$

and $\Delta = \begin{vmatrix} \sum wx^2 & \sum wx \\ \sum wx & \sum w \end{vmatrix} = \Sigma w \Sigma wx^2 - (\Sigma wx)^2.$

Hence for the slope,

$$s_b^2 = \frac{s^2}{\sum wx^2 - \{(\sum wx)^2/\sum w\}} \quad \text{and} \quad \text{SD}_b = \sqrt{(s_b^2)}; \tag{II.23}$$

and for the intercept,

$$s_a^2 = \frac{s^2 \sum wx^2}{\sum w \sum wx^2 - (\sum wx)^2} \quad \text{and} \quad \text{SD}_a = \sqrt{(s_a^2)}. \tag{II.24}$$

Equations (II.20), (II.21), (II.23) and (II.24) give the slope, intercept and the variances of the optimum weighted regression line. For the Lineweaver–Burk plot, we can equate $b = K_m/V$ and $a = 1/V$. In order to make accurate comparisons with other sets of data, it is necessary to have estimates of the variance or standard deviation of the two parameters V and K_m. By the same arguments outlined previously for the weights of the transformed variables, the standard deviation of V, s_V, is related to the standard deviation of the intercept, s_a ($a = 1/V$) by

$$s_V = V^2 s_a.$$

From the slope $b = K_m/V$, $K_m = b/a$ and since b and a are not stochastically independent an estimate of the variance of K_m, $s_{K_m}^2$, is given by the equation (Elmore *et al.*, 1963)

$$s_{K_m}^2 = \left(\frac{\delta K_m}{\delta a}\right)^2 s_a^2 + \left(\frac{\delta K_m}{\delta b}\right)^2 s_b^2 + 2\frac{\delta K_m}{\delta a}\cdot\frac{\delta K_m}{\delta b}\operatorname{cov}(a,b). \tag{II.25}$$

Since $\operatorname{cov}(a,b) = -s^2\bar{x}/\Sigma w(x-\bar{x})^2$ then

$$s_{K_m}^2 = \frac{b^2}{a^4}s_a^2 + \frac{s_b^2}{a^2} + \frac{2b}{a^3}\frac{s^2 \sum wx}{[\sum w \sum wx^2 - (\sum wx)^2]}.$$

The weighting factors, w_i, can be obtained by determining the velocity, v_i, at each value of $[S_i]$, a number of times and hence obtaining a mean value v_i and its variance $s^2_{v_i}$. An alternative, but less accurate, procedure is to determine the variance of a number of velocity determinations at high and low substrate concentrations and use interpolation to determine the variance at intermediate points. A third method, which can be applied to continuous product/time plots, is to use a polynomial regression analysis procedure (Elmore *et al.*, 1963) to determine the initial velocity at $t = 0$. In this case the standard deviation of the velocity is an estimate of the error in extrapolation to $t = 0$. However, it may be that the standard deviation of the initial velocity, v, based on the repeatability of individual kinetic runs, is greater than the error in determining the initial velocity by extrapolation to $t = 0$. In this case, the standard deviation of the worst case must always be used.

Fitting experimental data in reciprocal form yields satisfactory results only if the proper weighting factors are used. Note that since the proper weight contains the fourth power of the velocity, any 'rogue' points, i.e. ones that for some reason, usually human error, are not members of a set of points normally distributed about a mean value, will weight the curve unduly in their favour. In this case, a computer program based on the weighted least squares formulae should be cycled to remove the worst point, determine the new values of K_m and V, determine if they are significantly different from the values obtained with the set including the worst point and if they are, remove the worst point and repeat the process. Values of K_m and V are accepted only after a suitable statistical test has been passed.

II.2.3. Non-linear least squares regression. An alternative procedure for dealing with non-linear equations is to fit the data directly to the equation using a non-linear optimisation procedure. The method to be presented is based on an article by Wentworth (1965*a*, *b*) and is generally applicable; thus one could fit data directly to the Michaelis–Menten equation (equation (II.1)), any of the two-substrate rate equations (equation (6.12)), the Adair equation (equation (8.12)), pH–V equation (equation (4.5)), etc. In each case it is necessary to provide initial estimates of the various parameters to be fitted. If the variances of the data points are essentially constant then an unweighted fit can be carried out; otherwise a weighted non-linear regression must be used. Since the data are being fitted directly to a non-linear function, it is necessary to provide data points over a sufficient range that the shape of the curve is completely described.

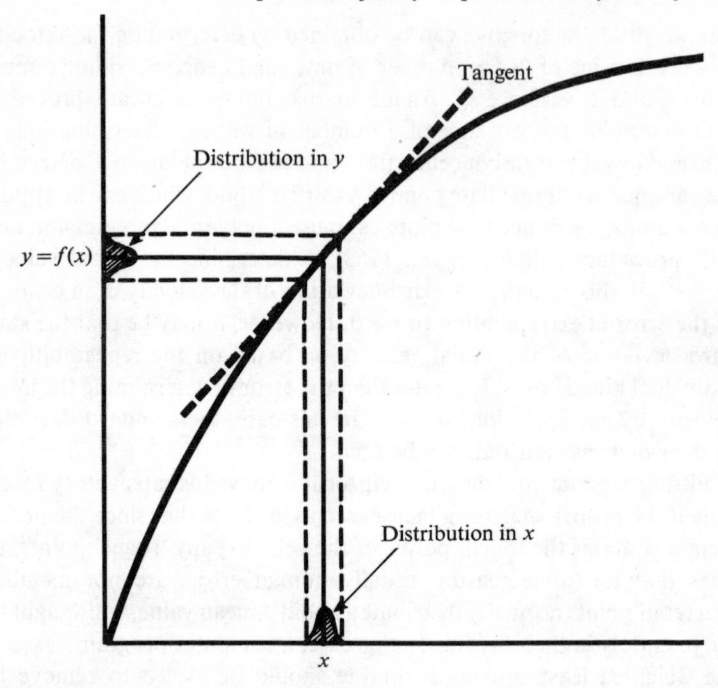

Fig. II.4. Schematic representation of the basis of non-linear optimisation techniques. Small errors in both x and y, indicated by the Gaussian curves, can be represented by the tangent to the curve.

Most non-linear optimisation procedures utilise the fact that small errors in the values of the variables x and y can be represented by the tangent to the curve, i.e. by the first term in a Taylor series expansion (fig. II.4). For the Michaelis–Menten equation (equation (II.1)), which relates the two variables $[S_i]$, v_i and the two parameters, V and K_m, assume that m pairs of observations of $[S_i]$, v_i are obtained and that both $[S_i]$ and v_i are subject to error. The uncertainty in the tru‹ ›sition of any point $[S_i]$, v_i is no longer a vertical error bar, as was the case when $[S_i]$ was assumed to be free from error, but is now an area, most likely an ellipse with the principle axis in the vertical direction since $[S_i]$ is likely to have the least error. Because the number of observations is finite and both $[S_i]$ and v_i are subject to error, it will not be possible to determine the true values of V and K_m but only estimates, with also an estimate of their standard deviations. This is achieved by a weighted least squares adjustment of the points $[S_i]$, v_i, $i = 1, m$ by minimising the sum of squares of the weighted residuals:

$$s = \sum (w_{s_i} \Delta[S_i]^2 + w_{v_i} \Delta v_i^2) = \text{minimum},$$

where $\Delta[S_i]$ and Δv_i are the deviations from the calculated values of the variables $[\bar{S}_i]$, $\bar{v}_i$ and $w_{s_i} = 1/\sigma_{s_i}^2$, $w_{v_i} = 1/\sigma_{v_i}^2$:

$$\Delta[S_i] = ([S_i] - [\bar{S}_i]),$$

$$\Delta v_i = (v_i - \bar{v}_i).$$

We can define a function $F_i([S_i], v_i, V, K_m)$

$$F_i = v_i - \frac{V[S_i]}{K_m + [S_i]} \tag{II.26}$$

as the difference between the calculated value of v_i, at particular estimates of V and K_m, and the experimental value of v_i. The function $F_i^0([S_i]$, v_i, V^0, $K_m^0)$, where the superscript 0 refers to the initial estimates of V and K_m, can be expressed as a Taylor series to give the adjusted values of V and K_m:

$$F_i([\bar{S}_i], \bar{v}_i, V, K_m) = F_i^0([S_i], v_i, V^0, K_m^0) - F_{v_i} \Delta v_i - F_{s_i} \Delta[S_i] - F_{V_i} \Delta V - F_{K_{m_i}} \Delta K_m \quad i = 1, 2, ..., m \tag{II.27}$$

where

$$F_{v_i} = \left(\frac{\delta F_i}{\delta v}\right)_{[S_i], V^0, K_m^0}, \qquad F_{s_i} = \left(\frac{\delta F_i}{\delta [S]}\right)_{v_i, V^0, K_m^0},$$

$$F_{V_i} = \left(\frac{\delta F_i}{\delta V}\right)_{[S_i], v_i, K_m^0} \quad \text{and} \quad F_{K_{m_i}} = \left(\frac{\delta F_i}{\delta K_m}\right)_{[S_i], v_i, V^0}$$

The values of $[\bar{S}_i]$, $\bar{v}_i$ are the calculated values that cause the condition equations, using the adjusted values of V and K_m, to be zero

$$F_i([\bar{S}_i], \bar{v}_i, V, K_m) = 0 \quad i = 1, 2, \cdots, m$$

i.e. $$F_i = \bar{v}_i - \frac{V[\bar{S}_i]}{K_m + [\bar{S}_i]} = 0 \quad i = 1, 2, \cdots, m.$$

hence from equation (II.27),

$$F_{v_i} \Delta v_i + F_{s_i} \Delta[S_i] + F_{V_i} \Delta V + F_{K_{m_i}} \Delta K_m = F_i^0([S_i], v_i, V^0, K_m^0)$$

where

$$F_i^0 = v_i - \frac{V^0[S_i]}{K_m^0 + [S_i]}$$

and $\Delta V = V^0 - V, \qquad \Delta K_m = K_m^0 - K_m.$

(V and K_m are the adjusted values of V^0 and K_m^0 after one iteration.) This eventually leads, in this case, to two simultaneous equations:

$$\sum \frac{F_{V_i} \cdot F_{V_i}}{L_i} \cdot \Delta V + \sum \frac{F_{V_i} \cdot F_{K_{m_i}}}{L_i} \cdot \Delta K_m = \sum \frac{F_{V_i} \cdot F_i^0}{L_i} \tag{II.28}$$

$$\sum \frac{F_{K_{m_i}} \cdot F_{V_i}}{L_i} \cdot \Delta V + \sum \frac{F_{K_{m_i}} \cdot F_{K_{m_i}}}{L_i} \cdot \Delta K_m = \sum \frac{F_{K_{m_i}} \cdot F_i^0}{L_i}, \tag{II.29}$$

where

$$L_i = \frac{F_{v_i}^2}{w_{v_i}} + \frac{F_{s_i}^2}{w_{s_i}}.$$

In coefficient form equations (II.28) and (II.29) are

$$b_{11}\,\Delta V + b_{12}\,\Delta K_m = c_1$$

and $b_{21}\,\Delta V + b_{22}\,\Delta K_m = c_2,$

which can be solved for ΔV and ΔK_m; hence

$$\Delta V = \frac{c_1 b_{22} - c_2 b_{12}}{b_{11} b_{22} - b_{21} b_{12}} \quad \text{and} \quad \Delta K_m = \frac{b_{11} c_2 - c_1 b_{12}}{b_{11} b_{22} - b_{21} b_{12}}.$$

The refined values of V and K_m are

$$V = V^0 - \Delta V \quad \text{and} \quad K_m = K_m^0 - \Delta K_m$$

and these *new* values of V and K_m can now be used as starter values for the second iteration. A number of iterations are carried out until a suitable statistical test is passed. This can be the difference between a parameter 'sigma external' (σ_{ext}) before and after each iteration:

$$\sigma_{\text{ext}}^2 = \frac{s}{m-2}$$

($m - 2$ is the number of degrees of freedom where there are m pairs of observations and two parameters to be evaluated).

s is given by

$$s = \sum \frac{F_i^0 \cdot F_i^0}{L_i} - \sum \frac{F_{V_i} \cdot F_i^0}{L_i} \cdot \Delta V - \sum \frac{F_{K_{m_i}} \cdot F_i^0}{L_i} \cdot \Delta K_m.$$

If the initial estimates of V and K_m are poor, then during the iteration process the parameters will fail to converge to unique values and new estimates of V^0 and K_m^0 should be tried. This is not likely to occur, however, if a linear regression is used initially to provide the starting values of V^0 and K_m^0.

II.3. Computer programs for enzyme kinetics

The procedures outlined in section II.2 are summarised in the following computer program listing of *KMVM*. This program carries out a weighted linear and non-linear least squares regression analysis of data using the Lineweaver–Burk and Michaelis–Menten equations respectively and determines K_m and V and their standard deviations. It can be used for inhibition studies or with reactions involving two or more substrates, in which case the parameters K_m and V will be functions of the concentrations of inhibitors, other substrates etc. which are held constant. The program consists of four segments; *MAIN* and the subroutines *WLSQ*, *WORST* and *KMFI*. *MAIN* reads the data, inverts it for the double reciprocal plot, computes the various sums Σwx, Σwy etc. and calls *WLSQ*. *WLSQ* determines the initial values of K_m and V by a weighted least squares linear regression of $1/v$ against $1/[S]$. These values are then used by *KMFI* which carries out a non-linear fit to the Michaelis–Menten equation. *MAIN* then calls *WORST* which determines the worst point from the weighted linear regression. This point is then temporarily removed and the cycle of computing K_m and V is repeated. The new values of K_m and V are checked to see if they are significantly different from the previous values. If there is no significant difference between the two sets of values then the first set of values is accepted. If there is a significant difference between the two sets of values, then the worst point is permanently removed and the process is repeated. Final acceptance of the parameters K_m and V is determined by passing a t-test for the number of points remaining. This is summarised in the following program listing.

```
 1   C   *************************** KMVM ***********************************
 2   C   *                                                                  *
 3   C   ********************* MAIN SEGMENT *********************************
 4   C   *                                                                  *
 5   C   * THIS PROGRAM WAS CRIGINALLY ONE DEVELOPED BY ELMORE,KINGSTON     *
 6   C   * AND SHIELDS J.CHEM. SOC.2070,(1963). IT WAS LATER MODIFIED       *
 7   C   * BY D.V.ROBERTS AND D.T.ELMORE TO INCLUDE NON-LINEAR FITTING.     *
 8   C   * THE PROGRAM DETERMINES THE KINETIC PARAMETERS KM AND V FOR      *
 9   C   * AN ENZYME CATALYSED REACTION. IT CONSISTS OF A MAIN SEGMENT     *
10   C   * AND THREE SUBROUTINES. THE DATA REQUIRED IS A VELOCITY,AN       *
11   C   * ESTIMATE OF ITS STANDARD DEVIATION AND THE SUBSTRATE            *
12   C   * CONCENTRATION, FOR OPTIMUM RESULTS THE NUMBER OF POINTS         *
13   C   * SHOULD PREFERABLY EXCEED ABOUT 8. THE SUBROUTINE WLSQ FIRSTLY   *
14   C   * OBTAINS AN ESTIMATE OF KM AND V BY FITTING THE DATA USING       *
15   C   * A LEAST SQUARES ANALYSIS TO THE LINEWEAVER-BURK EQUATION.       *
16   C   * THE STANDARD DEVIATIONS OF THE VELOCITIES ARE USED AS           *
17   C   * WEIGHTING FACTORS. THESE INITIAL VALUES OF KM AND V ARE         *
18   C   * THEN USED AS STARTER VALUES FOR A NON-LINEAR REGRESSION         *
19   C   * ANALYSIS (LEAST SQUARES) TO THE HYPERBOLA EQUATION IN           *
20   C   * SUBROUTINE KMFI.(FOR NON-LINEAR ANALYSIS SEE WENTWORTH          *
21   C   * J.CHEM.ED.(1965),44, 92  AND 162 ).THE WORST POINT IS           *
22   C   * DETERMINED USING SUBROUTINE WORST AND TEMPORARILY REMOVED       *
23   C   * AND THE COMPUTATIONS ARE THEN REPEATED LESS THIS POINT.         *
24   C   * THE NEW VALUES OF KM AND V ARE THEN COMPARED WITH THE           *
25   C   * ORIGINAL ONES AND IF THERE IS NO SIGNIFICANT DIFFERENCE         *
26   C   * (T-TEST),THE ORIGINAL VALUES ARE ACCEPTED. IF REMOVING THE      *
27   C   * WORST POINT PRODUCES VALUES OF KM AND V THAT SIGNIFICANTLY      *
28   C   * DIFFERENT THEN THE WORST POINT IS DROPPED AND THE PROCESS IS    *
29   C   * REPEATED WITH THE NEXT WORST POINT. A 95% T-TEST IS APPLIED     *
30   C   * THE FIRST CARD OF THE DATA HAS THE NUMBER OF POINTS IN I2       *
31   C   * FORMAT IN COLS. 1-2. THE REMAINDER OF THE CARD CAN BE USED      *
32   C   * FOR DESCRIPTIVE PURPOSES. IF COLS.1-2 ARE ZERO THEN THIS        *
33   C   * EXITS THE PROGRAM. THE NEXT CARDS ARE THE DATA CARDS AND        *
34   C   * HAVE THE SUBSTRATE CONCENTRATION, S ,THE VELOCITY, V ,AND       *
35   C   * THE STANDARD DEVIATION OF THE VELOCITY, SV ,PUNCHED IN          *
36   C   * E10.4 FORMAT WITH ONE SPACE BETWEEN THE NUMBERS.                *
37   C   *                                                                  *
38   C   ********************************************************************
39       DIMENSION X(60),Y(60),W(60),S(60),V(60),SV(60),T(48),BUF(69)
40       DATA T/12.706,4.303,3.183,2.776,2.571,2.447,2.365,2.306,2.262,
41      12.228,2.201,2.179,2.160,2.145,2.132,2.120,2.110,2.101,2.093,2.086,
42      22.080,2.074,2.069,2.064,2.060,2.056,2.052,2.048,2.045,2.042,2.040,
43      32.037,2.035,2.032,2.030,2.028,2.026,2.024,2.023,2.021,2.020,2.018,
44      42.017,2.015,2.014,2.013,2.011,2.011/
45   13  CONTINUE
46   95  FORMAT (1H1)
47       WRITE(3,95)
48   15  FORMAT (I2,1X,69A1)
49   C   READ THE FIRST CARD, N = NO. OF DATA POINTS IN I2 FORMAT
50   C   THE REMAINDER OF THE CARD CAN BE DECRIPTIVE.
51       READ(1,15)N,(BUF(J),J=1,69)
52       IF (N)89,89,120
53   14   FORMAT(1X,69A1)
54   120 WRITE(3,14)(BUF(J),J=1,69)
55   16  FORMAT (3(E10.4,1X))
56   C   READ THE DATA CARDS, SUBSTRATE CONCENTRATION, VELOCITY AND
57   C   STANDARD DEVIATION OF THE VELOCITY.
58       READ (1,16)(S(I),V(I),SV(I),I=1,N)
```

```
 59    17    FORMAT(1X,'       S              V              SD OF V        X              Y
 60         1      W')
 61          WRITE(3,17)
 62          J=0
 63          SW=0.0
 64          SWX=0.0
 65          SWY=0.0
 66          SWX2=0.0
 67          SWY2=0.0
 68          SWXY=0.0
 69    C     CALCULATES THE RECIPROCALS OF [S] AND V AND ALSO THE WEIGHT
 70    C     W. DETERMINES THE VARIOUS SUMS OF SQUARES ETC. FOR A
 71    C     WEIGHTED LINEAR LEAST SQUARES ANALYSIS.
 72          DO 18 I=1,N
 73          X(I)=1.0/S(I)
 74          Y(I)=1.0/V(I)
 75          W(I)=V(I)**4/SV(I)**2
 76          SW=SW+W(I)
 77          SWX=SWX+W(I)*X(I)
 78          SWY=SWY+W(I)*Y(I)
 79          SWX2=SWX2+W(I)*X(I)*X(I)
 80          SWY2=SWY2+W(I)*Y(I)*Y(I)
 81          SWXY=SWXY+W(I)*X(I)*Y(I)
 82    18    CONTINUE
 83    19     FORMAT (6(1X,1PE10.4,1X))
 84    C      WRITE OUT THE DATA THAT HAS BEEN READ IN AS A RECORD.
 85          WRITE(3,19)(S(I),V(I),SV(I),X(I),Y(I),W(I),I=1,N)
 86    C      CALLS SUBROUTINE WLSQ WHICH COMPUTES KM & V USING A WEIGHTED
 87    C      LEAST SQUARES REGRESSION BASED ON THE LINEWEAVER-BURK PLOT.
 88    20    CALL WLSQ (N,SW,SWX,SWY,SWX2,SWY2,SWXY,ORD,SLOPE,SD,KM,DKM,VMAX,
 89         1DVMAX,DORD,DSLOPE,KMO,VO)
 90           IF (J)10,11,10
 91    C     CALLS SUBROUTINE KMFI,WHICH CARRIES OUT A NON-LINEAR LEAST SQUARES FIT
 92    C     TO THE M-M EQUATION,USING THE VALUES OBTAINED FROM THE
 93    C     LINEAR LEAST SQUARES FIT AS STARTING VALUES.
 94    C     CALLS SUBROUTINE WORST. THIS DETERMINES THE WORST POINT
 95    C     REMOVES IT,RECALCULATES THE VALUES OF KM,V AND CHECKS IF
 96    C     THEY ARE SIGNIFICANTLY DIFFERENT FROM THE PREVIOUS VALUES.
 97    11    CALL KMFI (KMO,VO,V,S,SV,N)
 98          CALL WORST (N,X,Y,W,ORD,SLOPE,SW,SWX,SWX2,SWY,SWY2,SWXY,YWOBS,
 99         1XWOBS,WWOBS,KM,DKM,VMAX,DVMAX,DORD,DSLOPE,SD,S,V,SV)
100          J=J+1
101          N=N-1
102          GO TO 20
103    C      DETERMINE THE T (TOBS) FOR THE WEIGHTED LEAST SQUARES FIT
104    C      AND COMPARE WITH TABLE OF T VALUES.
105    10    TOBS=ABS(YWOBS-ORD-SLOPE*XWOBS)*SQRT(WWOBS)/SD
106    12    FORMAT (/1X,2HT=,F7.3////)
107          WRITE(3,12)TOBS
108          IF (TOBS-T(N-2))121,121,11
109    121   CONTINUE
110          GO TO 13
111    89    CONTINUE
112    88    FORMAT (//////1X,18HALL DATA PROCESSED)
113          WRITE(3,88)
114          END
```

```
 1  C       ***************SUBROUTINE WLSQ*********************************
 2  C       *                                                              *
 3  C       * THIS SUBROUTINE COMPUTES THE SLOPE AND INTERCEPT OF THE       *
 4  C       * LINEAR LEAST SQUARES REGRESSION ANALYSIS OF THE LINE          *
 5  C       * WEAVER-BURK PLOT. S2=VARIANCE OF THE LINE,SD= STANDARD        *
 6  C       * DEVIATION OF THE LINE. DSLOPE = STANDARD DEVIATION OF         *
 7  C       * THE SLOPE. DORD = STANDARD DEVIATION OF THE INTERCEPT         *
 8  C       * VO = INITIAL ESTIMATE OF VMAX FOR THE DIRECT FIT USING        *
 9  C       * KMFI. KMO = INITIAL ESTIMATE OF KM. DKM=SD OF KM,             *
10  C       * DVMAX=SD OF VMAX.                                             *
11  C       *                                                              *
12  C       ****************************************************************
13          SUBROUTINE WLSQ (N,SW,SWX,SWY,SWX2,SWY2,SWXY,ORD,SLOPE,SD,KM,DKM,
14         1VMAX,DVMAX,DORD,DSLOPE,KMO,VO)
15          REAL KM,NUM,KMO
16          NUM=SW*SWXY-SWX*SWY
17          DENOM=SW*SWX2-SWX*SWX
18          SLOPE=NUM/DENOM
19          ORD=(SWY-SLOPE*SWX)/SW
20          S2=(SWY2-SWY*SWY/SW-SLOPE*NUM/SW)/FLOAT(N-2)
21          SD=SQRT(S2)
22          DSLOPE=SD*SQRT(SW/DENOM)
23          DORD=SD*SQRT(SWX2/DENOM)
24          KM=SLOPE/ORD
25          VMAX=1.0/ORD
26          VO=VMAX
27          KMO=KM
28          DKM=SD*VMAX*SQRT((KM*KM*SWX2+2.0*KM*SWX+SW)/DENOM)
29          DVMAX=DORD/(ORD*ORD)
30          RETURN
31          END
```

```
 1  C      **************SUBROUTINE KMFI*************************************
 2  C      *                                                                *
 3  C      * THIS SUBROUTINE TAKES THE INITIAL ESTIMATES FOR KM AND V       *
 4  C      * AND USES THEM AS STARTER VALUES FOR A NON-LINEAR LEAST         *
 5  C      * SQUARES REGRESSION ANALYSIS ON THE HYPERBOLA EQUATION          *
 6  C      * V=V(MAX)*S/(KM+S) ACCORDING TO THE METHOD OF WENTHWORTH        *
 7  C      * J.CHEM.ED.1965,44, 92  AND 162.THE SUBROUTINE REITERATES       *
 8  C      * UNTIL THERE IS NO FURTHER SIGNIFICANT IMPROVEMENT IN THE       *
 9  C      * PARAMETERS. CONTROL THEN RETURNS TO THE MAIN SEGMENT           *
10  C      * WHERE THE WORST POINT IS REMOVED AND THE PROCESS IS RE-        *
11  C      * PEATED.                                                        *
12  C      *                                                                *
13  C      ******************************************************************
14         SUBROUTINE KMFI (KMO,VO,V,S,SV,N)
15         REAL KMO,L,KM
16         DIMENSION V(60),SV(60),S(60),F(60),FK(60),FS(60),L(60),
17        1KM(10),VM(10),SIGMAV(10),SIGMAK(10),SIGMAX(10),DELTAK(10),
18        2DELTAV(10),FV(60)
19  10     FORMAT(//1X,18HSTARTER VALUE KMO=,1PE11.4,2X,17HSTARTER VALUE VO=,
20        11PE11.4)
21         WRITE(3,10)KMO,VO
22  102    FORMAT(//1X,105HTHE FOLLOWING VALUES ARE OBTAINED BY APPLYING THE
23        1LEAST SQUARES METHOD DIRECTLY TO THE MICHAELIS EQUATION)
24         WRITE(3,102)
25  104    FORMAT(/1X,76HAN ARBITRARY STANDARD DEVIATION OF 1% IS ATTACHED TO
26        1 SUBSTRATE CONCENTRATION)
27         WRITE(3,104)
28         KM(1)=KMO
29         VM(1)=VO
30         M=0
31  13     CONTINUE
32         B11=0.0
33         B12=0.0
34         B21=0.0
35         B22=0.0
36         C1 =0.0
37         C2 =0.0
38         C3 =0.0
39         M=M+1
40         DO 38 I=1,N
41         P=VM(M)*S(I)
42         Q=KM(M)+S(I)
43         QQ=Q**2
44  C      F(I)=V(I)-VM(M)*S(I)/(KM(M)+S(I)). THIS DETERMINES THE
45  C      DEVIATIONS OF THE CALCULATED VALUE OF V(I) USING THE
46  C      INITIAL ESTIMATES OF VM(M) & KM(M)
47  C      FV(I)= PARTIAL DERIVATIVE OF F(I) WITH RESPECT TO VM.
48  C      FK(I)= PARTIAL DERIVATIVE OF F(I) WITH RESPECT TO KM.
49  C      FS(I)= PARTIAL DERIVATIVE OF F(I) WITH RESPECT TO S
50  C      L(I)= SUM OF RECIPROCALS OF THE WEIGHTS.
51         F(I)=V(I)-P/Q
52         FV(I)=-S(I)/Q
53         FK(I)=P/QQ
54         FS(I)=-VM(M)*KM(M)/QQ
55         L(I)= (FS(I)**2)*((S(I)*0.01)**2)+SV(I)**2
56         B11=B11+FV(I)**2/L(I)
57         B12 = B12+FV(I)*FK(I)/L(I)
58         B22=B22+FK(I)**2/L(I)
```

```
59            C1 = C1 + FV(I)*F(I)/L(I)
60            C2 =C2+FK(I)*F(I)/L(I)
61            C3 =C3+ F(I)**2/L(I)
62            B21=B12
63     38     CONTINUE
64     C       DELTAK(M) AND DELTAV(M) ARE THE ADJUSTMENTS TO THE INITIAL
65     C       VALUES OF KM AND V.
66     C       SIGMAX = SIGMA EXTERNAL. TESTS IF THERE IS NO SIGNIFICANT
67     C       CHANGE.
68     C       SIGMAK =SD OF KM BY DIRECT FIT PROCEDURE.
69     C       SIGMAV =SD OF V  BY DIRECT FIT PROCEDURE.
70            DENOM=B11*B22-B12*B21
71            DELTAK(M)=(B11*C2-C1*B12)/DENOM
72            DELTAV(M)=(C1*B22-B12*C2)/DENOM
73            KM(M+1)=KM(M)-DELTAK(M)
74            VM(M+1)=VM(M)-DELTAV(M)
75            SIGMAX(M)=((C3-C2*DELTAK(M)-C1*DELTAV(M))/FLOAT(N-2))
76            SIGMAK(M+1)=SQRT(B11/DENOM)
77            SIGMAV(M+1)=SQRT(B22/DENOM)
78            IF(M-1)120,13,120
79     120    IF(M-9)121,121,199
80     121    IF(SIGMAX(M-1)-1.0001*SIGMAX(M))199,199,13
81     199    CONTINUE
82     44     FORMAT(/1X,2HM=,I2,3X,15HSIGMA EXTERNAL=,1PE11.4)
83            M1=M-1
84            WRITE(3,44)(J,SIGMAX(J),J=1,M1)
85     39     FORMAT(//4X,2HKM,9X,8HSD OF KM,7X,4HVMAX,7X,10HSD OF VMAX,3X,1HM/)
86            WRITE(3,39)
87     40     FORMAT(/1X,1PE11.4,2X,1PE11.4,2X,1PE11.4,2X,1PE11.4,2X,I2)
88            DO 50 K=1,M
89            J=K-1
90            WRITE(3,40)KM(K),SIGMAK(K),VM(K),SIGMAV(K),J
91     50     CONTINUE
92            RETURN
93            END
```

```
 1  C      **************SUBROUTINE WORST*******************************
 2  C      *                                                            *
 3  C      * THIS SUBROUTINE DETERMINES THE WORST POINT AND REMOVES IT  *
 4  C      * CONTROL RETURNS TO THE MAIN SEGMENT WITH N REDUCED TO N-1  *
 5  C      * THE PARAMETERS KM AND V ARE RECALCULATED AND CHECKED TO    *
 6  C      * SEE IF THERE IS ANY SIGNIFICANT DIFFERENCE FROM THE FIRST  *
 7  C      * ESTIMATES OF KM AND V.                                     *
 8  C      *                                                            *
 9  C      **************************************************************
10         SUBROUTINE WORST (N,X,Y,W,ORD,SLOPE,SW,SWX,SWX2,SWY,SWY2,SWXY,
11        1YWOBS,XWOBS,WWOBS,KM,DKM,VMAX,DVMAX,DORD,DSLOPE,SD,S,V,SV)
12         REAL KM
13         DIMENSION X(50),Y(50),W(50),YCAL(50),DYCAL(50),WDYCAL(50),S(50),
14        1V(50),SV(50)
15  C       CALCULATES THE VALUE OF Y (YCAL) FROM THE SLOPE AND ORDINAL
16  C       INTERCEPT AT EACH VALUE OF THE VARIABLE X. DETERMINES THE
17  C       DEVIATION (DYCAL) OF EACH POINT FROM THE LINE, AND THE
18  C       WEIGHTED DEVIATION (WDYCAL).
19         DO 32 I=1,N
20         YCAL(I)=ORD+SLOPE*X(I)
21         DYCAL(I)=Y(I)-YCAL(I)
22         WDYCAL(I)=ABS(SQRT(W(I))*DYCAL(I))
23  32     CONTINUE
24  C       DETERMINES WHICH POINT (XWOBS,YWOBS) HAS THE WORST WEIGHTED
25  C      DEVIATION FROM THE LINE.
26         M=1
27         TEST=WDYCAL(1)
28         MM=2
29  33     IF (WDYCAL(MM)-TEST)34,34,120
30  120    TEST=WDYCAL(MM)
31         M=MM
32         MM=MM+1
33         IF (MM-N)33,33,35
34  34     MM=MM+1
35         IF (MM-N)33,33,35
36  35     YWCAL=YCAL(M)
37         YWOBS=Y(M)
38         XWOBS=X(M)
39         WWOBS=W(M)
40         WWDY=WDYCAL(M)
41         WDY=DYCAL(M)
42  C       THE NUMBER OF THE WORST POINT IS X(M),Y(M). THIS IS REMOVED
43  C       BY MOVING ALL THE FOLLOWING POINTS DOWN BY ONE AND THUS
44  C       OVERWRITING THE POINT X(M),Y(M) BY X(M+1),Y(M+1).
45         DO 36 J=M,N
46         X(J)=X(J+1)
47         Y(J)=Y(J+1)
48         W(J)=W(J+1)
49         S(J)=S(J+1)
50         V(J)=V(J+1)
51         SV(J)=SV(J+1)
52  36     CONTINUE
53  C       THE SUBSTRATE CONCENTRATION AND VELOCITY OF THE WORST POINT
54  C       ARE DETERMINED FROM THE RECIPROCALS OF 1/XWOBS AND 1/YWOBS.
55         SWORST=1.0/XWOBS
56         VWORST=1.0/YWOBS
57  201    FORMAT(//1X,111HTHE FOLLOWING VALUES ARE OBTAINED BY APPLYING THE
58        1WEIGHTED LEAST SQUARES METHOD TO THE LINEWEAVER-BURK EQUATION)
```

```
59            WRITE(3,201)
60      21    FORMAT(//4X,2HKM,9X,8HSD OF KM,7X,4HVMAX,7X,10HSD OF VMAX/)
61            WRITE(3,21)
62      22    FORMAT (/1X,1PE11.4,2X,1PE11.4,2X,1PE11.4,2X,1PE11.4)
63            WRITE(3,22)(KM,DKM,VMAX,DVMAX)
64      23    FORMAT(//4X,3HORD,8X,9HSD OF ORD,6X,5HSLOPE,6X,11HSD OF SLOPE/)
65            WRITE(3,23)
66      24    FORMAT(/1X,1PE11.4,2X,1PE11.4,2X,1PE11.4,2X,1PE11.4)
67            WRITE(3,24)(ORD,DORD,SLOPE,DSLOPE)
68      37    FORMAT (//1X,11HWORST POINT/)
69            WRITE(3,37)
70      38    FORMAT (1X,2HS=,1PE10.4,4X,2HV=,1PE10.4)
71            WRITE(3,38)SWORST,VWORST
72      39    FORMAT (1X,2HX=,1PE10.4,4X,2HY=,1PE10.4)
73            WRITE(3,39) XWOBS,YWOBS
74      40    FORMAT (//1X,15HDEV. FROM LINE=,1PE11.4,3X,19HWT. DEV. FROM LINE
75           11PE10.4)
76            WRITE(3,40) WDY,WWDY
77      41    FORMAT (//1X,3HSD=,1PE14.8/)
78            WRITE(3,41)SD
79      C       THE VARIOUS SUMS, SWX, SWY ETC ARE ALTERED BY REMOVING THE
80      C       VALUES ASOCIATED WITH THE WORST POINT AND THE NEW VALUES
81      C       OF KM AND V ARE DETERMINED LESS THE WORST POINT. THIS
82      C       OCCURS AFTER RETURNING TO MAIN AND WLSQ. THE VALUES OF
83      C       KM AND V ARE CHECKED TO SEE IF THEY ARE SIGNIFICANTLY
84      C       DIFFERENT FROM THE VALUES DETERMINED WITH THE WORST POINT
85      C       IF THEY ARE THEN THE PROCESS IS REPEATED BY REMOVING THE NEXT
86      C       WORST POINT ETC UNTIL A SUITABLE TOBS VALUE IS OBTAINED.
87            SW=SW-WWOBS
88            SWX=SWX-WWOBS*XWOBS
89            SWY=SWY-WWOBS*YWOBS
90            SWX2=SWX2-WWOBS*XWOBS*XWOBS
91            SWY2=SWY2-WWOBS*YWOBS*YWOBS
92            SWXY=SWXY-WWOBS*XWOBS*YWOBS
93            RETURN
94            END
```

Appendix III Analogue computer units

III.1. Introduction

The basis of an analogue computer is the operational amplifier. The attachment of various input impedance networks and a feedback impedance converts the operational amplifier into a summer, inverter or integrator. The operational amplifier (fig. III.1) has the following distinguishing characteristics:

(*a*) a high negative gain voltage amplification, A, usually of the order of 10^8 at DC;

(*b*) a high input impedance and for any input voltage the current flowing into the amplifier is negligible, typically 10^{-9} A;

(*c*) a low output impedance so that the output is not loaded down by anything attached to it.

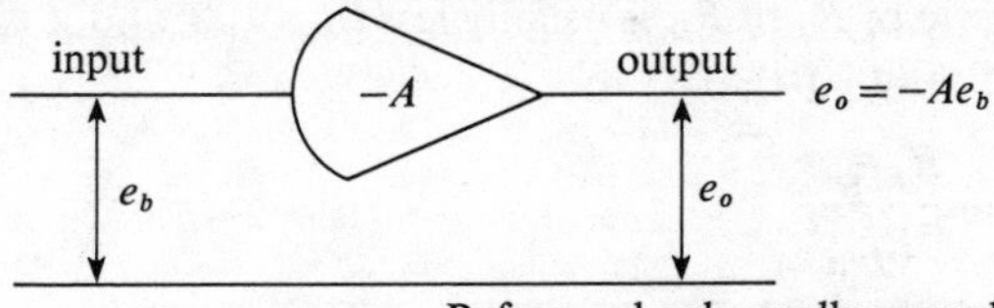

Fig. III.1. Diagrammatic representation of an operational amplifier.

III.2. Inverter

An inverter is produced by the addition of resistive input and output impedances.

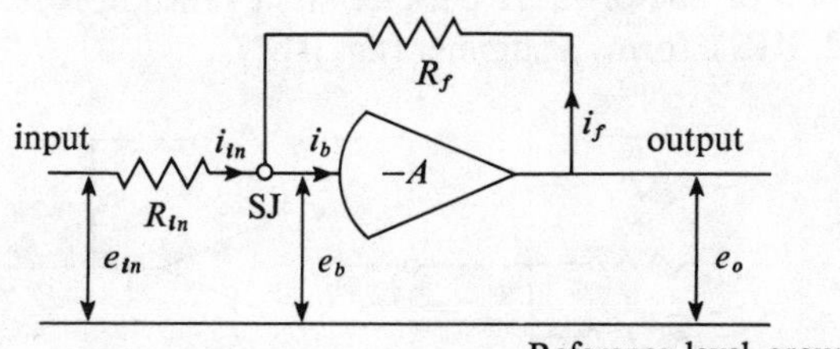

Fig. III.2. Simple inverter circuit.

In figure III.2, the input voltage, e_{in}, the voltage at the summing junction (SJ), the base voltage, e_b, and the output voltage, e_o, are all referred to a common reference level, such as ground. The output voltage is related to the voltage at the summing junction by the gain of the amplifier, thus

$$e_o = -Ae_b.$$

Using Kirchoff's laws, the nodal current equation at the summing junction SJ is

$$i_b = i_{in} + i_f$$

or, from Ohm's Law,

$$i_b = \frac{e_{in} - e_b}{R_{in}} + \frac{e_o - e_b}{R_f}.$$

Since i_b is effectively zero, 10^{-9} A, it can be neglected and replacing e_b by $-e_0/A$ we obtain

$$\frac{e_{in}}{R_{in}} + \frac{e_o}{AR_{in}} = -\frac{e_o}{AR_f} - \frac{e_o}{R_f}$$

or $$e_o = -\frac{R_f e_{in}/R_{in}}{1 + (1/A)\{(R_f/R_{in}) + 1\}}.$$

Since the ratio of R_f to R_{in} is usually less than 100, and A is much greater than 1 (typically 10^8!),

$$e_o = -\frac{R_f e_{in}}{R_{in}}.$$

The input–output relationship is dependent solely on the ratio of the feedback to input resistances. For an inverter, $R_f = R_{in}$, and hence the output voltage has the same amplitude as the input voltage but is of opposite sign.

III.3. Summer

The addition of one or more parallel input resistances to the previous circuit (fig. III.2), forms a summer (fig. III.3).

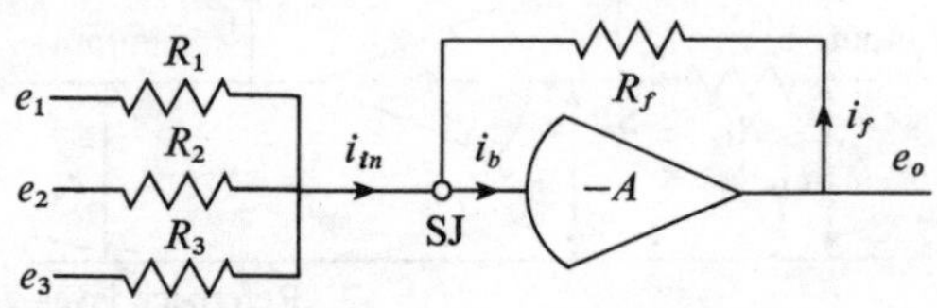

Fig. III.3. Summing amplifier circuit.

The summing junction nodal equation is

$$i_1 + i_2 + i_3 + i_f - i_b = 0.$$

Using Ohm's Law,

$$\frac{e_1 - e_b}{R_1} + \frac{e_2 - e_b}{R_2} + \frac{e_3 - e_b}{R_3} + \frac{e_o - e_b}{R_f} - i_b = 0.$$

Since both i_b and e_b are effectively zero ($e_b = -e_o/A$, but $e_o \sim 0$–$100\ V$ and $A \sim 10^8$), then

$$e_o = -\left[\frac{R_f e_1}{R_1} + \frac{R_f e_2}{R_2} + \frac{R_f e_3}{R_3}\right].$$

Usually the computer unit has four inputs; two have input resistances equal to the feedback resistance and the other two inputs have resistances that are one-tenth R_f, hence

$$e_o = -[e_1 + e_2 + 10e_3 + 10e_4]$$

or

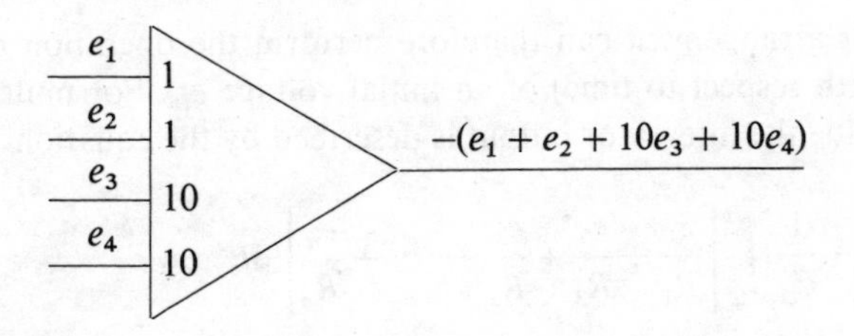

III.4. Integrator

In this case the feedback impedance is a capacitor; the amplifier circuit for a single input becomes as shown in fig. III.4.

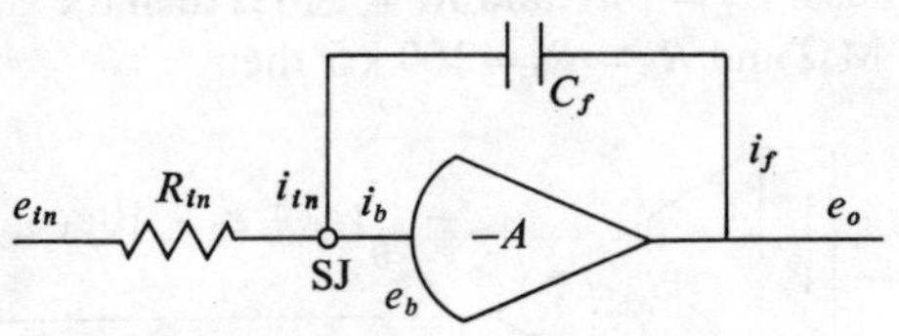

Fig. III.4. Simple integrator circuit.

The relation between capacitance, voltage drop and current for a capacitor with *no initial charge* is,

$$e = \frac{1}{C}\int_0^t i\,dt.$$

Thus the voltage drop across the capacitor can be expressed as

$$e_o - e_b = \frac{1}{C_f}\int_0^t i_f\,dt$$

or $$i_f = C_f\,\frac{d(e_o - e_b)}{dt}.$$

The current summation at the summing junction, SJ ($e_b = 0$, $i_b = 0$) is now

$$\frac{e_{in}}{R_{in}} + C_f\frac{de_o}{dt} = 0,$$

whose solution is

$$e_0 = -\frac{1}{R_{in}C_f}\int_0^t e_{in}\,dt.$$

This circuit arrangement can therefore perform the operation of integration (with respect to time) of an initial voltage e_{in}. For multiple resistive inputs, the integrator output is described by the equation

$$e_o = -\frac{1}{C_f}\int_0^t\left[\frac{e_1}{R_1} + \frac{e_2}{R_2} + \frac{e_3}{R_3} + \cdots + \frac{e_n}{R_n}\right]dt,$$

i.e. the output voltage is the integral of the algebraic sum of the input voltages (assuming $R_1 = R_2 = R_3 = \cdots = R_n$). The gain of the integrator is determined by the time constant of the input/output networks, i.e. $1/R_1C_f$, $1/R_2C_f$ etc. These gains are usually adjusted to be the same as the summer. Thus if $C_f = 1\ \mu$F and $R_1 = 1\ \text{M}\Omega$ then the gain would be 1. If $R_1 = R_2 = 1\ \text{M}\Omega$ and $R_3 = R_4 = 100\ \text{k}\Omega$ then

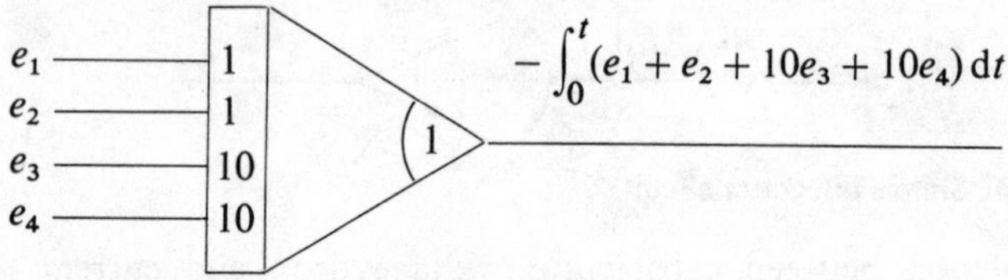

e_1 — 1
e_2 — 1
e_3 — 10
e_4 — 10
(1)
$$-\int_0^t (e_1 + e_2 + 10e_3 + 10e_4)\,dt$$

The feedback capacitor, C_f, can usually be altered and this changes the gains of the various inputs. This additional gain is signified by the number in the apex of the computer symbol. If C_f in the previous example were altered from 1 μF to 0.1 μF, then all the gains would be increased by a factor of 10, hence

Input	Gain
e_1	1
e_2	1
e_3	10
e_4	10

Integrator gain: 10

$$-10\int_0^t (e_1 + e_2 + 10e_3 + 10e_4)\,dt$$

III.5. Multiplier

There are several electronic methods for multiplying two or more variables, and since the electronic circuitry is complicated and varies with manufacturers, these devices are not described here.

Bibliography

Achs, M. J. & Garfinkel, D. (1968). *Comp. Biomed. Res.*, **2**, 92.

Adair, G. S. (1925). *J. Biol. Chem.*, **63**, 529.

Alberty, R. A. (1953). *J. Am. Chem. Soc.*, **75**, 1928.

Andrews, P. R., Smith, G. D. & Young, I. G. (1973). *Biochemistry*, **12**, 3492.

Arrhenius, S. (1889). *Z. Phys. Chem.*, **4**, 226.

Baird, J. B. & Elmore, D. T. (1968). *FEBS Lett.*, **1**, 343.

Bak, T. A. (1963). *Contributions to the Theory of Chemical Kinetics.* W. A. Benjamin Inc., New York.

Baker, B. R. (1967). *Design of Active-Site Directed Irreversible Enzyme Inhibitors.* J. Wiley & Sons. Inc., New York.

Barwell, C. J. & Hess, B. (1970). *Hoppe-Seyler's Z. Physiol. Chem.*, **351**, 1531.

Bell, R. M. & Koshland, D. E. (1971). *Science*, **172**, 1253.

Bender, M. L., Begue-Canton, M. L., Blakeley, R. L., Brubacher, L. J., Feder, J., Gunter, C. R., Kezdy, F. J., Kilheffer, J. V., Marshall, T. H., Miller, C. G., Roeske, R. W. & Stoops, J. K. (1966). *J. Am. Chem. Soc.*, **88**, 5890.

Berezin, I. V., Kazanskaya, N. F. & Klyosov, A. A. (1971). *FEBS Lett.*, **15**, 121.

Berezin, I. V., Varfolomeyev, S. D. & Martinek, K. (1970). *FEBS Lett.*, **8**, 173.

Bethell, M. R., Smith, K. E., White, J. S. & Jones, M. E. (1968). *Proc. Nat. Acad. Sci., USA*, **60**, 1442.

Blow, D. M., Birktoft, J. J. & Hartley, B. S. (1969). *Nature, Lond.*, **221**, 337.

Briggs, G. E. & Haldane, J. B. S. (1925). *Biochem. J.*, **19**, 338.

Brot, F. E. & Bender, M. L. (1969). *J. Am. Chem. Soc.*, **91**, 7187.

Brown, A. J. (1902). *Trans. Chem. Soc.*, **81**, 373.

Brown, F. A., Hastings, J. W. & Palmer, J. D. (Ed.) (1970). *The Biological Clock: Two Views.* Academic Press, New York & London.

Bruice, T. C., Brown, A. & Harris, D. O. (1971). *Proc. Nat. Acad. Sci. USA*, **68**, 658.

Bünning, E. (1964). *The Physiological Clock*. Springer-Verlag, Berlin.

Cafferata, R. L. & Freundlich, M. (1969). *J. Bacteriol.*, **97**, 193.

Carnahan, B., Luther, H. A. & Wilkes, J. O. (1969). *Applied Numerical Methods*. John Wiley, New York.

Chance, B. (1943). *J. Biol. Chem.*, **151**, 553.

Chance, B. (1963). In *Techniques of Organic Chemistry*, Ed. S. L. Friess, E. S. Lewis & A. Weissberger, vol. 8 part 2, p. 728. Interscience, New York.

Chance, B., Estabrook, R. W. & Ghosh, A. (1964*a*). *Proc. Nat. Acad. Sci., USA*, **51**, 1244.

Chance, B., Hess, B. & Betz, A. (1964*b*). *Biochem. Biophys. Res. Commun.*, **16**, 182.

Changeux, J-P., Gerhart, J. C. & Schachman, H. K. (1968). *Biochemistry*, **7**, 531.

Charlesworth, A. S. & Fletcher, J. R. (1967). *Systematic Analogue Computer Programming*. Pitman Ltd, London.

Chase, T. & Shaw, E. (1967). *Biochem. Biophys. Res. Commun.*, **29**, 508.

Cleland, W. W. (1963). *Biochim. Biophys. Acta*, **67**, 104.

Cleland, W. W. (1967). *Adv. Enzymol.*, **29**, 1.

Cohen, G. N. (1969). *Curr. Top. Cell. Reg.*, **1**, 183.

Cohlberg, J. A., Pigiet, V. P. & Schachman, H. K. (1972). *Biochemistry*, **11**, 3396.

Curtis, A. R. & Chance, E. M. (1974). *CHEK and CHEKMAT: Two Chemical Reaction Kinetics Programs*. Her Majesty's Stationery Office, London.

Dalziel, K. (1957). *Acta Chem. Scand.*, **11**, 1706.

Dalziel, K. (1968). *FEBS Lett.*, **1**, 346.

Dalziel, K. & Engel, P. C. (1968). *FEBS Lett.*, **1**, 349.

Datta, P. (1969). *Science*, **165**, 556.

Datta, P. & Gest, H. (1964). *Nature, Lond.*, **203**, 1259.

Davies, G. E. & Stark, G. R. (1970). *Proc. Nat. Acad. Sci., USA*, **66**, 651.

Dixon, M. (1953). *Biochem. J.*, **55**, 170.

Dowd, J. E. & Riggs, D. S. (1965). *J. Biol. Chem.*, **240**, 863.

Dynnik, V. V. & Sel'kov, E. E. (1973). *FEBS Lett.*, **37**, 342.

Eadie, G. S. (1952). *Science*, **116**, 688.

Easterby, J. S. (1973). *Biochem. Biophys. Acta*, **293**, 552.

Eckfeldt, J., Hammes, G. G., Mohr, S. C. & Wu, C-W. (1970). *Biochemistry*, **9**, 3353.

Eigen, M. (1954). *Discuss. Faraday Soc.*, **17**, 194.

Eisenberg, H. & Reisler, E. (1970). *Biopolymers*, **9**, 113.

Eisenberg, H. & Tomkins, G. M. (1968). *J. Mol. Biol.*, **31**, 37.

Elmore, D. T., Kingston, A. E. & Shields, D. B. (1963). *J. Chem. Soc.*, 2070.

Elmore, D. T., Roberts, D. V. & Smyth, J. J. (1967). *Biochem. J.*, **102**, 728.

Elmore, D. T. & Smyth, J. J. (1968*a*). *Biochem. J.*, **107**, 97.

Elmore, D. T. & Smyth, J. J. (1968*b*). *Biochem. J.*, **107**, 103.

Eyring, H. (1935*a*). *J. Chem. Phys.*, **3**, 107.

Eyring, H. (1935*b*). *Chem. Rev.*, **17**, 65.

Fischer, E. H., Pocker, A. & Saari, J. C. (1970). *Essays Biochem.*, **6**, 23.

Foster, R. J. & Niemann, C. (1953). *Proc. Nat. Acad. Sci., USA*, **39**, 999.

Frieden, C. (1959). *J. Biol. Chem.*, **234**, 809.

Frieden, C. (1967). *J. Biol. Chem.*, **242**, 4045.

Frieden, C. (1970). *J. Biol. Chem.*, **245**, 5788.

Garfinkel, D. (1968). *Comp. Biomed. Res.*, **2**, 31.

Garfinkel, D., Frenkel, R. A. & Garfinkel, L. (1968). *Comp. Biomed. Res.*, **2**, 68.

Garfinkel, D., Garfinkel, L., Pring, M., Green, S. B. & Chance, B. (1970). *Ann. Rev. Biochem.*, **39**, 473.

Gear, C. W. (1969). *Inf. Process.*, **68**, 187.

Gerhart, J. C. (1970). *Curr. Top. Cell Reg.*, **2**, 275.

Gerhart, J. C. & Pardee, A. B. (1962). *J. Biol. Chem.*, **237**, 891.

Gerhart, J. C. & Schachman, H. K. (1965). *Biochemistry*, **4**, 1054.

Ghosh, A. & Chance, B. (1964). *Biochem. Biophys. Res. Commun.*, **16**, 174.

Gibson, Q. H. & Milnes, L. (1964). *Biochem. J.*, **91**, 161.

Goldman, R. & Katchalski, E. (1971). *J. Theor. Biol.*, **32**, 243.

Gutfreund, H. (1955). *Discuss. Faraday Soc.*, **20**, 167.

Gutfreund, H. & Sturtevant, J. M. (1956*a*). *Biochem. J.*, **63**, 656.

Gutfreund, H., & Sturtevant, J. M. (1956*b*). *Proc. Nat. Acad. Sci., USA*, **42**, 719.

Hammes, G. G., Porter, R. W. & Stark, G. R. (1971). *Biochemistry*, **10**, 1046.

Hanes, C. S. (1932). *Biochem. J.*, **26**, 1406.

Harrison, B. S. & Hammes, G. G. (1973). *Biochemistry*, **12**, 1395.

Hartley, B. S., Brown, J. R., Kauffman, D. L. & Smillie, L. B. (1965). *Nature, Lond.*, **207**, 1157.

Hartley, B. S. & Kilby, B. A. (1952). *Biochem. J.* **50**, 672.

Hartridge, H. & Roughton, F. J. W. (1923). *Proc. R. Soc. Lond., Ser. A*, **104**, 376.

Hatfield, G. W., Ray, W. J. & Umbarger, H. E. (1970). *J. Biol. Chem.*, **245**, 1748.

Hess, B., Boiteux, A. & Kruger, J. (1969). *Adv. Enzyme Regul.*, **7**, 149.

Hess, B. & Wurster, B. (1970). *FEBS Lett.*, **9**, 73.

Heyde, E. (1973). *Biochim. Biophys. Acta*, **293**, 351.

Higgins, J. (1964). *Proc. Nat. Acad. Sci., USA*, **51**, 989.

Higgins, J. (1967). *Ind. Eng. Chem.*, **59**, 18.

Hill, A. V. (1913). *Biochem. J.*, **7**, 471.

Hill, L. R. & Brew, K. (1975). *Adv. Enzymol.*, **43**, 411.

Hinberg, I. & Laidler, K. J. (1972*a*). *Can. J. Biochem.*, **50**, 1334.

Hinberg, I. & Laidler, K. J. (1972*b*). *Can. J. Biochem.*, **50**, 1360.

Hofstee, B. H. J. (1952). *Science*, **116**, 329.

Hood, J. J. (1878). *Phil. Mag.*, **6**, 371.

Hood, J. J. (1885). *Phil. Mag.*, **20**, 323.

Hunter, A. & Downs, C. E. (1945). *J. Biol. Chem.*, **157**, 427.

Hurst, R. O. (1969). *Can. J. Biochem.*, **47**, 941.

Jameson, G. W., Roberts, D. V., Adams, R. W., Kyle, W. S. A. & Elmore, D. T. (1973). *Biochem. J.* **131**, 107.

Josephs, R. (1971). *J. Mol. Biol.*, **55**, 147.

Kaplan, H. & Laidler, K. J. (1967). *Can. J. Chem.*, **45**, 539.

Kibby, M. R. (1969). *Nature, Lond.*, **222**, 298.

King, E. L. & Altman, C. (1956). *J. Phys. Chem.*, **60**, 1375.

Kirschner, M. W. (1971). PhD Thesis, University of California, Berkeley.

Kirschner, M. W. & Schachman, H. K. (1971). *Biochemistry*, **10**, 1919.

Kirtley, M. E. & Koshland, D. E. (1967). *J. Biol. Chem.*, **242**, 4192.

Klotz, I. M. (1946). *Arch. Biochem.*, **9**, 109.

Knorre, W. A. (1968). *Biochem. Biophys. Res. Commun.*, **31**, 812.

Knowles, J. R. (1972). *Acc. Chem. Res.*, **5**, 155.

Knowles, J. R. & Preston, J. M. (1968). *Biochim. Biophys. Acta*, **151**, 290.

Koshland, D. E. Jr (1973). *Sci. Am.*, **229**(4), 52.

Koshland, D. E., Némethy, G. & Filmer, D. (1966). *Biochemistry*, **5**, 365.

Krupka, R. M. & Laidler, K. J. (1961). *J. Am. Chem. Soc.*, **83**, 1445.

Kryloff, N. & Bogoliuboff, N. (1947). *Introduction to Non-Linear Mechanics*. Princeton University Press.

Kuchel, P. W., Nichol, L. W. & Jeffrey, P. D. (1974). *J. Theor. Biol.*, **48**, 39.

Kuchel, P. W., Nichol, L. W. & Jeffrey, P. D. (1975). *J. Biol. Chem.*, **250**, 8222.

Kuchel, P. W. & Roberts, D. V. (1974). *Biochim. Biophys. Acta*, **364**, 181.

Lienhard, G. E. (1972). *Ann. Rep. Med. Chem.*, **7**, 249.

Lienhard, G. E. (1973). *Science*, **180**, 149.

Lindblad, P. & Degn, H. (1967). *Acta Chem. Scand.*, **21**, 791.

Lineweaver, H. & Burk, D. (1934). *J. Am. Chem. Soc.*, **56**, 658.

Lotka, A. J. (1910). *J. Phys. Chem.*, **14**, 271.

Lotka, A. J. (1920). *Proc. Nat. Acad. Sci., USA*, **6**, 410.

McClure, W. R. (1969). *Biochemistry*, **8**, 2782.

McCracken, D. D. & Dorn, W. S. (1964). In *Numerical Methods and Fortran Programming*, p. 330. J. Wiley & Sons Inc., New York.

Manning, G. B. & Campbell, L. L. (1961). *J. Biol. Chem.*, **236**, 2953.

Meighen, E. A., Pigiet, V. & Schachman, H. K. (1970). *Proc. Nat. Acad. Sci., USA*, **65**, 234.

Melo, A. & Glaser, L. (1965). *J. Biol. Chem.*, **240**, 398.

Michaelis, L. & Menten, M. L. (1913). *Biochem. Z.*, **49**, 333.

Michaelis, L. & Rothstein, M. (1920). *Biochem. Z.*, **110**, 217.

Monod, J. & Jacob, F. (1961). *Cold Spring Harbor Symp. Quant. Biol.*, **26**, 389.

Monod, J., Wyman, J. & Changeux, J-P. (1965). *J. Mol. Biol.*, **12**, 88.

Morales, M. & McKay, D. (1967). *Biophys. J.*, **7**, 621.

Neet, K. E., Nanci, A. & Koshland, D. E. (1968). *J. Biol. Chem.*, **243**, 6392.

Nichol, L. W., Jackson, W. J. H. & Winzor, D. J. (1967). *Biochemistry*, **6**, 2449.

Nichol, L. W., Kuchel, P. W. & Jeffrey, P. D. (1974). *Biophys. Chem.*, **2**, 354.

Nichol, L. W., O'Dea, K. & Baghurst, P. A. (1972). *J. Theor. Biol.*, **34**, 255.

Nichol, L. W., Smith, G. D. & Ogston, A. G. (1969). *Biochim. Biophys. Acta*, **184**, 1.

Orr, G. W. (1973). PhD thesis, Queen's University of Belfast.

Page, M. I. & Jencks, W. P. (1971). *Proc. Nat. Acad. Sci., USA*, **68**, 1678.

Palmer, G. & Massey, V. (1968). In *Biological Oxidations*, ed. T. P. Singer, p. 263. Interscience, New York.

Prigogine, I. (1947). *Etude thermodynamique des Phenomenes irreversibles*. De Soer Liège.

Pye, E. K. (1969). *Can. J. Bot.*, **47**, 271.

Pye, K. & Chance, B. (1966). *Proc. Nat. Acad. Sci., USA*, **55**, 888.

Rabin, B. R. (1967). *Biochem. J.*, **102**, 22C.

Roberts, D. V., Adams, R. W., Elmore, D. T., Jameson, G. W. & Kyle, W. S. A. (1971). *Biochem. J.*, **123**, 41P.

Roberts, D. V. & Elmore, D. T. (1974). *Biochem. J.*, **141**, 545.

Rodiguin, N. M. & Rodiguina, E. N. (1964). *Consecutive Chemical Reactions: Mathematical Analysis and Development* (English edn), ed. R. F. Sneider. Van Nostrand, New York.

Rosenbusch, J. P. & Weber, K. (1971). *J. Biol. Chem.*, **246**, 1644.

Santi, D. V., McHenry, C. S. & Perriard, E. R. (1974). *Biochemistry*, **13**, 467.

Scarano, E., Geraci, G. & Rossi, M. (1967). *Biochemistry*, **6**, 192.

Schrödinger, E. (1944). *What is Life*? Cambridge University Press, London.

Schwert, G. W. & Eisenberg, M. A. (1949). *J. Biol. Chem.*, **179**, 665.

Schwert, G. W., Neurath, H., Kaufman, S. & Snoke, J. E. (1948). *J. Biol. Chem.*, **172**, 221.

Sel'kov, E. E. (1968). *Eur. J. Biochem.*, **4**, 79.

Sel'kov, E. E. (Ed.) (1971). *Oscillatory Processes in Biological and Chemical Systems, All-Union Symposium on Oscillatory Processes*, Vol. II, Puschino-no-oka, Moscow. (In Russian with English summaries.)

Smith, G. D., Roberts, D. V. & Kuchel, P. W. (1975). *Biochim. Biophys. Acta*, **377**, 197.

Spangler, R. A. & Snell, F. M. (1967). *J. Theor. Biol.*, **16**, 381.

Stewart, J. A. & Ouellet, L. (1959). *Can. J. Chem.*, **37**, 751.

Sund, H., Pilz, I. & Herbst, M. (1969). *Eur. J. Biochem.*, **7**, 517.

Tanizawa, K., Ishii, S-I. & Kanaoka, Y. (1968). *Biochem. Biophys. Res. Commun.*, **32**, 893.

Tanizawa, K., Ishii, S-I. & Kanaoka, Y. (1970). *Chem. Pharm. Bull.*, **18**, 2346.

Teipel, J. & Koshland, D. E. (1969). *Biochemistry*, **8**, 4656.

Theorell, H. & Chance, B. (1951). *Acta Chem. Scand.*, **5**, 1127.

Thomson, W. (1876*a*). *Proc. Roy. Soc. Lond.*, **24**, 269.

Thomson, W. (1876*b*). *Proc. Roy. Soc. Lond.*, **24**, 271.

Umbarger, H. E. (1956). *Science*, **123**, 848.

Valentine, R. C. (1968). *Proceedings of the 4th Regional Conference on Electron Microscopy, Rome*, **2**, 3.
Van't Hoff, J. H. (1887). *Z. Phys. Chem.*, **1**, 481.
Vaughan, R. J. & Westheimer, F. H. (1969). *Anal. Biochem.*, **29**, 305.
Von Euler, H., Josephson, K. & Myrbäck, K. (1924). *Z. Physiol. Chem.*, **134**, 39.
Waley, S. G. (1953). *Biochim. Biophys. Acta*, **10**, 27.
Weber, K. (1968*a*). *Nature, Lond.*, **218**, 1116.
Weber, K. (1968*b*). *J. Biol. Chem.*, **243**, 543.
Wentworth, W. E. (1965*a*). *J. Chem. Ed.*, **42**, 96.
Wentworth, W. E. (1965*b*). *J. Chem. Ed.*, **42**, 162.
Wilhelmy, L. (1850). *Ann. Physik. u. Chem., Poggendorff's*, **81**, 499.
Wilkinson, G. N. (1961). *Biochem. J.* **80**, 324.
Witt, H. T., Rumberg, B., Schmidt-Mende, P., Siggel, U., Skerra, B., Vater, J. & Weikard, J. (1965). *Angew. Chem. Int. Ed.*, **4**, 799.
Wolfenden, R. (1972). *Acc. Chem. Res.*, **5**, 10.
Wong, J. T-F. & Hanes, C. S. (1962). *Can. J. Biochem. Physiol.*, **40**, 763.
Woolf, B. (1932). In *Allgemaine Chemie der Enzyme*, ed. J. B. S. Haldane & K. Stern, p. 119. Steinkopf, Leipzig & Berlin. (See also Haldane, J. B. S. (1957). *Nature, Lond.*, **179**, 832.)
Wu, C-W. & Hammes, G. G. (1973). *Biochemistry*, **12**, 1400.
Yamazaki, I., Yokota, K. & Nakajima, R. (1965). *Biochem. Biophys. Res. Commun.*, **21**, 582.
Yates, R. A. & Pardee, A. B. (1956). *J. Biol. Chem.*, **221**, 757.
Yielding, K. L. & Tomkins, G. M. (1961). *Proc. Nat. Acad. Sci., USA*, **47**, 983.

Index